林良益

Santa Clara Spring. 1994.

THE SYNTHESIS APPROACH TO DIGITAL SYSTEM DESIGN

THE KLUWER INTERNATIONAL SERIES IN ENGINEERING AND COMPUTER SCIENCE

VLSI, COMPUTER ARCHITECTURE AND DIGITAL SIGNAL PROCESSING

Consulting Editor
Jonathan Allen

Latest Titles

VLSI Design of Neural Networks, U. Ramacher, U. Ruckert
ISBN: 0-7923-9127-6
Synchronization Design for Digital Systems, T. H. Meng
ISBN: 0-7923-9128-4
Hardware Annealing in Analog VLSI Neurocomputing, B. W. Lee, B. J. Sheu
ISBN: 0-7923-9132-2
Neural Networks and Speech Processing, D. P. Morgan, C.L. Scofield
ISBN: 0-7923-9144-6
Silicon-on-Insulator Technology: Materials to VLSI, J.P. Colinge
ISBN: 0-7923-9150-0
Microwave Semiconductor Devices, S. Yngvesson
ISBN: 0-7923-9156-X
A Survey of High-Level Synthesis Systems, R. A. Walker, R. Camposano
ISBN: 0-7923-9158-6
Symbolic Analysis for Automated Design of Analog Integrated Circuits, G. Gielen, W. Sansen,
ISBN: 0-7923-9161-6
High-Level VLSI Synthesis, R. Camposano, W. Wolf,
ISBN: 0-7923-9159-4
Integrating Functional and Temporal Domains in Logic Design: The False Path Problem and its Implications, P. C. McGeer, R. K. Brayton,
ISBN: 0-7923-9163-2
Neural Models and Algorithms for Digital Testing, S. T. Chakradhar, V. D. Agrawal, M. L. Bushnell,
ISBN: 0-7923-9165-9
Monte Carlo Device Simulation: Full Band and Beyond, Karl Hess, editor
ISBN: 0-7923-9172-1
The Design of Communicating Systems: A System Engineering Approach, C. J. Koomen
ISBN: 0-7923-9203-5
Parallel Algorithms and Architectures for DSP Applications, M. A. Bayoumi, editor
ISBN: 0-7923-9209-4
Digital Speech Processing: Speech Coding, Synthesis and Recognition A. Nejat Ince, editor
ISBN: 0-7923-9220-5
Sequential Logic Synthesis, P. Ashar, S. Devadas, A. R. Newton
ISBN: 0-7923-9187-X
Sequential Logic Testing and Verification, A. Ghosh, S. Devadas, A. R. Newton
ISBN: 0-7923-9188-8
Introduction to the Design of Transconductor-Capacitor Filters, J. E. Kardontchik
ISBN: 0-7923-9195-0

THE SYNTHESIS APPROACH TO DIGITAL SYSTEM DESIGN

edited by

Petra Michel
Ulrich Lauther
Peter Duzy

Siemens AG

foreword by
A. Richard Newton

Kluwer Academic Publishers
Boston/Dordrecht/London

Distributors for North America:
Kluwer Academic Publishers
101 Philip Drive
Assinippi Park
Norwell, Massachusetts 02061 USA

Distributors for all other countries:
Kluwer Academic Publishers Group
Distribution Centre
Post Office Box 322
3300 AH Dordrecht, THE NETHERLANDS

Library of Congress Cataloging-in-Publication Data

The Synthesis approach to digital system design / edited by Petra Michel, Ulrich Lauther, Peter Duzy.
p. cm. -- (The Kluwer international series in engineering and computer science. VLSI, computer architecture, and digital signal processing)
Includes bibliographical references and index.
ISBN 0-7923-9199-3 (alk. paper)
1. Digital electronics--Data processing. 2. Electronic circuit design--Data processing. 3. Computer-aided design. I. Michel, Petra. II. Lauther, U. III. Duzy, Peter. IV. Series.
TK7868.D5S96 1992
621.381--dc20 91-46109
CIP

Printed on acid-free paper.

Printed in the United States of America

Contents

Contents

Foreword

Over the past decade there has been a dramatic change in the role played by design automation for electronic systems. Ten years ago, integrated circuit (IC) designers were content to use the computer for circuit, logic, and limited amounts of high-level simulation, as well as for capturing the digitized mask layouts used for IC manufacture. The tools were only aids to design—the designer could always find a way to implement the chip or board manually if the tools failed or if they did not give acceptable results. Today, however, design technology plays an indispensable role in the design of electronic systems and is critical to achieving time-to-market, cost, and performance targets.

In less than ten years, designers have come to rely on automatic or semi-automatic CAD systems for the physical design of complex ICs containing over a million transistors. In the past three years, practical logic synthesis systems that take into account both cost and performance have become a commercial reality and many designers have already relinquished control of the logic netlist level of design to automatic computer aids.

To date, only in certain well-defined areas, especially digital signal processing and telecommunications, have higher-level design methods and tools found significant success. However, the forces of time-to-market and growing system complexity will demand the broad-based adoption of high-level, automated methods and tools over the next few years.

Research in the area of high-level design automation has been ongoing, world wide, for more than twenty years now. Many of the techniques used today have been adapted from earlier approaches, made practical by changes in hardware implementation technologies, design requirements, and the power of modern computers and workstations. However, today's researchers are certainly adding their share of new design techniques and algorithms to the broad body of knowledge already assembled. A significant portion of the overall challenge of making high-level design automation practical involves sorting through the existing collection of techniques and finding the right combination of methods and tools for a particular problem domain.

I am very pleased that the many talented contributors to this book have taken the time to document this collection of state-of-the-art techniques and algorithms used in high-level description, synthesis and verification, and have presented them in a way that new students in the field will find them understandable on their own as well as in the context of the CALLAS design system. By using a complete, high-level design system as a vehicle for the text, the authors have captured a number of the essential aspects of high-level design methods as well as the basic algorithms and data structures that support them. The text is an excellent starting point for the developers of new high-level design aids and for potential users of such systems as well.

A. Richard Newton
Berkeley, California
October, 1991

Acknowledgements

Many discussions must precede the development of a book like this. While the contributors mentioned throughout this book have played a major part in its writing, their many other colleagues in our department have been closely involved as well. Especially, we would like to acknowledge the contributions of those colleagues who are directly involved in synthesis, verification, hardware-description languages, and testing in the Systems Design Automation Group in Corporate Research and Development of Siemens AG.

But in addition to those colleagues from Siemens, we would like to acknowledge the contributions made by one other person in particular. A. Richard Newton was involved in the development of this book throughout all stages of its evolvement. He participated in the many discussions on the taxonomy of synthesis, he influenced our work in the various fields of system design, and he reviewed most of the chapters in this book. All contributors appreciated his valuable advice.

Special thanks go to the following people for their help in certain chapters of this book: To Birgit Lutter for her efforts in helping with the figures in Chapter 1; to Wendell Baker for helpful discussions and suggestions concerning readability and contents of Chapter 6; to Franz Rammig for his really quick response when we asked him to write Chapter 9, and to Michael Hohenbichler for his help in editing this chapter. All of us appreciated the suggestions of Joyce McLean regarding the correct use of English language. Certainly everybody relied on one of the authors, Wolfgang Glunz, to prepare the camera-ready copy of the book. Special thanks go to Carl Harris for his patience when we delayed the date for the final manuscript month after month and to Egon Hörbst for his understanding when for some time everybody seemed to be engaged exclusively with this book.

Our final thanks go to our families who supported us while we were working through nights and early mornings.

Petra Michel
October, 1991

THE SYNTHESIS APPROACH TO DIGITAL SYSTEM DESIGN

Chapter 1

Introduction to Synthesis

Ulrich Lauther

The rapid technological development of the last thirty years has provided means for the design and fabrication of larger and larger electronic systems. This was not possible without equally rapid progress in the development of design methodologies and tools for automated design, which support the designer in the application of these methodologies. Whereas in the early days, design automation was mainly concerned with design verification (circuit and logic simulation, layout verification), later on also with the layout process (physical design starting from a netlist of functional blocks and a corresponding library of predesigned cells), a rapid development of design automation tools can be observed in the late eighties and early nineties that allow for the specification or description of a system at various levels of abstraction in the behavioral domain and for the automated or even automatic implementation of a design starting from this input.

The terms "specification" and "description" are not used interchangeably in this book. Loosely speaking, a *specification* describes behavior in terms of results, whereas a *description* defines behavior in terms of procedure. This distinction gives rise to the informal notion of describing "what" (specification) as opposed to "how" (*implementation*), and hence to the notion that specification languages are "non-procedural" in nature. A description, therefore, represents one way of implementing the function. A *model* is an abstraction of an implementation. Models, in general, are used to capture a specific aspect, e.g., behavior or structure, of a system or subsystem in order that one can reason about it.

By automation of design verification and physical design a tremendous reduction of design effort has been achieved in the past. This is illustrated in Figure 1.1 (modified from [Nomu85]). The use of modern equipment for schematic entry, the application of simulation for verification of gate-level design descriptions and fully automatic placement and routing starting from flat netlists resulted in a reduction of design effort of roughly 50%. A further re-

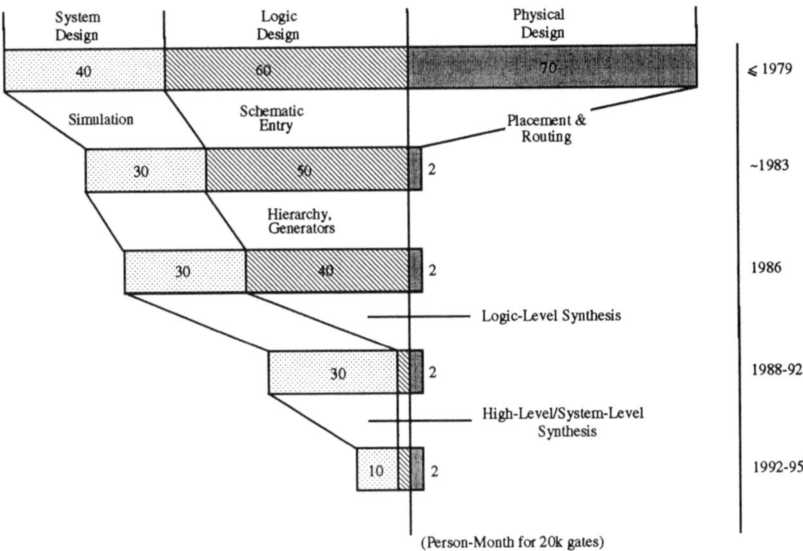

Figure 1.1: Development of effort spent in various design steps over the years.

duction was achieved by supporting hierarchical design and by the availability of module generators, e.g., generators for programmable logic arrays (PLAs). The next big step was the introduction of logic-level synthesis and optimization techniques, allowing for automatic translation of truth tables into minimized networks of logic gates. Another significant reduction of design effort and design time will result from synthesis methods, tools, and systems that transform a very high level design description—or even specification—into an adequate implementation. From the well-known folktheorem, stating that a programmer (or a designer), on the average, can produce 10 lines of correct code (or draw 10 transistors, place 10 cells) a day, independent of the abstraction level of the programming (or design) language used, it becomes clear that moving up to higher levels of abstraction will result in a further reduction of design cost and time-to-market. Another driving force behind the move to higher entry points into the design process is the need to come up with implementations which are correct the first time, thus eliminating the need for time consuming engineering changes and redesign[1] cycles. In principle, synthesis techniques promise *correctness by construction*, and even if the designer needs to take part in the design process, working at higher levels of abstraction will be less prone to error than dealing with the nitty-gritty details of lower levels.

In the meantime, these methods have reached such a degree of maturity that writing a text dedicated to this methodology seems timely.

The design of electronic systems can be tackled at various levels of abstraction, dealing with design data of different domains, requiring specific tools to support or to automate associated design steps. Therefore, to set the stage for the main body of this book, first a taxonomy of design automation is developed in this chapter and then—to ease the introduction of specific methods and tools—an ideal design flow from a high-level specification down to fabrication is sketched.

The focus is on the various forms of synthesis; the understanding of the term "synthesis" as it is used in this text is explained and the terms used for the subtasks and variants of synthesis are defined.

1.1 The Y-Chart—Levels and Domains of Description

For the development of a taxonomy of design automation in the field of electronic systems, the Y-chart, first introduced by Gajski and Kuhn [GaKu83], is a most valuable reference point. The basic idea underlying the Y-chart is that each element of an electronic system can be described within three different domains. These domains are the *behavioral domain*, the *structural domain*, and the *physical/geometrical domain*. In the behavioral domain, the (intended) behavior of the system is specified, ideally without any reference to the way this behavior is achieved by an implementation. In the structural domain, one deals with the system as a hierarchy of functional elements and their interconnections. In the physical domain, the structure of the design is mapped into space, without any reference to its functionality.

The domains are represented by the three axes of a "Y" which converge at a common center point (see Figure 1.2). Within each domain, elements can be described at various levels of abstraction. These levels are represented as points along the three axes, with higher (more abstract) levels at the periphery, lower levels near the center of the chart.

The lowest level of the design hierarchy a chipdesigner ever has to deal with is the *circuit level*. The structural elements belonging to this level are transis-

[1] "engineering change" and "redesign" are common euphemisms for fixing errors in specification, description, or implementation of a design.

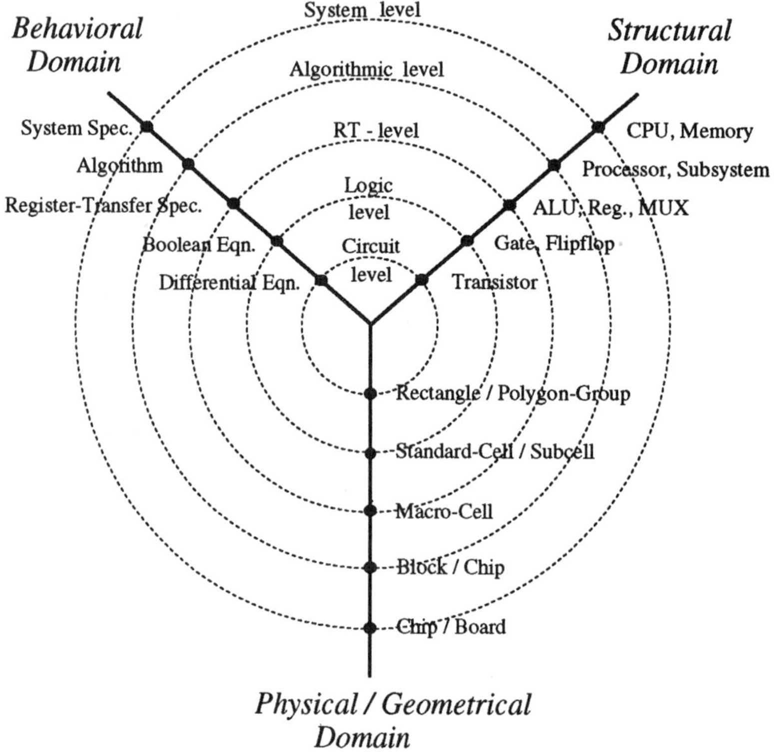

Figure 1.2: Domains and levels of description.

tors, resistors, capacitors, etc. Within the behavioral domain, the transistor is described by differential equations that specify dependencies between voltages applied to its terminals and resulting currents between the terminals. Similar equations result if circuit-level elements are connected into networks. Within the structural domain, the transistor is used to build more complex functional units, e.g., gates. Here, a gate is described by the interconnection pattern between transistors; the transistor itself is just a set of terminals, whose inner connectivity is expressed by the common transistor symbols. Within the physical/geometrical domain, the transistor is represented by a set of polygons or rectangles which specify the regions on the chip to which specific process steps (e.g., etching, diffusion, metallization, etc.) have to be applied during fabrication in order to achieve the behavior described within the behavioral domain for this individual transistor.

Moving up within the behavioral domain, the next level of abstraction,[2] the *logic level*, is that of Boolean equations, at which voltage levels have been abstracted to logic values "true" and "false" or "1" and "0". In Chapter 9, abstraction mechanisms as they are used for this purpose in the context of simulation are dealt with in detail.

The next level is that of operations on sets of logic values which have been grouped into "words" and whose 0-1-patterns are interpreted as numerical values to which arithmetic and logic operations are applied. Since activities in systems at this level of abstraction are typically described by transfers of values between registers, this level is called *register-transfer level* (RT-level).

To describe the behavior of a (sub)system, i.e., to describe how the system processes input values supplied at its input-ports in order to come up with the required answers at its outputs, operations are grouped into algorithms, thus the term *algorithmic level.*

In many cases, a complex system is composed from subsystems. Each subsystem can be described by a sequential process (though the implementation may use some fine-grain parallelism); these processes run concurrently, using various communication methods for interaction. This level of abstraction is called *system level.*

The corresponding design units on the structural and physical axis are shown in Figure 1.2.

So far, only the various abstraction levels for dealing with *data* within the behavioral domain have been considered. Similar levels of abstraction apply to time and control. Time is considered to be continuous at the circuit level, discrete (clock cycles) at the logic through algorithmic level, and even more abstract models of time, e.g., operation cycles, are used at the system level. There is no notion of control at the circuit level. At the logic level, control exists, but is not clearly distinguished from other activity in the circuit. At the register-transfer level, control takes the form of possibly abstract states; at the algorithmic level it is abstracted to control structures, e.g., loops, branching, procedure calls. Finally at the system level, it takes the form of synchronization and communication between processes.

The Y-chart provides a convenient framework for the description or definition of design tasks. They can be expressed as transitions between points on the axes of the chart, graphically represented by directed arcs. For example, a transition from a network of modules (subsystem on the structural axis) to the point representing the chip on the physical/geometrical axis corresponds to the task of placement and routing of macro-cells, where starting from a netlist of

[2]The exact number and definition of abstraction levels depends on the specific design task at hand; in simulation, for instance, the switch level, a level between circuit and logic level, is often used.

functional blocks (modules), the full layout of a chip (placement of macro-cells and geometry of interconnections) is generated. Reconstructing a network of transistors (structural domain) from polygones (physical domain) is the task of an extraction program. Device recognition and parameter extraction correspond to a further transition from the structural axis to the respective point on the behavioral axis.

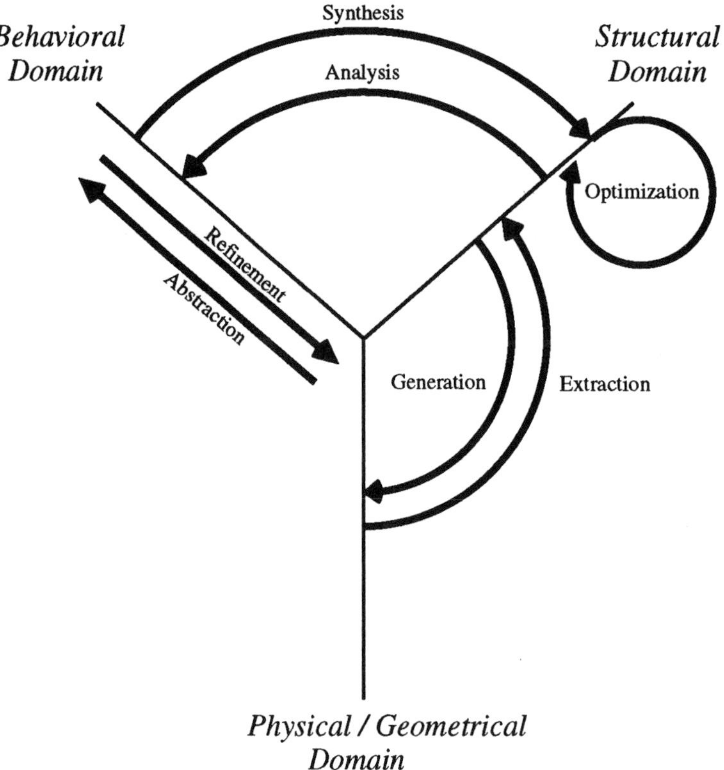

Figure 1.3: Transitions in the Y-chart defining design steps.

Based on these transitions, the terms "generation," "extraction," "synthesis," and "analysis" can be defined as they are used in this book (see Figure 1.3). Transitions from the structural domain to the physical domain are called *generation*, reverse transitions are called *extraction*, those from the behavioral to the structural domain *synthesis*, and transitions in the opposite direction are called *analysis*. The latter is typically used to support verification. To validate the implementation, the actual behavior found by analysis is compared with the initially described intended behavior.

For movements within one axis of the Y-chart the following definitions are used: Steps represented by arcs directed towards the center are referred to as *refinement*, those of the opposite direction are called ***abstraction***. Arcs whose head- and tailnodes are identical (self-loops) symbolize a transformation within one domain and one level, called ***optimization***. In optimization, the basic functionality remains constant, but the quality of the design (expressed by an objective function in terms of, e.g., performance, area, and power consumption) is improved.

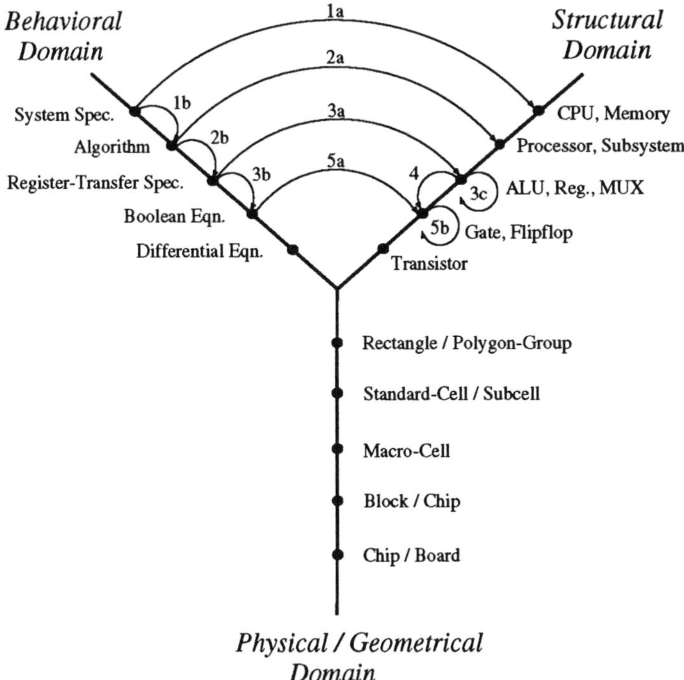

Figure 1.4: Transitions in the Y-chart defining various synthesis and optimization steps. 1: System-level synthesis; 2: High-level synthesis; 3: RT-level synthesis and optimization; 4: Technology mapping; 5: Logic-level synthesis and optimization.

In most cases design steps provide two types of results. The behavior of a system specified at some level of abstraction is transformed into a structural description at the same level. This is the synthesis contribution of a design step. The structural description thus synthesized refers to abstract objects described at the next lower level of abstraction. For these lower level objects behavior is either defined or implied. This is the refinement component of the design step. The behavioral descriptions of the lower-level objects provide the starting point for the next synthesis step.

The levels of abstraction and domains of description introduced above have, of course, to be taken with a grain of salt. Very often, design descriptions contain both behavioral *and* structural elements, or can be interpreted both ways. Thus, the definition of design steps as transitions in the Y-chart, as intuitive as it seems at a first glance, can become difficult, or at least look rather far-fetched. A refined approach to classification is based on the observation that each step in the design process adds information to the overall (multi-level and multi-domain) system description. Such a classification is shown in Table 1.1.

Notion	Adds	Based on
Synthesis	Structural information	Behavioral/Structural information
Analysis	Behavioral information	Structural information
Generation	Physical information	Structural information
Structural extraction	Structural information	Physical/geometrical information
Behavioral extraction	Behavioral information	Physical/geometrical information
Refinement	More detailed information	More abstract information from same domain
Abstraction	More abstract information	More detailed information from same domain
Optimization	Improved version	Former version from same domain/level

Table 1.1: Design steps in terms of information added to the overall system description.

1.2 Design Flow and Taxonomy of Synthesis

The ultimate goal of design automation for electronic systems is to fully automate the transformation of a specification given at the highest level of abstraction in the behavioral domain into a description at the lowest level in the physical domain, e.g., the mask geometry which provides the interface to fabrication. A software system that can provide this transformation may rightfully be called a *silicon compiler*.[3] To attempt this transformation in one step is not practical. A structural representation provides a convenient bridge between behavior and layout. But one intermediate step is not sufficient. To achieve subproblems of manageable complexity, the design process has to be broken down into many subtasks.

The design flow of a hypothetical silicon compiler consisting of a "pipeline" of synthesis tasks serves as an example for such a break down into subtasks and—at the same time—as a vehicle to introduce synthesis steps. In this chapter, a top-down view of the design process is taken, mostly for didactic reasons. In reality, support for the lower levels of design is further advanced and more often used in practice. The main body of the book follows, therefore, a bottom-up approach to design automation.

1.2.1 System-Level Synthesis

At the highest level of abstraction in the behavioral domain, the system is specified by its functionality (e.g., instruction set of a computer) and a set of constraints to be met (e.g., speed, power consumption, fabrication cost). The result of the first synthesis step is a partitioning of the system into subsystems, a set of communicating concurrent processes, (Figure 1.4, 1a) together with a behavioral description at the algorithmic level (Figure 1.4, 1b) for each of these subsystems. Currently, both partitioning and specification of subsystems are performed manually; nevertheless, some subtasks can be automated. Methods used in this field are presented in Chapter 7. The transition from a system-level specification to one or more subsystem descriptions at the algorithmic level is called *system-level synthesis.* As can be seen in Chapter 7, this term is somewhat fuzzy, often used as well for optimization steps at the higher levels of abstraction.

1.2.2 High-Level Synthesis

A behavioral description at the algorithmic level provides the starting point for *high-level synthesis*, often also referred to as *algorithmic-level synthesis*. The

[3] This term is often used in a much narrower sense, e.g., for module generators, for layout systems based on procedural layout description languages, etc.

term ***algorithmic description*** is used throughout this book as a synonym for "behavioral description at the algorithmic level." The algorithmic description defines a precise procedure for the computational solution of a problem. It specifies behavior in terms of operations and computation sequences on inputs to produce the required outputs. Basic elements of the description are very similar to those of general programming languages. Examples include arithmetic and logic operations applied to variables, and control structures, such as branches, loops, and procedure calls. Mapping (Figure 1.4, 2a,b) of variables to registers or signals, operators to functional units, and control structures to actions of a controller is, in principle, a straightforward process, but there are many different ways to solve this task. Basically, three different subtasks can be distinguished: functional units of appropriate types and number have to be selected, operations have to be assigned to time slots, and operations need to be assigned to specific functional units. These tasks are called ***resource allocation, scheduling***, and ***resource assignment***, respectively. Obviously, there is a time/area tradeoff: allocating more resources allows for more parallel execution of operations, thus giving higher performance at higher hardware cost (chip area). The feasible region (subject to constraints given by the user) of the space spanned by time and area is often called ***design space.*** To find an implementation corresponding to the optimum point within this space is one task of scheduling, resource allocation, and resource assignment.

In Chapter 6 a thorough discussion of the relevant problems of high-level synthesis and their solution is provided.

1.2.3 Register-Transfer Level Synthesis

The result of high-level synthesis is typically an initial description at the register-transfer level of a data path and a controller. Operations have already been assigned to control steps, but there is still freedom in the implementation of data path and controller. This is the starting point for *register-transfer level synthesis.*

Concerning the data path, resource allocation and assignment can be improved based on more detailed knowledge about physical characteristics of alternate implementations of the various blocks (resynthesis). Even the initial assignment of operations to control steps can be modified by structural changes (register relocation) to optimize performance (retiming) or area. In the strict sense (as defined in Section 1.1) this is not a synthesis step per se but rather an optimization step (see Figure 1.4, 3c); both, input and output are descriptions in the structural domain.

Controller synthesis, on the other hand, is a true synthesis step involving a transition from behavior to structure. Input of this step is an abstract description of a finite state system in terms of states and state transitions. Typical tasks include the selection of an appropriate controller architecture, state assignment, input/output encoding, and decomposition of controllers into smaller interacting finite state machines (Figure 1.4, 3a,b).

There is considerable overlap between high-level synthesis and register-transfer level synthesis. This is due partly to the fact that many systems do not provide the full pipeline of synthesis steps described here but start at the register-transfer level. As they mature, one tries to extend their scope, for example, by inclusion of resource sharing, a concept similar to scheduling and allocation in high-level synthesis. (Compatible operations in mutually exclusive branches can share functional units.)

In Chapter 5 methods for register-transfer level synthesis are explained in detail.

1.2.4 Logic-Level Synthesis

As a result of the previous steps, the system to be designed is broken down into blocks of combinational logic and storage elements. The behavior of combinational logic is described by Boolean functions. Their optimization and mapping to a gate-level hardware structure is the task of *logic-level synthesis* (Figure 1.4, 5a,b). There are a number of different representations for Boolean functions, some of them reflecting the style of the hardware implementation. For instance, equations in sum of products form look very much like the description of a PLA. In any case, as is demonstrated in Chapter 3, the translation from the behavioral domain to the structural domain is straightforward at the logic level. Actually, a gate-level schematic can be understood both as an implementation or as a specification of the intended behavior. Thus, the main point in logic-level synthesis is optimization, specifically *logic minimization*, aiming at minimal area (measured as number of literals).

If two-level logic is the target architecture, the combinational blocks can be handed over to a PLA generator. Thus, for the first time the physical/geometrical axis of the Y-chart is reached. In the case of multi-level logic one further step is needed.

1.2.5 Technology Mapping

The result of logic synthesis—in the case of multi-level logic—is a network of abstract gates. To implement the abstract network in hardware, groups of abstract gates are mapped to matching physical library cells of a given target technology, thus the terms *library mapping* or *technology mapping* (Figure 1.4, 4). There are many ways to cover the network with cells, resulting in different area and delay values. Optimization strategies for this problem are dealt with in Chapter 4.

Technology mapping, however, is not restricted to the logic level. Some systems perform technology mapping right after register-transfer level optimizations, mapping functional units to macro-blocks which are described both in the structural and the physical domain. Advantages and drawbacks are presented, again, in Chapter 4.

This concludes the synthesis pipeline. Further steps deal with the transition from the structural to the physical domain at lower levels of abstraction, i.e., layout generation, which is not covered by this book.

1.3 Entry Points and User Interfaces

Today, the full pipeline of synthesis and optimization steps described above is not available in a single system. As a consequence, various entry points into the (fictive) pipeline exist and adequate user interfaces (graphical or textual languages) must be provided. In the past, verification of the logic-level design by simulation and nearly completely automated physical design—placement and routing of predesigned cells—have been the main application fields of design automation in digital system design. Both steps start from a structural description at the logic or register-transfer level. The natural interface at these levels is a netlist, typically in a graphical representation, i.e., a schematic.

As the entry point moves to higher levels of abstraction, other interfaces are needed. At the logic level, truth tables or Boolean equations are adequate means. The latter can be represented graphically (circuit diagrams) or in textual form.

At the register-transfer level, the data flow between storing elements within a (sub)system is described. As in (imperative) software programming, assignments to variables and operations on their values are specified. Typically, in a register-transfer level description, the designer implies a strict mapping between variables in the description to registers in the hardware implementation and assigns operations to control steps. Thus, the transition from this behavioral description to the structural domain is a straightforward one-to-one mapping.

Descriptions at the algorithmic level look very much like those at the register-transfer level. But, according to the larger flexibility of high-level synthesis, the close links between variables and registers and between operations and control-steps are gone. Registers and operations belong to the resources high-level synthesis systems have to manage. *Hardware description languages* (HDLs), introduced in Chapter 2, are used at all levels of abstraction discussed above, both in the structural and the behavioral domain, thus playing a crucial role in synthesis.

Input languages are not the only user interface. The user needs access to results of individual synthesis steps. Not all problems in synthesis are understood well enough to exclude the user. Rather, he or she needs to inspect results, to modify parameters, to reformulate input descriptions in an attempt to achieve better solutions, or even to change synthesis results directly or to integrate manually designed pieces of hardware. Thus, synthesis is generally an interactive process, to be supported by adequate interactive tools.

1.4 Validation

Interactive design is an error prone process. Automatic design is not foolproof either, since synthesis software may contain "bugs." Therefore, it is essential to include methods for validation into a synthesis system. Basically, there are two complementary approaches to the problem, application of *formal methods* within the synthesis system, and validation by *simulation*. Formal methods can be used either concurrently with interactive design, thus guaranteeing that all transformations the user attempts are correct, or a posteriori, to prove consistency between implementation and specifications, or to prove certain properties of an implementation. The former approach is presented in Chapter 8. The latter, being in principle independent of the way an implementation was produced and feasible only for small designs, is only weakly related to the synthesis approach and thus outside the scope of this text.

Even if a design could be proved to be a correct implementation of the specification, the specification might be erroneous, i.e., specify another functionality than the user had in mind. Simulation provides an answer to this problem. There are more arguments for the necessity of simulation in a synthesis environment, which can be found in Chapter 9, where synthesis related aspects of simulation are presented.

1.5 Testing

An activity often overlooked by designers is *testing*. To consider testing issues only after the design has been completed leads to delays, designs which are hard to test, and more area overhead than necessary. Synthesis provides the opportunity to take care of testing and testability from the very beginning and in an integrated way. Optimization methods used in synthesis tend to remove redundancies and thus improve testability implicitly. If additional hardware is needed to achieve sufficient testability, the generation and inclusion of these pieces of hardware can be integrated into the synthesis process at various levels of abstraction, relieving the user of this unloved task and giving better overall results, because the additional hardware can be included into the optimization steps. These and other synthesis related aspects of testing are covered in Chapter 10.

Chapter 2

Hardware Description Languages and their Relevance to Synthesis

Andreas Hohl

2.1 Introduction

Hardware Description Languages (HDL) are used to describe various aspects of a hardware system to be designed. Many languages exist for this purpose, e.g., AHPL [Hill74], CAP/DSDL [Ramm80b], CDL [Chu74], CONLAN [Pilo83], DDL [Diet74], ELLA [MoPT85], ISP [Siew74a], PMS [Siew74b], and VHDL [VHDL87]. As most of them are derived from general programming languages, e.g., ADA, ALGOL, C, PASCAL, PL/1 [Comp74, Comp77, Hohl90], a close relationship is evident between the use of HDLs in hardware design and of general programming languages in software engineering.

In Figures 2.1a and 2.1b, a simple VHDL[1] model, or program, and an ADA program are shown to demonstrate the similarity of an HDL and a general programming language. HDLs, such as VHDL, which are used to describe a hardware system on different levels of abstraction are referred to as *multi-level HDLs*. In hardware design different HDL descriptions represent the design on the respective levels of abstraction. The descriptions themselves can be validated by simulation (cf. Chapter 9). Synthesis systems are used to transform an HDL description on one level of abstraction to a lower level.

---

[1]Very High Speed Integrated Circuits-HDL.

```
entity hello_world is
 -- VHDL version
end hello_world;
architecture behavior
          of hello_world is
begin
 assert false
 report "Hello world.";
end behavior;
```

(a) VHDL

```
with text_io;

procedure hello_world is
-- ADA version
begin
 text_io.put("Hello world.");
end hello_world;
```

(b) Ada

Figure 2.1: Hello world.

The following sections, therefore, focus on the relevant aspects of HDLs for the synthesis process. For this purpose, an example of a simple piece of hardware—a bar-code reader—is introduced to present the major characteristics of HDLs, especially VHDL.

2.2 VHDL Example - Bar-Code Preprocessor

A processor for fast bar-code recognition is shown in Figure 2.2.

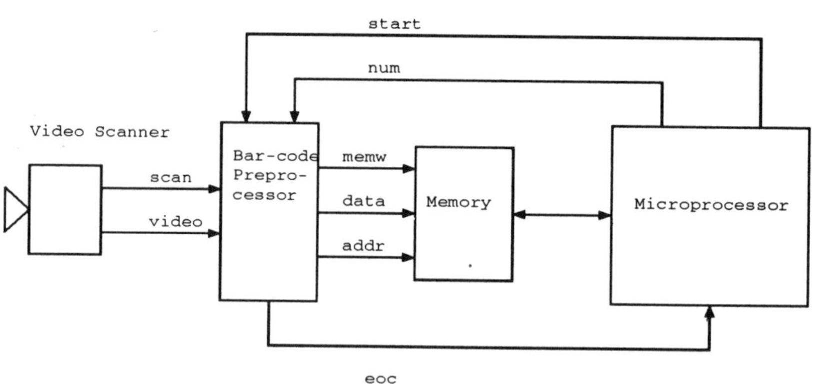

Figure 2.2: Fast bar-code processor.

The hardware system bar-code processor [März90] is composed of the modules: video scanner, bar-code preprocessor, memory, and microprocessor. The bar-codes, sampled by the video scanner, are recognized in real-time by the preprocessor, stored into a memory, and decoded by a microprocessor.

```
package barcodetypes IS
constant  fa       : bit := '0';
constant  tr       : bit := '1';
constant  wh       : bit := '0';
constant  bl       : bit := '1';
end barcodetypes;
library callas;
use callas.callas.all, work.barcodetypes.all;

-- This algorithm describes a preprocessor for fast barcode readers
entity bc IS
  port(
    scan      : in  bit;                         -- The scan signal of a ccd or laser scanner
    video     : in  bit;                         -- The video signal  "    "    "    "
    start     : in  bit;                         -- Start of conversion signal of a microproc.
    num       : in  unsigned(7 downto 0);        -- Given number of black-white transitions
    eoc       : out bit;                         -- End of conversion signal to a microproc.
    memw      : out bit;                         -- Write signal for a memory
    data      : out unsigned(7 downto 0);        -- Data to be written to a memory
    addr      : out unsigned(7 downto 0)         -- Address signals for a memory
      );
end bc;
architecture be of bc is
begin
  p: process
    constant  null8   : unsigned(7 downto 0) := "00000000";
    constant  overfl  : unsigned(7 downto 0) := "11111111";
    variable  white   : unsigned(7 downto 0);  -- Counts the width of white bars
    variable  black   : unsigned(7 downto 0);  -- Counts the width of black bars
    variable  hnum    : unsigned(7 downto 0);  -- Stores the given number of black-white transitions
    variable  actnum  : unsigned(7 downto 0);  -- Counts the number of black-white transitions
    variable  flag    : bit;                   -- Flag for recognizing a black-white transition
    variable  hmemw   : bit;
  begin
      wait for cycles(1);
      eoc <= tr;
      memw <= hmemw;
      while (start = fa) loop
        wait for cycles(1);
      end loop;
      hnum := num;
      loop  -- Repeat until barcode  pattern has been recognized
        wait for cycles(1);
        memw <= hmemw;
        eoc <= fa;
        while (scan = fa) loop
          wait for cycles(1);
        end loop;
        flag   := wh;
        white := null8;
        black := null8;
        actnum := null8;
        loop    -- Repeat until a counter (black or white) overflows
                -- or the barcode pattern has been recognized
          wait for  cycles(1);
          memw <= hmemw;
          if video = wh then
              white := white + 1;
              if flag = bl then
                actnum := actnum + 1;
                hmemw := fa;
              else
                hmemw := tr;
              end if;
              flag  := bl;
              white := null8;
              data <= black;
          else
              -- same for video = bl
          end if;
          addr <= actnum;
          exit when (white = overfl) OR (black = overfl);
        end loop;
        exit when (actnum = hnum) AND (white = overfl);
      end loop;
  end process;
end be;
```

Figure 2.3: VHDL model of bar-code preprocessor.

In the following, the focus is on the design of the preprocessor, which has the tasks to measure the width of the black and white bars, to count the number of bars, and finally to check whether the bar-code is correct. In Figure 2.3 the algorithm of the bar-code preprocessor is specified in VHDL. To ease the understanding of the VHDL model, a schematic overview of the various parts of a VHDL model is given in Figure 2.4.

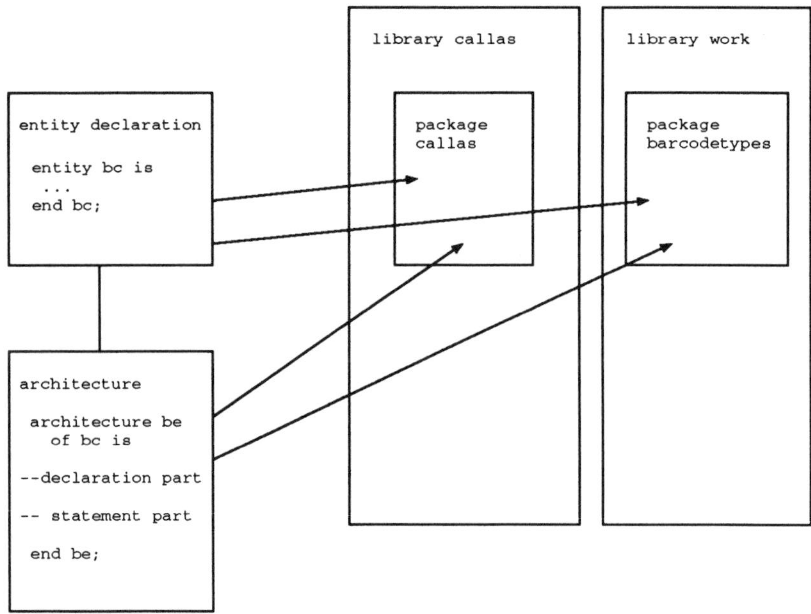

Figure 2.4: VHDL partitioning of fast bar-code preprocessor.

The fast bar-code preprocessor is described by a *VHDL entity*, consisting of an *entity declaration* and an *architecture*. The declaration identifies the model with the entity name, the entity interface, the *port list*, and other optional parts (cf. Section 2.3). The architecture itself describes an implementation of the entity. The preprocessor is described in this example at the algorithmic level, i.e., if the algorithm is executed, it shows the required behavior. VHDL supports this description style by the *process statement*, nested in the architecture part. The process statement contains a *declaration part* and a *statement part*, in which the respective statements are executed sequentially in the order of their occurrence in an infinite loop. The bar-code model contains, as known from common programming languages, *constant* and *variable declarations*, *loop*, *conditional*, *exit*, and *assignment statements*. In contrast to programming languages, an HDL has to support timing behavior and the description of concurrent actions. Therefore, in VHDL it is possible to distinguish between variables, as known from programming languages, and signals. In case of a signal assignment, a waveform containing a value and a time is assigned, i.e., the signal assumes a specified value after a specified time. In addition, the *wait statement* in VHDL allows for the specification of a wait-for-time or wait-for-event condition. Furthermore, VHDL provides various *concurrent statements* (cf. Section 2.3) to describe concurrency within a hardware system. For example, the hardware system "bar-code processor" can be described using a top entity "system," for which all nested entities (bar-code preprocessor, microprocessor etc.) are executed in parallel, and which communicate via the ports specified in the entity interface.

Entities can use global declarations from packages, similar to the ADA [ADA83] package concept. The use clause makes these declarations visible within an entity declaration and an architecture. Both entity declaration and architecture, as well as *package declarations* and *package bodies*—the implementation part of a package—are referred to as *design units.* Libraries are a container for design units and the *library clause* specifies the libraries to be used for a VHDL model.

A high-level synthesis system requires a description similar to the one shown in Figure 2.3 as an input in order to synthesize a corresponding hardware realization.

2.3 VHDL Hardware Description Language

In this chapter, the IEEE Standard hardware description language VHDL is used as an example to point out the features and characteristics of modern HDLs. The basic structure of a VHDL model, i.e., the entity declaration and the architecture, were introduced in the previous section. In this section VHDL is presented in more detail and additional language constructs are introduced, followed by examples. VHDL has its origin in the programming language ADA [ADA83]. The development of the language VHDL started in 1980, and was sponsored by the US Department of Defense (DoD) with the intention to create a standard HDL in the VHSIC[2] project. The language concepts of ADA, which was also initiated by the DoD, had to be extended by the following features to handle hardware design:

[2]Very High Speed Integrated Circuit.

- Description of hardware types (e.g., signals, bit strings)
- Description of timing behavior
- Data driven control structures (in addition to algorithmic control structures)

As introduced along with the bar-code example, a VHDL model consists of at least one design entity, which is separated into an entity declaration and at least one architecture body. In Figure 2.5 the basic structure of an entity is shown.

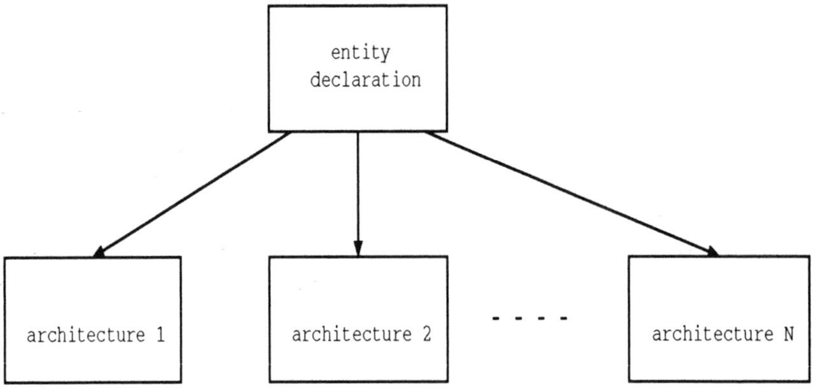

Figure 2.5: Basic structure of a VHDL entity.

The entity declaration contains the interface of the model to the environment, *attribute* descriptions, specifications and further descriptions (i.e., *assertions*) common to all architectures. An architecture represents one possible implementation of the model. Different implementations can result from different design variants or from different levels of abstraction in the design process. VHDL itself is a multi-level description language and provides a *behavioral*, a *structural* and a *data flow* view of a hardware model, whereas an arbitrary mixture of these views in a model or architecture is optional. In this context, the term "behavioral" stands for an algorithmic description of the function of the hardware system, enhanced optionally with some timing information. A structural description contains a list of connected blocks, where each block represents a subsystem. The third kind of view is used to describe the flow of data within a hardware system. A typical example is a register-transfer description, where data is exchanged between registers and similar complex objects. As an example a full adder (cf. Figures 2.6, 2.7, 2.8) is used to illustrate the different views in VHDL. The names of the architectures are chosen to reflect the different views, e.g., "behavioral_view," although an arbitrary name can be assigned to an architecture.

```
entity full_adder is
 port( a, b, cin: in bit; -- input ports
       sum, cout: out bit); -- output ports
end full_adder;
```

Figure 2.6: VHDL entity declaration of a full adder.

Each VHDL model (cf. Figure 2.6) starts with the keyword **entity**, followed by the name of the entity and an optional port declaration. Three input ports of type *bit* and two output ports of the same type are described. "Bit" is a predefined enumeration type ('0', '1'). The *end statement* contains the keyword **end** followed by a repetition of the entity name, and indicates the end of the entity declaration. Each statement in VHDL is concluded by a semicolon. Comments start with "- -" and extend to the end of a line.

```
architecture behavioral_view of full_adder is
 begin
 process -- a process contains several statements
         -- and will be executed in a sequential manner.
  variable i               : integer; -- index variable
  constant sum_vector      : bit_vector (0 to 3)  := "0101"
  constant carry_vector    : bit_vector (0 to 3)  := "0011"
  -- sum_vector and carry_vector are derived from
  -- the function table of the full adder.
  begin
   i := 0;
   if a = '1' then i := i + 1; end if;
   if b = '1' then i := i + 1; end if;
   if cin '1' then i := i + 1; end if;
   sum <= sum_vector(i) after 20 ns;
   -- signal assignment, similar to an
   -- ordinary variable assignment
   cout <= carry_vector(i) after 20 ns;
   wait on a, b, cin;
   -- process is sensitive to signals a, b, cin
 end process;
end behavioral_view;
```

Figure 2.7: Architecture body containing a behavioral view of a full adder.

The architecture body is divided into a declarative and a statement part (similar to an entity declaration), headed by the keyword **architecture**, the name of the architecture and the name of the corresponding entity. The body of the architecture "behavioral_view" (Figure 2.7) contains a process statement, which is the language construct provided for the description of the behavior of a hardware component or even a hardware system. The declarative part of the architecture is empty in this example. A process statement may also contain a declarative part with an equivalent structure. The process statement itself is part of the statement part of the architecture, and additionally contains its own statement part. The end statement concludes the architecture body.

```
architecture dataflow_view of full_adder is
 signal s : bit;
 begin
  s <= a xor b after 5 ns;
  sum  <=  s xor cin after 5 ns;
  cout <= (a and b) or (s and cin) after 10 ns;
end dataflow_view;
```

Figure 2.8: Architecture body containing a data flow view of a full adder.

The architecture "dataflow_view" (cf. Figures 2.8) contains a local declaration of a signal "s" and a *block statement* as the architecture body. All statements within a block statement are initiated in parallel. The signal assignment statements, which describe a data flow view of a full adder, are based on the corresponding Boolean equations. A signal assignment is executed in two steps. The expression on the right hand side of the equation sign is evaluated after each change of any variable term involved,[3] and the resulting value will be assigned to the signal after the specified delay.[4] Any changes of the expression value in this time interval have (in general) no effect on a previous assignment. The exception occurs if the delay of the youngest assignment is less than or equal to the previous one and the previous value assignment has not yet been executed.

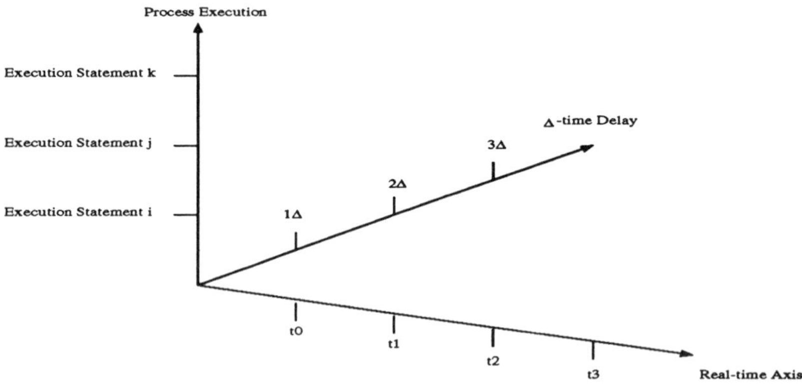

Figure 2.9: VHDL, three dimensions of time.

[3]This is an example for a data-driven control structure, i.e., the execution of certain statements in a VHDL model depends on a specific change of variable or signal values.

[4]For example, after 10 nanoseconds (ns).

To gain a clearer understanding, the time model of VHDL is discussed in the following. The underlying simulation model is that of an event driven simulation, i.e., a real-time axis (cf. Figure 2.9) reflects the advance of time in terms of discrete events.

Besides the real-time axis, a delta-time delay provides the deterministic handling of concurrency (multiple drivers of a signal) and zero-delay in case of signal assignment. This means that any signal assignment is split into an *initiation activity*, the first step of a signal assignment execution, and an *execution activity*, the second step as described above. The execution takes place at least one delta-cycle after the initiation. Uncertainties caused by multiple assignments to a signal have to be solved by the definition of a *resolution function* provided by the user. A short example, shown in Figure 2.10, clarifies this correlation.

```
begin     -- block statement, where all nested
          -- statements are initiated concurrently
 sig_a <= sig_b;  -- signal assignment with zero-delay
 sig_b <= sig_a;  -- signal assignment with zero-delay
end;
```

Figure 2.10: Concurrent signal assignment.

If the block statement is initiated at time t_0, both signal assignments are executed. The semantics of VHDL asks for both expressions on the right-hand side to be evaluated and the assignments of the values to take place within simulation cycle $t_0 + \Delta$. This means that signals sig_a and sig_b will change their values. This mechanism makes sure that the execution of the VHDL program is independent of a particular implementation by a simulator.

Additionally, VHDL distinguishes between an *inertial* and a *transport delay*. The first one takes the physical aspects, the inertia (e.g., a change of an output demands a minimum hold-time of the corresponding input signals, otherwise there will be no effect), into account and is the default delay. The latter one, indicated by the keyword **transport**, is used to describe changes without considering inertia.

The third axis (see Figures 2.9 and 2.11) of the time model reflects the execution of statements within processes. For example, a process describes an algorithm without any "real" delay in terms of time, that is to say nanoseconds. Variable assignments within a process, for instance, consume no time. Therefore, no global variables exist in a VHDL model. Otherwise, what happens in case of multiple assignments to a variable at the same time has to be defined to avoid a simulator-dependent execution of a VHDL program. In general, the underlying time model of any HDL describes a real-time behavior and has to define the language semantics for the description of concurrent activities in an HDL model.

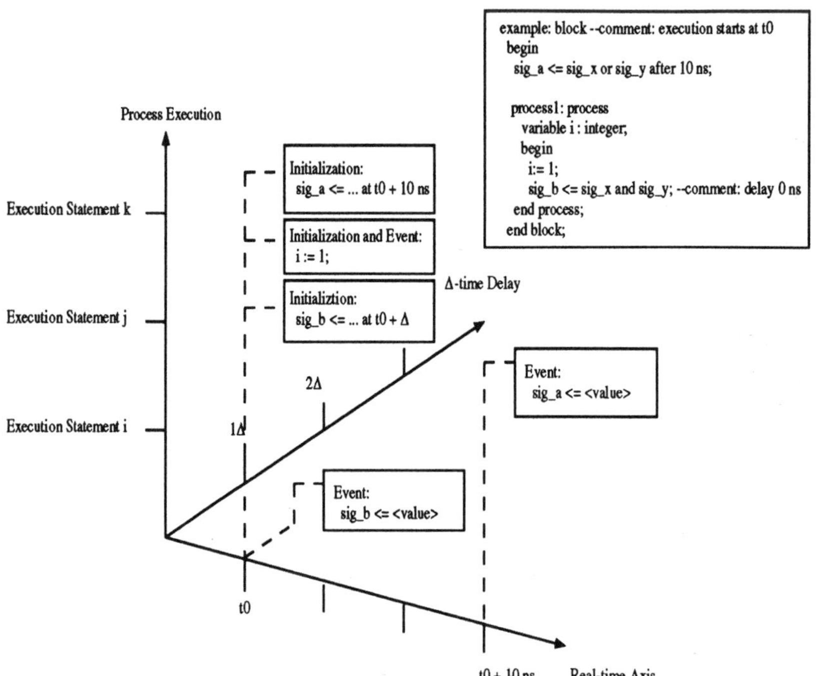

Figure 2.11: VHDL time model.

```
architecture structural_view of full_adder is
component half_adder
 port ( i1, i2:  in bit;
        carry: out bit;
        sum:   out bit);
end component;
component or_gate
 port ( i1, i2:  in bit;
        o:       out bit);
end component;
signal x, y, z: bit;
begin
 s1: half_adder port map (a,b,x,y);
 s2: half_adder port map (y,cin,z,sum);
 s3: or_gate port map (x,z,cout);
end structural_view;
```

Figure 2.12: Architecture body containing a structural view of a full adder.

The structural description of the full adder in the architecture "structural_view" (Figure 2.12) corresponds closely to the hardware representation of a full adder. Mainly, the architecture describes a netlist and its components.

External and internal signals are wired as described in the *port map clause.*

In the following, additional language constructs introduced by some examples are presented.

Configurations. Configurations (cf. Figure 2.13) are used to describe how design entities are linked to form a complete hardware model. A configuration declaration defines which VHDL entity is instantiated by a *component instantiation statement.* For example, the configuration declaration presented in Figure 2.13 means that for the two half-adder components of the architecture "structural_view" the entity "ha_123" is to be taken from the library "work." The binding (i.e., connection of ports etc.) of a component instance is left unspecified, and is performed by additional configuration specifications. The feature of configuration declaration and specification is useful in the case of complex multi-level descriptions. Furthermore, it is mandatory to allow for a separate compilation of design units which have a dependency as in the full-adder example, where entity "full_adder" uses entity "half_adder" as a component.

```
library work;
configuration config_full_adder of full_adder is
use work.all;
 for structural_view
     for s1, s2: half_adder
      use entity ha_123;
     end for;
     for s3: or_gate
      use entity or_xy;
     end for;
 end for;
end config_full_adder;
```

Figure 2.13: VHDL configuration.

Design entities are stored in libraries and can be used via the library clause and the use clause as shown in Figure 2.13. The *for clauses* attach the components "half_adder" and "or_gate" to corresponding library elements.

Subprograms and packages. Procedures and functions are used as subprograms, similar to procedures and functions in ADA or PASCAL. A package (cf. Figure 2.14) is also used in the same way as in ADA. It consists of a package declaration which can be referred to by other entities and a realization part, the *package body.* The package global_values from Figure 2.14 contains some constants to be used for example in *delay clauses* and the description of 7-valued logic. To describe hardware systems, a simple 2-valued logic "type bit is ('0', '1')," a predefined type in VHDL, is insufficient. Hence, multi-valued logic can be described in a package.

```
package global_values is
 constant trisingtime:   time := 10 ns;
 constant tfallingtime:  time := 15 ns;
 constant tchangingtime: time := 12 ns;
 type bit_7 is ('0', '1', 'X', 'L', 'H', 'Y', 'Z');
 function bitval(value:bit_7) return bit;
 function bit_7val(value:bit) return bit_7;
 type bit_7vector is array(natural range <>) of bit_7;
 function resolve(sources:bit_7vector) return bit_7;
end global_values;
--
--
--
package body global_values is
 function bitval(value:bit_7) return bit is
  constant bits: bit_vector := "0100000";
 begin
  return bits(bit_7'pos(value));
 end;
 -- further declarations and implementations
end global_value;
```

Figure 2.14: VHDL package.

Types. The following type categories are provided by VHDL (see Figure 2.15): enumeration, integer, real, physical types (for example, type time), floating point types, structured types, access types, file types.

```
type logic is ('X', '0', '1', 'Z');
type unsigned is array (natural range <>) of logic;
type subroutine_modes is (expand, share);
type time is range implementation_defined
       -- e.g. represented by 32 Bits or by 64 Bits.
      units
       fs;
       ps = 1000 fs;
       ns = 1000 ps;
       us = 1000 ns;
       ms = 1000 ms;
       sec = 1000 ms;
       min = 60 sec;
       hr = 6o min;
     end units;
type sub_real is -100.0 to +100.0;
type coordinate is
      record
       x : integer;
       y : integer;
     end record;
subtype counter_value is integer range 0 to 15;
```

Figure 2.15: VHDL type declarations.

Some of them are predefined, such as, integer or real. From the user's point of view, it is worthwhile to mention that these predefined types have implementation specific constraints, e.g., rounding, and overflow properties. Additionally, VHDL provides mechanisms well known from general programming languages to declare user defined types.

Declarations. VHDL offers several kinds of entities that are declared explicitly or implicitly, identified by names and associated with a declaration scope: types, subtypes, objects (signal, variable), files, interfaces, aliases, attributes, components, entities, configurations, subprograms, packages. The concept of scope and visibility of identifiers is similar to concepts from general programming languages. In the VHDL examples of this chapter various kinds of declarations are presented.

Attribute and configuration specifications. Specification attributes (Figure 2.16) can be associated with various objects in VHDL. As mentioned earlier, a configuration specification describes further binding information for a given component instantiation. In Figure 2.16 it is emphasized that the half-adders used in the full-adder example are instantiated by the entity "han," architecture "cell," which are stored in the library "rt." The *generic map* defines a specific delay value, "tha_delay," and the port map specifies the binding of the ports. The configuration specification is nested in an architecture, whereas a configuration declaration is a separate design unit.

```
entity full_adder is
port(....);

 attribute author of full_adder:entity is "Mr. D.";
 ...
end full_adder;
library rt;
architecture structural_view of full_adder is
 ...
 for all:half_adder
  use entity rt.han(cell)
       generic map(delay => tha_delay);
       port map(X => A,..., O => Sum);
 end for;
 ...
```

Figure 2.16: VHDL attribute and configuration specification.

Expressions. As known from programming languages, expressions may contain various kinds of operators such as arithmetic operators (+, -, /, etc.), logical operators (AND, OR, EXOR, etc.), function calls (e.g., bitval, cf. Figure 2.14), relational operators (<, >, =, etc.), and others.

Sequential and concurrent statements. VHDL provides various sequential statements: wait statement, assertion statement, signal assignment statement, variable assignment statement, procedure call statement, if statement, case statement, loop statement, next statement, exit statement, return statement, and null statement. These statements are used to describe algorithms for the execution of a subprogram or process, which show the required behavior of a hardware system, e.g., the bar-code preprocessor. Sequential statements execute in the order of their appearance, i.e., as they are written down.

Concurrent statements are: process statement, concurrent procedure call, concurrent assertion statement, concurrent signal assignment statement, and component instantiation statement, and generate statement. They are used to define interconnected blocks and processes describing the overall behavior or structure of a VHDL model. They are executed asynchronously with respect to their environment, and thus are a crucial feature for the description of the concurrency of modules within a hardware system. In Figure 2.17 a VHDL model of the fast bar-code processor is shown, in which all modules (processor, bar-code preprocessor, memory, video scanner) are instantiated by the corresponding component instantiation statements, i.e., all modules, components, are concurrently activated and communicate via their ports.

```
entity barcode_system is
-- identification block
  -- creation date: 91/07/12
  -- author: andreas hohl
  -- function: test bench of bar-code preprocessor
-- end identification block
end barcode_system;
architecture test_body of barcode_system is
-- architecture declarative part
  -- component component_name generic_list port_list
component processor
 generic ( -- generic parameter );
 port ( -- port list );
end component;
component bc -- this is the bar-code preprocessor
 port ( -- interface to i2c-bus
        scan: in bit;
        video: in bit;
        ..... );
end component;
component memory
-- ....
end component;
component video_scanner
-- ....
end component;
begin -- architecture body
 processor1: processor port map ( -- ports );
 memory1: memory port map ( -- ports );
 bc1: bc port map ( scan, video, start, num,
                    eoc, memw, data, addr );
 video_scanner1: video_scanner port map ( scan, video );
end test_body;
```

Figure 2.17: VHDL model of the bar-code processor.

Multi-valued logic. In contrast to other HDLs, VHDL has no built-in multi-valued logic but offers the possibility of describing any form of logic. A minimum logic set for describing hardware contains the values '0', '1', 'X' (undefined state), and 'Z' (high impedance state). Other multi-valued logic sets contain 7 or 9 different values, taking different signalstrengths into account. In general, a package providing the declaration of a multi-valued logic, including operators and functions declaration, has to be defined for describing hardware systems in VHDL.

2.4 Relevance of HDLs to Synthesis

An obvious application of HDLs in connection with synthesis is as an input for synthesis, among other input forms such as logic diagrams, Boolean equations, block diagrams or truth tables, which can also be described by HDLs. For its efficient use in synthesis several points have to be discussed:

1. Multi-level descriptions
2. Formal semantics of HDLs
3. Synthesis specific language requirements
4. Validation aspect
5. Synthesis subset definition and policy of use

The first two aspects, multi-level description capability, and formal semantics for HDLs are, in general, evident:

- The design process requires HDL models of various levels of abstraction to handle complex hardware designs and to be efficient. Multi-level HDLs are therefore a necessity.
- The definition of a formal semantics of HDLs for various purposes, e.g., simulation, synthesis, etc., is mandatory to achieve tool independence, i.e., the HDL description has to express "what the designer says," the tools must not interpret "what the designer means."

The third issue, synthesis specific language requirements, can in some cases be seen as synthesis-tool-dependent. The fourth issue, validation aspect, is a more tool-independent requirement. Although synthesis is often considered as a method for correctness by construction, the designer wants to validate the intermediate HDL models by simulation. The fifth issue, definition of subsets for synthesis of an HDL, is tool-dependent again. Often, there are too many possibilities for the description of one hardware model with an HDL. Consider for example the various ways to specify a finite state machine in VHDL, e.g., with a case statement or nested "if then else" statements. The chosen VHDL subset and a tool-dependent policy of use restricts this variety.

2.4.1 Multi-Level Descriptions

A hardware design process passes through different levels of abstraction and involves different views, i.e., various HDL models are generated manually or automatically. Discussing HDLs as an input for synthesis, the different levels of abstractions that are used as the starting point for the different synthesis systems have to be considered. Therefore, for a **standard** HDL it is necessary to cover the whole range of possible entry levels for synthesis tools.

As mentioned before, if a synthesis tool generates an intermediate HDL model, it is very helpful for the designer to read the same HDL as the input HDL for the synthesis tool. It is much easier to compare simulation results from the higher-level model to the lower-level ones, if the models are written in the same HDL. Otherwise, an intelligent result checker, often the designer, has to perform this costly task.

2.4.2 Formal Semantics of HDLs

An HDL model describes a hardware system in terms of control and data abstraction, including the timing behavior. A synthesis system interprets the description and translates it into a lower level representation and the process results, eventually, in a hardware implementation. To achieve well-defined results, the semantics of an HDL have to be defined. Usually the semantics of an HDL are determined by operative semantics definition in a formal [Pilo83, YaIs89, IsYY90] or informal way, that is, formal in terms of an abstract simulation engine or informal by a description of the elaboration and execution simulation cycle in a natural language. Of course the latter case may result in different simulation results due to lack of preciseness of this description.

2.4.3 Synthesis-Specific Language Requirements

Synthesis requires specific description capabilities from an HDL. In general, these requirements are supported and can be satisfied by using a language in a "synthesis-oriented" way. If a language provides no explicit construct or feature to meet a requirement, the designer has to determine a *policy of use*. The following requirements can be found in the literature, although some of them are tool-dependent and therefore not generally agreed upon:

- Separation of control and data description
- Handling of don't care information
- Data types
- Specific description of timing

VHDL offers the possibility to separate the description of control and data information as postulated in [NaSp90]. The control handling is described in a process statement (Figure 2.18) with the label "control." In this case, the reader may ask if it is preferable to include this in the language. The data manipulation is realized within a block statement labeled "data." Both parts are executed concurrently.

```
entity hardware_unit is
 port( -- port list
      );
end hardware_unit;
architecture synthesis_view of hardware_unit is
-- architecture declarative part
begin -- architecture body
 control: process ( -- Sensitivity list,
                    -- similar to a wait statement
                    -- within the process statement,
                    -- i.e. activity starts if a signal
                    -- from this list changes its value.
                  )
          -- declaration and statement part
        end process;
 data: block -- concurrent activities
             -- declaration and statement part
       end block;
end synthesis_view;
```

Figure 2.18: VHDL example for separation of control and data abstraction.

Another requirement is the representation of don't care information. The designer wants to be able to specify a don't care and the synthesis tool may use this information for optimization (see Chapter 3). Don't cares can also be specified on higher levels than the bit level, e.g., in case of opcode handling of a CPU to be designed. In this case, a policy of use is mandatory, i.e., the HDL model has to express what to do. In case of don't care at the bit level the policy of use can be defined by the synthesis tool, e.g., to do optimization. The next requirement, data types, can be realized in various ways. On the one hand, an HDL may provide all requested data types as built-in data types for synthesis purposes. On the other hand, an HDL can provide some basic data types and a mechanism to build user-defined types. Especially for high-level synthesis, requirements exist [WhNe90] to support enumeration, vector, structure, and union type constructs. In this case, an enumeration is a finite set of symbols; a vector is a one-dimensional array of elements of one type; a structure defines an unordered composite of elements of various types; and a union is a composite of a number of mutually exclusive elements. With the exception of unions, VHDL provides all these type constructs. The description of timing behavior varies for different levels of abstraction [WaTh85]. Additionally, synthesis tools can use a specific timing description in a way similar to don't cares. Suppose, a signal assignment has to be delayed at least n time units and at most m time units, where $n < m$, but any value within this time interval is valid, a hardware realization can be synthesized which fits this constraint.

2.4.4 Validation Aspect

Fulfilling the requirements mentioned above is not a problem if the designer wants to use this description exclusively as an input for a synthesis tool. For example, VHDL provides the language features to separate control and data information as described above, allows definition of a multi-valued logic including a don't care value, provides adequate data types and supports the description of specific timing by using attributes. Problems occur if the HDL model, built for synthesis purposes, is used for validation also, namely the semantics of simulation. The handling of don't care information and the specific description of timing pose special problems.

Suppose, the designer uses a signal sig_state of type bit_vector(1 to 4) to represent the current state of a component. A constraint is made that in a certain case, the value is assumed to be "1??0". This description has a specific semantics for a synthesis tool, but the interpretation for simulation is not as easy. The synthesis semantics may be: The coding of this state must be a '1' at the leftmost bit and a '0' at the rightmost bit, but the coding of the other bits can be chosen arbitrarily, e.g., in a way that hardware is minimized. The simulation semantics may be: If a '?' occurs in a bitvector, any logic value of the user-defined multi-valued logic has to be considered during simulation. Therefore, the resulting value of a signal assignment, in this case, is not a single bitvector, but a set of bitvectors. If the chain of assignments is of a certain depth, the number of possible sets increases drastically and this results in a long simulation run.

The questions are which value or set of values are to be chosen for a don't care value for simulation, and which delay or set of delays are to be chosen in case of a specified timing interval. More generally, the definition of a formal semantics for these language features is necessary. Some publications address these issues [YaIs89, IsYY90, WhNe90].

2.4.5 Synthesis Subset Definition

The problem of subset definition of an HDL occurs if a given HDL is not designed for synthesis purposes exclusively. Language constructs provided for simulation may be hard to synthesize or can not be synthesized at all. For example, VHDL includes an I/O package providing files and subprograms for file access. Additionally, if a policy of use for synthesis is recommended or defined, the construction of synthesis tools is much easier. This is the reason for specifying subsets of a language and policies of use for synthesis. Until now, no general agreement on a synthesis subset for VHDL has been achieved, rather the capabilities of different synthesis tools define the subsets. Some citations from the literature may clarify this statement:

[LiGa89]:

> "For example, the language allows events that are defined as any change on a signal line. Such events are not easily realized in hardware since storage elements are triggered by positive or negative edge signal transitions but not both. Similarly, VHDL specifies delay with an after clause that does not distinguish among inputs. Efficient synthesis algorithms, on the other hand, require delay specifications for each input-output pair. Furthermore, guard expressions allow any combinations of signals, although designers know that only one signal (clock) is used for writing into storage elements."

[LMOI89]:

> "It [VHDL] is more powerful in its descriptive abilities than we can currently handle. While, at the same time, it is less precise then we would like. Features in VHDL, such as support of complex data structures, real number arithmetic, and file input and output make it convenient to model sophisticated designs, but require unrealistic capabilities of synthesis tools."

In [Meye89] D.D. Gajski says:

> "For synthesizable description, however, the commands that are used are not as important as the modeling style, you could use exactly the same subset of VHDL for a simulator and a synthesis system. But some constructs that are acceptable to a simulator will not be synthesizable because of the way they are written, not because of the commands they use."

What makes things worse, although VHDL is a hardware description language, its description range is not strictly limited to digital hardware system design. Firmware or software can be modeled as well. Especially at a high level of abstraction, system or algorithmic level, those latter aspects may occur more frequently and a synthesis tool does not have to implement those parts as a piece of hardware.

As an example of a synthesis subset, the VHDL subset supported by the CALLAS synthesis tool [KoGD90] is introduced briefly. In Table 2.1 an overview is given of the language constructs provided in VHDL, depending on the description domains. Constructs supported by CALLAS are marked with a (C). In some cases, additional semantics restrictions [Meye89] are imposed by the synthesis tool.

CALLAS synthesizes from algorithmic descriptions (Chapter 6), where structural information is used only to connect various behavioral parts.

In general, the tool starts with a VHDL process as a basic synthesis unit containing sequential VHDL statements and synthesizes the corresponding hardware equivalent. For an algorithmic description, logic operations such as **and** and **or**, relational operations such as $<$, $>$, and $=$, and arithmetic operations such as $+$, $-$, $*$, and $/$ need to be supported, but the use of operands and operations may be restricted. As with most of the high-level synthesis approaches, scalar types such as **integer** or enumeration types or predefined types such as **boolean** are provided in the CALLAS subset. For the specification or description of the control flow, the basic control constructs **if**, **case**, and loops, **while** and **for** loops and the **loop** and **exit** statement to model repeat loops, are included as well.

Specific features of algorithmic descriptions typically concern structural implications. VHDL attributes are used by the CALLAS system to control, for example, the handling of function or procedure calls.[5] An attribute specifies whether these have to be inline expanded or mapped to subprocessors (see Chapter 6).

The capture of timing in an algorithmic description is a crucial feature as well. VHDL allows for three different **wait** statements to suspend a process. Among these the **wait until <condition>** statement has been included in the subset to allow the specification of I/O protocols or waveforms in terms of system clock cycles.

A special synthesis package containing synthesis specific data types and operations is offered in CALLAS. A 4-valued logic ('X', '0', '1', 'Z') is provided within this package. A typical example of a CALLAS input description is shown in Figure 2.19.

Each process within the behavioral description is mapped to a corresponding hardware entity. Commercial synthesis systems generally support the transformation of a register-transfer level description to a hardware realization. Beside

[5] This is an example for a policy of use of the language by the CALLAS synthesis system.

the subset definition of each synthesis tool, a so-called "cookbook," is provided to inform the hardware designer about the synthesis strategy of the tool. Therefore, the results of each synthesis step depend heavily on the input description, i.e., how a VHDL model is written, even if the behavior described is the same. This is one of the problems to be solved by the next generation of synthesis tools.

Behavior	Structure	Data flow
entity declarations (C)	entity declarations	entity declarations (C)
architecture bodies (C)	architecture bodies	architecture bodies (C)
subprograms (C)	subprograms	subprograms (C)
ports (C)	ports	ports (C)
process statements (C)	block statements	block statements
variables (C)	signals	signals (C)
assignment (C)	components	assignment (C)
procedure call (C/not yet implemented)		
if, case, loop statements (C)		
next exit, return, null statements (C)		
package declaration	package declaration	package declaration
package bodies	package bodies	package bodies
type declarations (C)	type declarations	type declarations (C)
generics (C)	generics	generics (C)
wait statement (C)	config. declarations	disconnection specification
signals (C)	config. specification	register, bus signals
	generate statement	concurrent assignments (C), guards
aliases	aliases	aliases
attribute declaration	attribute declaration	attribute declaration
attribute specification	attribute specification	attribute specification
constant declaration (C)	constant declaration	constant declaration (C)
subtype declaration (C)	subtype declaration	subtype declaration (C)
concurrent assertions (C)	concurrent assertions	concurrent assertions (C)
concurrent proc. call	concurrent proc. call	concurrent proc. call
files	buses	
dynamic allocation		

Table 2.1: CALLAS VHDL synthesis subset.

```
entity callas_example is
 port (insig1 : in integer;
       insig2 : in bit;
       insig3 : in bit;
       inoutsig : inout bit;
       outsig : out integer
       );
end callas_example;
architecture behavior of callas_example is
signal local_signal : integer;
begin
 p1: process
      begin
       -- ...
      end process;
 p2: process
      begin
       -- ...
     end process;
-- ...
 pn: process
      begin
       -- ...
     end process;
end behavior;
```

Figure 2.19: CALLAS VHDL input description.

2.5 Outlook

To predict future trends in the field of HDLs and their relevance to synthesis, two aspects are to be highlighted: on the one hand, the future trends in research on HDLs, and on the other hand, the use of HDLs in hardware design in relationship to synthesis techniques and tools.

In research, the standard hardware description language VHDL is a major topic. From the user's point of view, a standard HDL is a benefit to overcome the Tower of Babel situation of HDLs. But the reader must realize that every HDL, including VHDL, is only a means and not a design methodology, which helps to overcome his/her design problems. This is very important because a designer has to focus on design methods. In the field of HDLs the next step to be expected is the support of object-oriented approaches [Marc87]. Also, a strong trend towards integration of different methods and tools can be identified which also will effect the development of HDLs [Newt91].

Today a hardware system is often understood as a digital hardware system and therefore existing HDLs mainly support this aspect. In reality systems contain analogue parts, mechanical parts, and software parts. These aspects are not yet well supported by existing techniques and tools. Therefore, HDLs and related methodologies and tools have to be enhanced to take these aspects into account. Possible approaches may address the issues of analogue and software parts within a system by embedding appropriate descriptions, e.g., software programs, within a digital hardware system description. A first step of integration is the solving of synthesis requirements for HDLs as pointed out in Section 2.4. Most HDLs have been developed for description and simulation purposes rather than for synthesis purposes.

The IEEE Standard HDL VHDL can be seen as a bridge between research work and the real world of hardware design. Therefore, there is a trend to design at higher levels of abstraction, rather than at register-transfer level or logic level, and VHDL acts as a means for this purpose.

2.6 Problems for the Reader

1. Give a short explanation (two sentences) of the term "HDL."
2. What are the different application fields of HDLs?
3. What is the relationship between HDLs and the design process?
4. Determine the relationship between general programming languages and HDLs.
5. What is a VHDL description? What are the main characteristics of the hardware description language VHDL?
6. Comment on the statement "HDL is only a means for the hardware designer, rather than a method."
7. What is the relationship between HDLs and synthesis?
8. What are the requirements of synthesis for HDLs?
9. What is the semantics of an HDL?
10. Explain the term "policy of use of an HDL."

Chapter 3

Logic-Level Synthesis

Michael Pilsl

3.1 Introduction

Logic-level synthesis (sometimes also called logic synthesis) closes the gap between high-level synthesis and conventional CAD tools for physical design. During the late 80s great progress was made in research and development, which has led to powerful commercial tools. These tools gain more and more importance in industry because they shorten the logic design process significantly and provide optimizations outperforming manual designs in many cases. As can be seen in this chapter, logic synthesis permits technology independent specifications of circuits. Thus it is quite easy to use another technology or manufacturer, without having to start from scratch. In this chapter some of the fundamental ideas and algorithms underlying logic-level synthesis tools are presented.

Technologies and design styles. As the designer faces a variety of technologies (CMOS, NMOS, ECL) and design styles (full-custom, semi-custom, module or layout generators, printed circuit board (PCB) design), one of the most useful advantages of logic synthesis is the abstraction of combinational logic to a technology independent specification. However, this implies that different synthesis strategies must be combined to take full advantage of the various technologies and design styles.

Optimization criteria. Before presenting synthesis strategies, the criteria for optimizations in logic synthesis tools are outlined. The main goal in logic-level synthesis is usually the minimization of area. However, to compete with human designers at least the same attention must be paid to the timing requirements of the circuit. The optimization task can be described as follows:

optimize a circuit in order to minimize its area, meeting given timing constraints at the same time. In addition, the performance of the synthesis tools and their flexibility with respect to different technologies and design styles must be considered. Therefore, synthesis algorithms are formulated as independent of specific technologies as possible, but as technology specific as needed. This way the same basic algorithms can be used for a variety of technologies and design styles. However, some inaccuracies in the estimation of area or timing of the circuit must be accepted. This chapter deals mainly with area optimization methods that can be formulated without considering technological aspects in detail.

Synthesis strategies. For the reasons mentioned above, strategies tailored to some specific target architectures and design styles have been developed. Mainly two different strategies have been established:

- Two-level logic minimization, targeting regular two-level implementations of logic such as programmable logic arrays (PLAs), or programmable logic devices (PLDs).
- Multi-level logic optimization, targeting cell based designs such as standard cell, gate array, or field programmable gate array (FPGA) designs.

In fact, two-level logic is a special case of multi-level logic; however, the restriction to two levels of logic leads to much more powerful optimization methods with respect to computation time and quality of results. Furthermore, two-level logic minimization is better understood than multi-level logic minimization and better heuristics to cope with the problem have already been developed. For the latter, relatively simple heuristic approaches are commonly used. Though results obtained so far are convincing and in many cases even better than manually optimized designs, in some cases results may be poor. This is especially true for regularly structured logic such as arithmetic units, e.g., adders, parity generators, or complete data paths. For this kind of logic other specialized synthesis tools (e.g., expert systems, layout synthesis tools) may be used.

Interfaces to high-level synthesis and library mapping. Logic synthesis tools start from a technology independent specification of logic blocks, generated either from high-level synthesis tools or provided directly by the user. Usually, the specifications are given in textual form using hardware description languages such as VHDL, or as truth tables, or logic equations. In addition, timing and area constraints can be specified. After the minimization step two-level logic can be directly implemented as a PLA module, or in a PLD device. In case of multi-level logic, library mapping procedures, such as described in Chapter 4, implement logic using elements of a cell library. Another way of implementing multi-level logic is the partitioning of multi-level logic into smaller fractions, which can be realized as individual complex cells using cell layout generation tools. In the latter case, certain technology-specific constraints, e.g., maximal number of transistors in series or in parallel, are used to guide the partitioning process. FPGA devices are available, which can realize relatively large multi-level logic modules. Therefore, FPGA devices are becoming more and more popular for the implementation of digital circuits.

3.2 Preliminaries and Definitions

In the following, binary Boolean algebra is introduced as the basic mathematical tool for dealing with combinational circuits.

Definition 3.1 (Binary Boolean algebra) *An algebraic system* $(B, *, +, ')$, *where* $*$ *(logical AND) and* $+$ *(logical OR) are binary operators and* $'$ *(logical complement or NOT operator) is an unary operator, is denoted as* <u>*binary Boolean algebra*</u> *if the following axioms are valid [Wake90]:*

Axiom 3.1 (Value domain)

$x = 0 \quad \textit{if} \quad x \neq 1 \qquad\qquad x = 1 \quad \textit{if} \quad x \neq 0$

Axiom 3.2 (Complement)

$\textit{If}\; x = 0, \quad \textit{then} \quad x' = 1 \qquad \textit{If}\; x = 1, \quad \textit{then} \quad x' = 0$

Axiom 3.3 (AND and OR operation)

$0 * 0 = 0 \qquad\qquad 1 + 1 = 1$

$1 * 1 = 1 \qquad\qquad 0 + 0 = 0$

$0 * 1 = 1 * 0 = 0 \qquad\qquad 1 + 0 = 0 + 1 = 1$

From these axioms the following set of theorems can be derived. Several of these theorems are used in this chapter to explain the logic-level synthesis task.

Theorem 3.1 (Constant)

$0 \in B \quad \textit{and} \quad 0 * x = 0 \qquad\qquad 1 \in B \quad \textit{and} \quad 1 + x = 1$

Theorem 3.2 (Associativity)

$(x * y) * z = x * (y * z) \qquad\qquad (x + y) + z = x + (y + z)$

Theorem 3.3 (Commutativity)

$x * y = y * x \qquad\qquad x + y = y + x$

Theorem 3.4 (Idempotency)

$x * x = x \qquad\qquad x + x = x$

Theorem 3.5 (Distributivity)

$x * (y + z) = x * y + x * z \qquad\qquad x + y * z = (x + y)(x + z)$

Theorem 3.6 (Identity)
$1 * x = x \qquad 0 + x = x$

Theorem 3.7 (Involution)
$x'' = x$

Theorem 3.8 (DeMorgan)
$(x * y)' = x' + y' \qquad (x + y)' = x' * y'$

Given a *variable*, a *literal* refers to the variable or its complement, e.g., the literals a and a' refer to the variable a. However, the literals a and a' are distinct, though they point to the same variable.

Definition 3.2 (Completely specified binary Boolean function)
A completely specified binary Boolean function f with n inputs and m outputs (multiple-output function) is defined by the mapping:

$$f : B^n \to B^m, \quad \textit{with} \quad B := \{0, 1\} \quad \textit{and} \quad n, m \in \mathbb{N}$$

In the special case of $m = 1$ the function $f : B^n \to B$ is said to be a *single-output function*. Obviously, each multiple-output function can be described by the set $(f_1, ..., f_k, ..., f_m)$ of single-output functions. Since this is not always the most suitable representation of multiple-output functions, some other representation methods are presented below. Further, no distinction is made between single-output and multiple-output functions in the following explanations, except for cases when this fact is relevant.

The set of all input combinations in B^n of completely specified functions can be split into two sets:
The set $F_k := f_k^{-1}(1) = \{x \in \{0,1\}^n | f_k(x) = 1\}$ is the *on-set* of the Boolean function $f_k, k = 1, ..., m$, containing all input combinations for which the output of the function f_k is defined to be $1 \in B$.
The set $F_k' := f_k^{-1}(0) = \{x \in \{0,1\}^n | f_k(x) = 0\}$ is the *off-set* of the Boolean function $f_k, k = 1, ..., m$, containing all input combinations for which the output of the function f_k is defined to be $0 \in B$.

Obviously, the following statements are valid for the on- and off-set of completely specified Boolean functions:

$$F_k \cup F_k' = B^n, \qquad k = 1, ..., m$$
$$F_k \cap F_k' = \emptyset, \qquad k = 1, ..., m$$

A Boolean function $(x_1, ..., x_n) \to f_k, k = 1, ..., m$ that maps only certain points $x \in \{0,1\}^n$ onto values $f_k(x) \in \{0,1\}$ but whose value is undefined for some other arguments is said to be an *incompletely specified Boolean function*.

The undefined values (*don't cares*) are indicated by a dash and can be interpreted as values corresponding to input combinations that will never appear in the foreseen application. During the minimization process these values may be arbitrarily added to the on-set or off-set in order to achieve the best optimization. More formally, an incompletely specified Boolean function is a mapping from the set $\{0,1\}^n$ onto the set $\{0,1,-\}^m$. Incompletely specified Boolean functions are described by their *on-* (F_k), *off-* (F_k'), and *don't-care* (D_k) *sets.* Given the sets $F_k, F_k', D_k \subseteq \{0,1\}^n$ with

$$F_k \cup F_k' \cup D_k = \{0,1\}^n \quad \text{and}$$

$$F_k \cap F_k' = \emptyset, F_k \cap D_k = \emptyset, F_k' \cap D_k = \emptyset,$$

the following functions are incompletely specified Boolean functions:

$$\begin{aligned} f_k &: F_k \cup F_k' \rightarrow \{0,1\}^m \\ f_k(x) &= 1 \quad \text{for} \quad x \in F_k \\ f_k(x) &= 0 \quad \text{for} \quad x \in F_k', \quad k = 1, ..., m \end{aligned}$$

where F_k is the on-set, F_k' is the off-set and D_k is the don't care set of the incompletely specified Boolean function f_k.

A Boolean function of n variables may be visualized as the set of vertices of n-dimensional cubes (n-cube). Each vertex represents one of 2^n possible input combinations of B^n and is marked (in a geometrical representation) as belonging to the on-set, the off-set, or don't care set. These vertices are identified by the logical *product* of n variables. For instance, in Figure 3.1, the 3-dimensional geometrical representation (3-space) of the Boolean function $f : (x_1, x_2, x_3) \rightarrow (f_1, f_2)$ with three inputs and two outputs, using two 3-cubes is shown. As can be seen, both output functions f_1 and f_2 are incompletely specified functions. In the geometrical representation the product $x_1 x_2' x_3$ identifies the vertex where $x_1 = 1, x_2 = 0, x_3 = 1$. This vertex, labeled with "2" in Figure 3.1 may also be viewed as the intersection of three orthogonal planes described by the identities $x_1 = 1, x_2 = 0$, and $x_3 = 1$.

Obviously, 2^n different product terms (*minterms*) with n variables exist, corresponding to the 2^n vertices of the n-cube. Thus an n-variable Boolean function can be described by the logical *sum* of those minterms that correspond to its on-set. This special *sum of product form* can be trivially derived from the geometrical representation and is denoted as *disjunctive canonical form.* A vertex of an n-cube may be viewed as an *0-subcube*. In general, a *k-subcube* of an n-cube is the set of 2^k vertices described by the specification of $n - k$

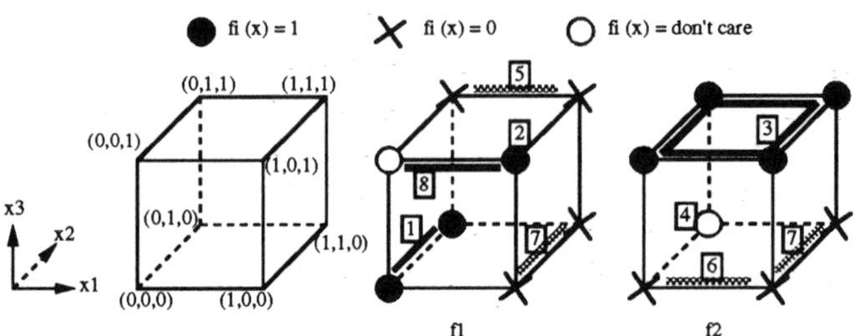

Figure 3.1: Geometrical representation.

variables. Again in 3-space for instance, the product $x_1'x_3'$ labeled with "1" in Figure 3.1 identifies all vertices of the 3-cube where $x_1 = 0$ and $x_3 = 0$ (000 and 010). This product term, written as vector (0-0) is said to *cover* the two vertices or in other words the two minterms $x_1'x_2x_3'$ and $x_1'x_2'x_3'$ are contained in $x_1'x_3'$. More formally, $c_i \supseteq c_j$ means that the cube c_i covers (*contains*) another cube c_j. Using subcubes, the on-set of a Boolean function can again be expressed as the logical sum of product terms. This form may be much more compact than the sum of minterms, where each term covers just one vertex. In fact, selecting a minimal set of (sub-) cubes to cover the on-set (and possibly a part of the don't care set) is what two-level logic optimization is all about.

A cube c is an *implicant* of a function f_k, if $c \subseteq F_k \cup D_k$, e.g., in Figure 3.1 the cube labeled with "8" is an implicant of the function f_1, containing the vector (101) from the on-set of f_1 and the vector (001) from the don't care set of f_1.

A cube or implicant c is *prime* if removing any literal from c causes $c \cap F_k' \neq \emptyset$, which means that c is not contained in any other implicant of f_k. Again, in Figure 3.1 the product $x_1x_2'x_3$, labeled with "2" is not prime, since it is contained in the implicant $x_2'x_3$, labeled with "8." However, it is obvious that this implicant $x_2'x_3$ is prime.

The set $C = \{c_1, c_2, .., c_k\}$ of cubes is said to be a *cover* of a Boolean function f, if the cubes contain all the vertices corresponding to the on-set and no vertices corresponding to the off-set of the Boolean function f. In other words, a cover represents the union of the on-set of f and possibly some points of the don't care-set of f.

A cover is a *prime cover* if all of its cubes are prime implicants, e.g., the two implicants $x_1'x_3'$ and $x_2'x_3$ form a prime cover of the function f_1, since they cover all vertices of the on-set of f_1 and both of them are prime.

An *essential prime implicant* of a cover C is a prime implicant that contains a minterm of C not contained in any other prime cube. An implicant is a *multiple-prime implicant* if it is a prime implicant for more than one output function f_i of the cover C.

3.2.1 Representation of Boolean Functions

Boolean functions may be represented in many ways. Some common representations are discussed below in more detail.

a) Truth tables:

As any function with a discrete and finite domain, a Boolean function $f = (f_1, ..., f_m)$ can be represented by a canonical table, containing a list of all input patterns of $x = (x_1, ..., x_n)$ and their corresponding output values $f_1(x_1, ..., x_n), ..., f_m(x_1, ..., x_n)$. If a value is not defined for a certain input argument (don't care) a dash ("–") is used to specify the output.
In Figure 3.2 the truth table for a decoder is shown, converting a BCD (Binary-Coded-Decimal) encoded digit into seven signals for a display unit. Only the binary vectors 0000 ... 1001 are used for representing a decimal digit (0 ... 9). Therefore, the outputs of the decoder are marked don't care for the input vectors 1010 ... 1111. However, truth tables are only useful for describing circuits with a small number of inputs (< 6), since the number of rows is 2^n for n input variables.

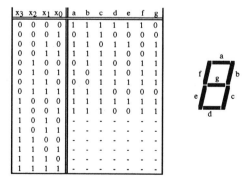

x_3	x_2	x_1	x_0	a	b	c	d	e	f	g
0	0	0	0	1	1	1	1	1	1	0
0	0	0	1	0	1	1	0	0	0	0
0	0	1	0	1	1	0	1	1	0	1
0	0	1	1	1	1	1	1	0	0	1
0	1	0	0	0	1	1	0	0	1	1
0	1	0	1	1	0	1	1	0	1	1
0	1	1	0	0	0	1	1	1	1	1
0	1	1	1	1	1	1	0	0	0	0
1	0	0	0	1	1	1	1	1	1	1
1	0	0	1	1	1	1	0	0	1	1
1	0	1	0	-	-	-	-	-	-	-
1	0	1	1	-	-	-	-	-	-	-
1	1	0	0	-	-	-	-	-	-	-
1	1	0	1	-	-	-	-	-	-	-
1	1	1	0	-	-	-	-	-	-	-
1	1	1	1	-	-	-	-	-	-	-

Figure 3.2: Truth table of a BCD to 7-segment decoder.

b) Sum of product forms:

As already mentioned, sum of product forms can be used to represent functions. However, sum of product forms can only express completely on-set F

	x1	x2	x3	f1	f2	
	0	-	0	1	0	1
3	-	-	1	0	1	
	1	0	1	1	0	2

off-set F'

	x1	x2	x3	f1	f2	
	-	0	0	0	1	6
7	1	-	0	1	1	
	-	1	1	1	0	5

□ Label of cube with respect to Figure 3.1

Figure 3.3: Matrix representation.

specified Boolean functions. For instance, the sum of product form $f = x_1x_2'x_3 + x_1'x_3'$ expresses the on-set of the Boolean function f_1 in Figure 3.1.

c) Matrix representations:

Extending the vectors denoting (sub-) cubes by one entry for each output of the function, a compact matrix-like notation for incompletely specified functions is obtained. For the example of Figure 3.1 the matrix representation of the on-set F and off-set F' of function f is shown in Figure 3.3. The first n columns (n inputs) of this matrix are called the *input part* and the last m columns (m outputs) are called the *output part* of the function. A "1" in the output part of, e.g., the on-set f_1, means that the associated cube is part of it and an "0" indicates that the cube is not part of the on-set of f_1. The matrix representation of Boolean functions is well suited for two-level logic minimization programs because it can be easily implemented as a data structure.

d) Factored forms:

Besides the usual sum of product representation, logic functions can be written as *factored forms.* This kind of representation is especially useful for multi-level logic, since it represents logic closer to its multi-level implementation. Often, the number of literals of functions in factored form is used as a measure for area during multi-level optimizations. Factored forms are defined recursively:

- a literal is a factored form
- a sum of factored forms is a factored form
- a product of factored forms is a factored form.

For example the sum of products $f = adf + aef + bdf + bef + cdf + cef + g$ may be represented as the factored form $f = (a + b + c)(d + e)f + g$. As can be seen, a factored form is a sum of products of sums of products of arbitrary depth.

e) Boolean networks:

A Boolean network [BRSW87] $G = (V, E)$ is a directed acyclic graph (DAG) such that for each vertex $v \in V$ an associated Boolean function f_i, and a Boolean variable y_i representing f_i exists. There is a directed edge $e \in E$ from $v_i \in V$ to $v_j \in V$ if f_j explicitly depends on y_i or y_i'. Furthermore, some of the variables in G may be classified as primary inputs or primary outputs. Boolean networks represent multi-level logic, where the functionality of each node may be expressed by one of the representations of logic functions discussed above. In Figure 3.4 a Boolean network is shown, in which each node is represented by a sum of products. In this case the variables $\{a, b, c, d, e, g, h, i, k, m\}$ are the primary inputs, $\{f_1, f_2\}$ are the primary outputs, and $\{z0, z3\}$ represent intermediate variables.

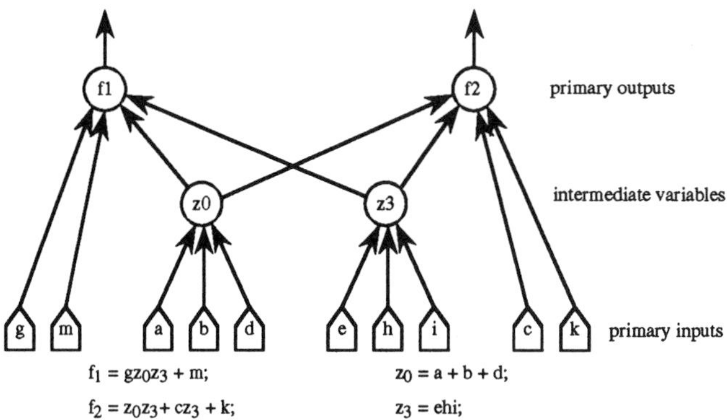

Figure 3.4: Example of a Boolean network.

f) Further representation methods for logic functions:

Another way of representing Boolean functions is known as *binary decision diagram* (BDD) introduced by Lee and Akers [Lee59, Aker78]. Like truth tables, BDDs provide a canonical representation of completely specified Boolean functions, but BDDs are based on graphs. Bryant [Brya86] has shown that most logic operations can be performed in a very efficient way using BDDs. A generalization of the BDD concept proposed by Karplus [Karp89] are *if-then-else-DAGs*. These methods will influence understanding and development of new minimization techniques. However, a detailed explanation of these interesting techniques is beyond the scope of this book. The interested reader is referred to the literature.

3.2.2 Manipulation of Boolean Functions

The *support* $sup(f)$ of a function f is the set of all variables used for the specification of the function f. The function $f = c_1 + c_2 + ... + c_n$, represented as a sum of cubes is said to be *algebraic minimal* or to be an *algebraic expression*, if $c_i \not\supseteq c_j$, for all i, j and $i \neq j$ (i.e., no cube c_i covers another cube c_j). Another way of saying this is that *function f is minimal with respect to single-cube containment.*

Let f and g be two arbitrary algebraic expressions of the form

$$f = c_1 + c_2 + ... + c_k, \qquad g = d_1 + d_2 + ... + d_n$$

then the product of f and g is given by

$$f * g = \sum_{i=1}^{k} \sum_{j=1}^{n} c_i * d_j$$

In general, the product $f * g$ is said to be a *Boolean product*. However, the product $f * g$ is named *algebraic product* if the functions f and g have disjoint supports ($sup(f) \cap sup(g) = \emptyset$). The result of an algebraic product is always algebraic minimal. Thus no further minimization step is required to transform the result into an algebraic expression.

Example:

$$f = (a + b) \qquad \Rightarrow sup(f) = \{a, b\}$$
$$g = (a' + c) \qquad \Rightarrow sup(g) = \{a, c\}$$
$$h = (c + d') \qquad \Rightarrow sup(h) = \{c, d\}$$

$sup(f) \cap sup(g) = \{a\}$ and $f * g = aa' + ac + a'b + bc$

Applying the axioms of Boolean algebra, the product bc can be written as $abc + a'bc$.
Since $abc + ac$ reduces to ac and $a'bc + a'b$ equals $a'b$:

$\Rightarrow f * g = ac + a'b + abc + a'bc = ac + a'b$
$\Rightarrow f * g$ is a Boolean product

$sup(f) \cap sup(h) = \emptyset$ and $f * h = ac + ad' + bc + bd'$

$\Rightarrow f * h$ is an algebraic product

The *quotient* q of two expressions f and g ($q = f/g$) is the largest sum (as many cubes as possible) of cubes q, so that

$$f = q * g + r$$

where r is the *remainder*. If the product $q * g$ is algebraic minimal, the division f/g provides the unique *algebraic quotient*. In the other case, the division f/g is one of several possible *Boolean quotients*.
The expression g is said to be an *algebraic divisor* of f if:

$$f = q * g + r, \quad q \neq \emptyset \quad \text{and} \quad sup(g) \cap sup(q) = \emptyset$$

The expression g is said to be an *algebraic factor* of f if:

$$f = q * g, \quad q \neq \emptyset \quad \text{and} \quad sup(g) \cap sup(q) = \emptyset$$

Example:

$$\begin{array}{lcllll} f & = & ab + ad + bc + cd & & & \\ g & = & ad + cd + b & & & \\ h & = & a + c & & & \\ f/h & = & b + d, \quad r = \emptyset; & \Rightarrow & h & \text{is an algebraic factor of } f \\ g/h & = & d, \quad r = b; & \Rightarrow & h & \text{is an algebraic divisor of } g \end{array}$$

Let us now consider the following two functions y and z:

$$\begin{array}{lcl} y & = & aehi + behi + dehi + m = ehi(a + b + d) + m \\ z & = & a + b + d \end{array}$$

As can be seen, the function z is completely contained within the function y. Therefore, the function z can be used to express the function y. This operation is termed *substitution*. More formally, the operation *substitute*(f, g, l) is the process of replacing a divisor g contained in a function f by a new literal l:

$$substitute(f, g, l): \qquad f :\rightarrow f_{new} = (f/g) * l + r$$

In the example *substitute*(y, z, x) evaluates to:

$$\begin{array}{rcl} y_{new} & = & (y/z) * x + r, \quad \text{where} \quad y/z = ehi, \quad \text{and remainder} \quad r = m \\ \Rightarrow y_{new} & = & ehix + m, \quad \text{with} \quad x = a + b + d \end{array}$$

The inverse operation of substitution is called *backsubstitution*. In that case, a literal, representing an intermediate variable of the function f, is expanded by the function associated with the intermediate variable.

A function f is is said to be *positive/negative* in a variable x_i, if x_i appears only in uncomplemented/complemented form in the sum of product form of f. If f is either positive or negative in x_i, f is said to be *unate* in x_i. If a function f is unate in each one of its variables, it is called *unate*. For instance, the function $f = ab' + cb'$ is unate because the variables a and c are used in their positive form and the variable b in its negative form. However, the function $g = ab' + cb$ is not unate, since the variable b is used both in its negative and positive form.

Given an algebraic representation of a logic function g, the *Shannon expansion* [Shan48] is defined as the decomposition of g into:

$$g = x_i g_{x_i} + x_i' g_{x_i'}$$

where g_{x_i} and $g_{x_i'}$ are called the *cofactors* of g with respect to the variable x_i. The cofactors g_{x_i} and $g_{x_i'}$ are obtained by:

$$g_{x_i} = g(x_i = 1), \quad \text{and} \quad g_{x_i'} = g(x_i = 0)$$

Assuming that the function $f = abc' + a'c + b$ is to be developed with respect to the variable a, the resulting equivalent representation of f is:

$$\begin{aligned} f &= a * f(a = 1) + a' * f(a = 0) \\ f &= a(bc' + b) + a'(c + b) \end{aligned}$$

Some heuristic minimization programs, e.g., PRESTO [Brow81] and ESPRESSO [BHMS84], use the *unate recursive paradigm* for function manipulation during the optimization process. This technique permits, for example, simplification of functions or efficient computation of the complement of a function. The paradigm recursively applies the Shannon expansion to a given function, until the cofactors are unate. The recursive Shannon expansion can be displayed in the form of a binary tree. The adjacent leaf nodes of the binary tree are then manipulated and merged again following rules which depend on the manipulation task (e.g., complementation or simplification).

In the example of Figure 3.5 the recursive Shannon expansion of the logic function $f = abcd + ab'c'd' + a'bc'd' + abc'd' + a'bcd' + b'cd + abc'd$ is shown. The variable for the recursive Shannon expansion (called *splitting variable*) is selected in a way to bring the two cofactors as close as possible to a unate function. A good strategy is to use the *most binate* variable, which is the variable appearing most often in its positive and negative form. In the example, variable b is selected as the first splitting variable. For each of the two cofactors obtained, again the most binate variable is selected for Shannon expansion. The recursion stops when a cofactor is unate.

3.3 Minimization of Two-Level Logic

3.3.1 Optimization Problem

The goal of two-level logic minimization is to find an area minimal implementation of a Boolean function in form of a PLA. It is sufficient to realize only the

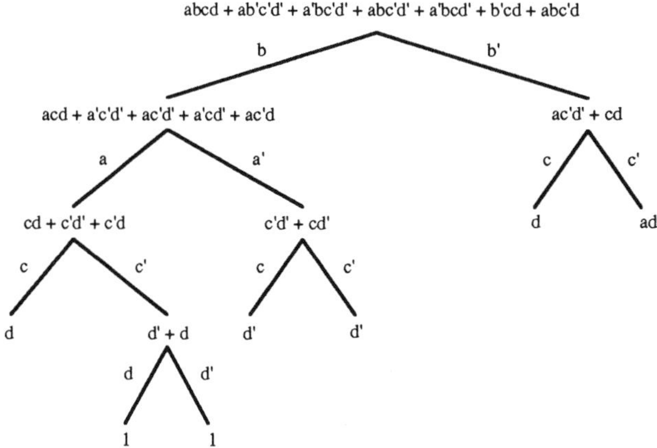

Figure 3.5: Example of recursive Shannon expansion.

cover of f (Section 3.2), where the don't care set D can be used for minimization purpose. The cover of a Boolean function can be represented by a sum of products form, which is directly related to its implementation as PLA. For instance, in Figure 3.6 a schematic layout of a typical PLA is presented. Each input is fed, complemented and uncomplemented, through the input-plane of the PLA. To form a product term, input lines are connected via transistors (horizontal lines). Again, in the output-plane, transistors combine the product terms to realize the sum of products for each output (vertical lines). In fact, the interconnected transistors of the PLA shown in Figure 3.6 realize the NOR function ($f_{NOR} = (a + b)'$) of their gate inputs. However, together with the inverting buffers at the inputs and outputs of the PLA, the sum of product form of the cover is implemented. Applying axioms of the Boolean algebra it is obvious that the sum $a + b = ((a + b)')'$ and the product term $ab = (a' + b')'$. Thus, a variable which appears uncomplemented in a product term correlates to a transistor connected with the appropriate complemented input. The realized covers for the functions f_1 and f_2 are obviously given by $f_1 = x_1'x_3' + x_2'x_3$

and $f_2 = x_1'x_2 + x_2'x_3$. In Figure 3.6 it is shown that the area of a PLA is dominated by the number of inputs and outputs (number of columns of the PLA) and the number of product terms (rows) of the PLA. Since the number of inputs and outputs is fixed, the goal of two-level minimization is to find a cover of the Boolean function f with minimal number of terms in its sum of product representation. If several minimal solutions exist, the sum of product form with the smallest number of literals is preferred in order to minimize the number of transistors.

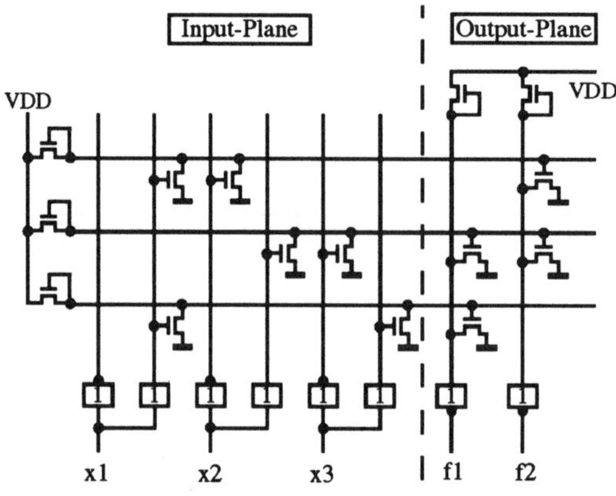

Figure 3.6: NOR-NOR schematic of a PLA.

3.3.2 Basics of Two-Level Logic Minimization

A cover C of the Boolean function f is said to be *irredundant*, if the set $C' = C \backslash \{c_i\}$ is no longer a cover of f, after removing the cube c_i from the cover C. In other words, if no *redundant* cube $c_i \in C$ exists that can be removed without changing the function f, f is irredundant. This does not imply that an irredundant cover C is a minimal representation of the function f. There may exist several irredundant covers for a function, which differ in their number of terms and literals. To obtain the minimal sum of product representation, the cover with the minimal number of terms and literals must be selected. This situation is depicted in Figure 3.7. Two prime and irredundant covers for the displayed function exist. However, only the cover C_2 is the minimal cover. Irredundancy of a function f is a stronger restriction than the single cube containment requested for algebraic expressions.

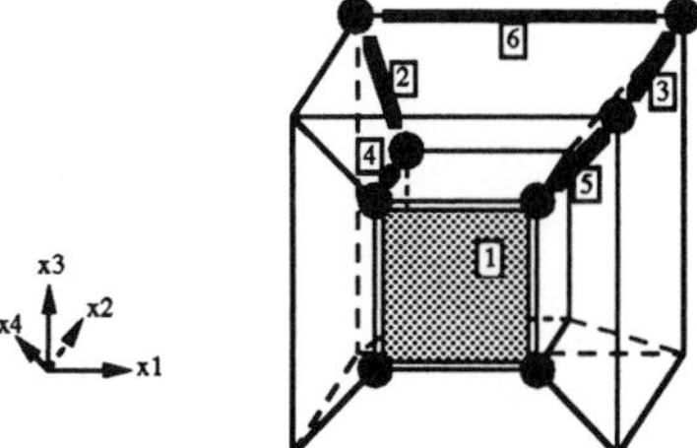

Two prime and irredundant covers exist:

$$C_1 = \text{“1”} + \text{“4”} + \text{“5”} + \text{“6”}$$
$$= x_2'x_4' + x_1'x_3x_4' + x_1x_2'x_3 + x_2x_3x_4$$

$$C_2 = \text{“1”} + \text{“2”} + \text{“3”}$$
$$= x_2'x_4' + x_1'x_2x_3 + x_1x_3x_4$$

Figure 3.7: Minimal cover versus irredundant cover.

All commonly used two-level logic optimization procedures rely on the following mechanisms:

1. Search for prime cubes or multiple prime cubes, considering the cubes of the don't care set and delete the smaller already covered cubes
2. Remove redundant cubes

An example will demonstrate these mechanisms. A Boolean function with three primary inputs and two primary outputs is given: $f : (x_1x_2x_3) \rightarrow (f_1f_2)$. The Boolean function can be visualized using three-dimensional cubes. Initially the Boolean function is given by the geometrical representation in Figure 3.8.

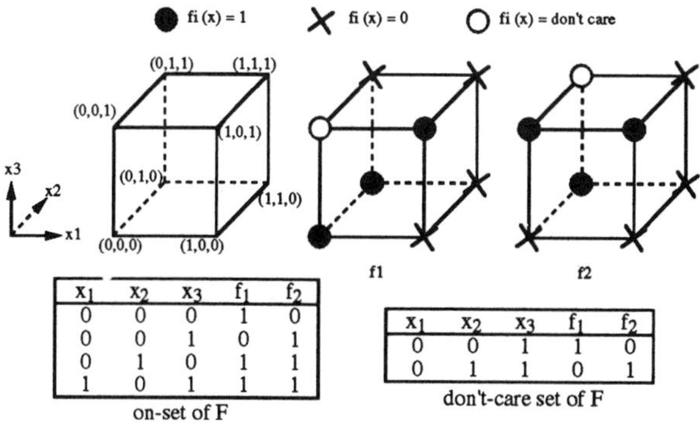

x_1	x_2	x_3	f_1	f_2
0	0	0	1	0
0	0	1	0	1
0	1	0	1	1
1	0	1	1	1

on-set of F

x_1	x_2	x_3	f_1	f_2
0	0	1	1	0
0	1	1	0	1

don't-care set of F

Figure 3.8: Initial Boolean function.

In sum of product form the two outputs are specified as follows:

$$f_1 = x_1'x_2'x_3' + x_1'x_2x_3' + x_1x_2'x_3; \qquad d_1 = x_1'x_2'x_3$$
$$f_2 = x_1x_2'x_3 + x_1'x_2'x_3 + x_1'x_2x_3'; \qquad d_2 = x_1'x_2x_3$$

After searching for prime cubes (without considering the don't care set) f reduces to (Figure 3.9):

$$f_1 = x_1'x_3' + x_1x_2'x_3; \qquad d_1 = x_1'x_2'x_3$$
$$f_2 = x_2'x_3 + x_1'x_2x_3'; \qquad d_2 = x_1'x_2x_3$$

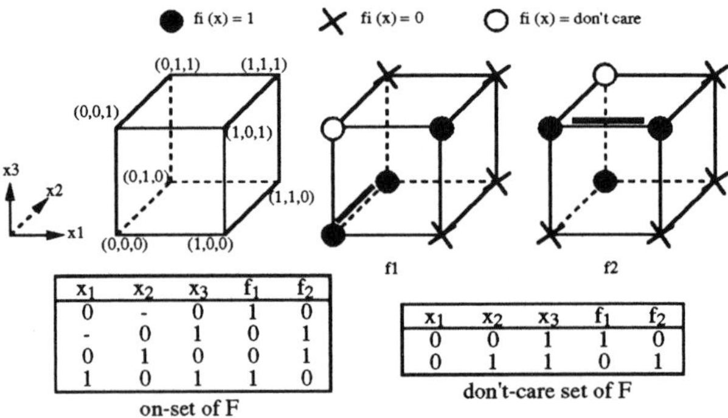

x1	x2	x3	f1	f2
0	-	0	1	0
-	0	1	0	1
0	1	0	0	1
1	0	1	1	0

on-set of F

x1	x2	x3	f1	f2
0	0	1	1	0
0	1	1	0	1

don't-care set of F

Figure 3.9: Prime cubes of the Boolean function.

Further minimization can be achieved using cubes of the don't care set to expand the prime cubes (Figure 3.10). This may result in a larger set of prime implicants. However, using cubes of the don't care set allows building of larger cubes covering more points of the Boolean space B. Furthermore, a richer set of prime implicants increases the probability of detecting multiple prime implicants.

$$\begin{aligned} f_1 &= x_1'x_3' + x_2'x_3 + x_1'x_2' \\ f_2 &= x_2'x_3 + x_1'x_2 + x_1'x_3 \end{aligned}$$

Looking at both outputs of the Boolean function, one multiple prime implicant can be detected (Figure 3.11). The detection of multiple prime implicants allows the usage of AND gates of the first stage for several OR gates of the second stage of the PLA.

In the sum-of-product form the multiple prime implicant is emphasized:

$$\begin{aligned} f_1 &= x_1'x_3' + \mathbf{x_2'x_3} + x_1'x_2' \\ f_2 &= \mathbf{x_2'x_3} + x_1'x_2 + x_1'x_3 \end{aligned}$$

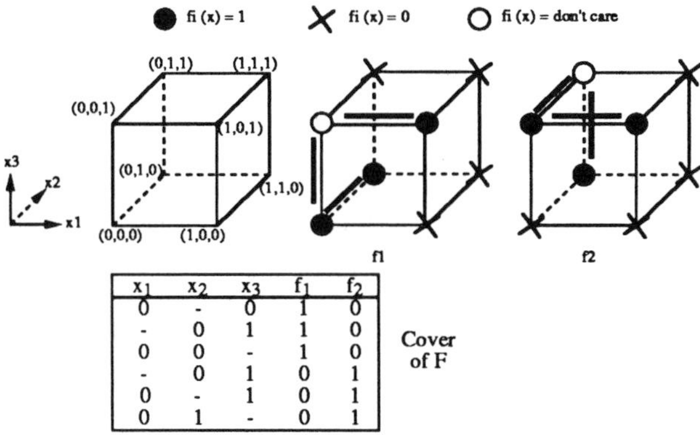

x_1	x_2	x_3	f_1	f_2
0	-	0	1	0
-	0	1	1	0
0	0	-	1	0
-	0	1	0	1
0	-	1	0	1
0	1	-	0	1

Figure 3.10: Prime implicants of the Boolean function.

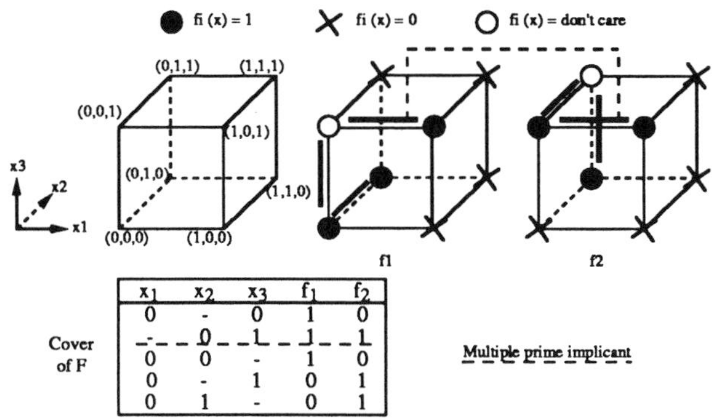

x_1	x_2	x_3	f_1	f_2
0	-	0	1	0
-	0	1	1	1
0	0	-	1	0
0	-	1	0	1
0	1	-	0	1

Figure 3.11: Multiple prime implicants of the Boolean function.

By this last step (Figure 3.12) the removal of redundant cubes is demonstrated. As can be seen, the given Boolean function is completely covered by three cubes only. Therefore, the two redundant cubes $x_1'x_2'$ $(00-)$ of f_1 and $x_1'x_3$ $(0-1)$ of f_2 can be removed from the cover of f. After removal of redundant cubes, the resulting cover of the function f is prime and irredundant.

$$\begin{aligned} f_1 &= x_1'x_3' + x_2'x_3 \\ f_2 &= x_2'x_3 + x_1'x_2 \end{aligned}$$

Finally, in Figure 3.13 the two-level realization of the minimized Boolean function is depicted. This result can be implemented by a PLA with a minimal number of AND and OR gates.

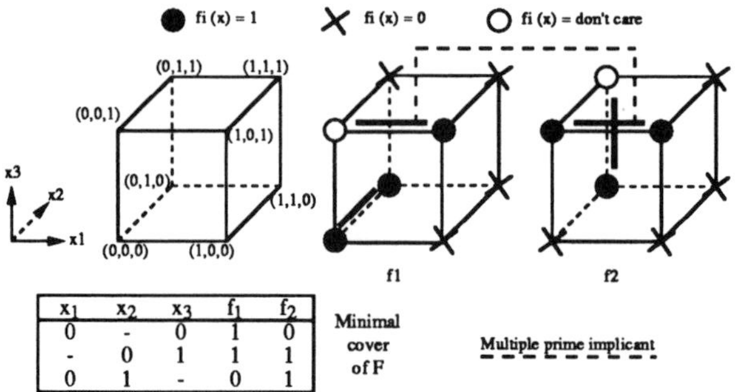

x_1	x_2	x_3	f_1	f_2
0	-	0	1	0
-	0	1	1	1
0	1	-	0	1

Figure 3.12: Prime and irredundant cover of the function.

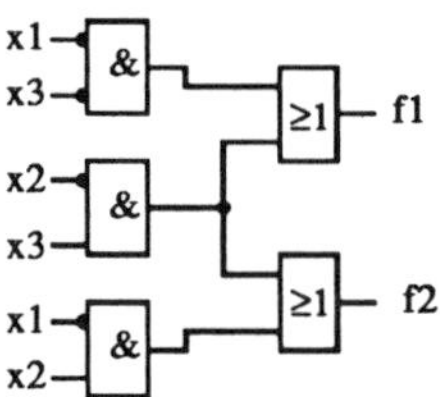

Figure 3.13: Two-level realization of the minimized Boolean function.

3.3.3 Exact Algorithm

For the problem of minimizing two-level logic, exact algorithms have been proposed by Quine and McCluskey [Quin55, McCl56]. Most of the earlier minimization algorithms are based on the Quine-McCluskey philosophy, requiring the following two steps:

1. Generation of all prime implicants
2. Extraction of a minimum prime cover

Unfortunately this method has several disadvantages. First, the method is restricted to completely specified Boolean functions, meaning that the truth table representation grows with 2^n for an n-input Boolean function. Further, it has been shown in [HoCO74] that the number of generated prime implicants may be as large as $3^n/n$. In addition, the second minimization step involves the solution of an NP-complete minimum covering problem. This restricts the Quine-McCluskey method to small Boolean functions with few outputs and up to 14–20 inputs. Larger problems require prohibitive amounts of CPU time and memory.

For these reasons efficient optimization techniques using various heuristic approaches have been introduced. Efficient heuristic two-level logic minimizers are available which are capable of minimizing large functions in reasonable CPU time.

3.3.4 Simple Minimization Heuristic

Based on the unate recursive paradigm a simple minimization heuristic performing fast single output minimization can be constructed. This simple heuristic is restricted to completely specified functions, i.e., the don't care set is empty or ignored. The technique is based on recursive Shannon expansion and can be stated as a recursive expression:

$$SIMPLIFY(f) = MERGE(x_i SIMPLIFY(f_{x_i}) + x_i' SIMPLIFY(f_{x_i'}))$$

The $MERGE$ operation merges two previously decomposed nodes into one. Two rules, which allow an optimization are checked. If none of the rules has fired, the merging is simply the reverse operation of the Shannon expansion.

First a variable g is introduced. This variable is initially empty and used to collect common or covered cubes, detected by one of the rules.

Rule 1 takes advantage of identical cubes in the two nodes:

R1: $c_k \in f_{x_i} \wedge c_k \in f_{x_i'} \quad \Rightarrow f_{x_i} := f_{x_i} \backslash \{c_k\}, f_{x_i'} := f_{x_i'} \backslash \{c_k\}, g := g \cup \{c_k\}$

Rule 2 uses the containment relation ($\supseteq$) between two cubes in the nodes to be merged:

R2: $c_k \in f_{x_i} \wedge c_j \in f_{x_i'} \wedge c_k \supseteq c_j \quad \Rightarrow f_{x_i'} := f_{x_i'} \backslash \{c_j\}, g := g \cup \{c_j\}$

Finally, the two nodes are merged and the result is returned:

return $(x_i f_{x_i} + x_i' f_{x_i'} + g)$

In Figure 3.14 the application of the MERGE operation on the decomposed function from the example in Figure 3.5 is presented. For each of the merged nodes the fired merging rule is indicated. This procedure leads to the minimized function:

$$f = abd + abc' + a'bd' + b'cd + ac'd'$$

The function f consists of only 5 product terms, thus 2 product terms have been saved. Though the results of this heuristic are not guaranteed to be prime

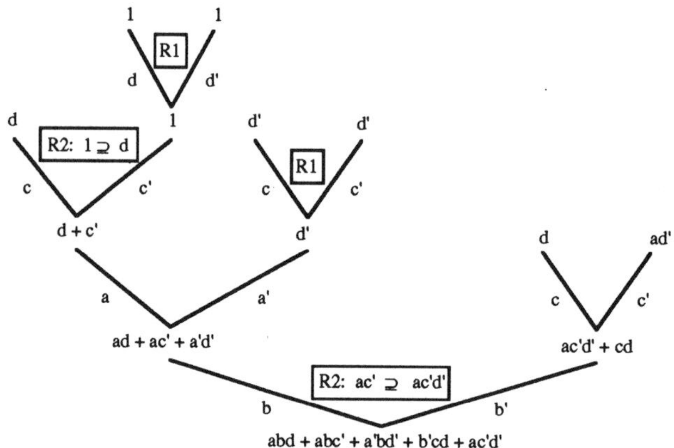

Figure 3.14: Merging of decomposed nodes.

and irredundant, in many cases the results are near optimum. In the example it can be seen that the result is prime but not irredundant. The product term abc' is redundant, because abc' is covered by the two terms abd and $ac'd'$.

3.3.5 Minimization Heuristic ESPRESSO

One of the first practical, heuristic minimization programs is MINI [HoCO74]. Later a program called PRESTO was introduced by D. Brown [Brow81] including ideas of Svoboda [SvWh79]. One of the most widely used heuristic two-level minimization programs is ESPRESSO [BHMS84]. ESPRESSO combines the best ideas found in MINI and PRESTO with some new heuristics. This tool performs two-level minimization executing the following steps:

1. *Complement*: Calculation of the complement considering the on-set and the don't care-set of the given incompletely specified Boolean function. The obtained off-set is needed in the following Expand step.

2. *Expand*: Expansion of implicants to prime implicants

3. *Essential Primes*: All essential prime implicants are detected and removed (but not deleted!) from the other implicants since they do not have to be considered during the following iteration loop.

4. *Irredundant Cover*: In this step an irredundant cover of the implicants is calculated. This cover is not guaranteed to be the absolute minimum because heuristics are used to solve the covering problem.

5. *Reduce*: Now the prime implicants are reduced. In principle this is the reverse process to the expand step (2), but another strategy is chosen to get a new set of implicants.
6. The steps 2, 4, and 5 are repeated until there is no further improvement.
7. *Lastgasp*: In this step one further reduce, expand, irredundant cover calculation is performed using a different strategy. If an improvement is obtained, the process continues with step 5, otherwise with step 8.
8. *Makesparse*: Now the essential prime implicants removed before are inserted again. Then the usage of the prime implicants is minimized. Finally, those implicants which are not used maximally are expanded one last time. This can lead to a further reduction of the number of implicants since other combinations of implicants may cover these implicants, which can then be removed.

The cost function of ESPRESSO which controls the iteration loops is a triple of three values: $Cost = (NPT, NLI, NLO)$, where NPT is the number of product terms of the cover, NLI is the number of literals in the input part of the cover and NLO represents the number of literals in the output part of the cover. In step 6 of the algorithm the costs of the current iteration are compared to the previous values. Only if none of the three values of the cost function have decreased, the iteration process is terminated.

3.4 Optimization of Multi-Level Logic

Multi-level logic implementations have become more and more important. This is due to the fact that this implementation style has several advantages compared to a two-level PLA implementation. There are many combinational logic blocks which are not well suited for two-level implementation. As an example, the minimum two-level implementations of specific Boolean functions may have up to 2^{n-1} product terms where n is the number of primary inputs of the function. This explosion of product terms is a serious problem especially for circuits that can be represented efficiently using EXOR operations like adders, parity generators or encryption/decryption systems. In addition, multi-level logic can often be implemented using less area and as a much faster circuit than an equivalent two-level (PLA) realization.

3.4.1 Synthesis Strategy

The starting point for multi-level logic optimization is a specification of the circuit by logic equations or truth tables. Usually, the synthesis process is split into two phases. In the first phase the overall structure of the circuit is manipulated to optimize it mainly in terms of area. This is the technology independent phase where the circuit is expressed as a Boolean network. The previously described algebraic and Boolean manipulation methods are used for these transformations. Since the representation of the design is technology independent, the constraints for the design can only be reflected in a very simplistic fashion, namely in the cost functions guiding the optimizations. The second phase of logic synthesis maps the design onto cells of a user specified target library (library- or technology-mapping), and performs technology-dependent optimizations taking the given constraints into account. In the following sections the technology independent phase of logic synthesis is presented. Methods for library mapping and technology-dependent optimizations are explained in Chapter 4.

3.4.2 Optimization Problem

The quality of multi-level circuits depends on many factors such as timing characteristics, chip area, power dissipation, driving capabilities and so on. Bounds for these parameters are derived from the environment of the logic module and from the specification of the circuit. The problem is how to synthesize circuits automatically, satisfying these correlated constraints. In practice, it is most important to meet the timing specification of the circuit. However, multi-level logic optimization is commonly more concerned with optimizing the area of circuits, than meeting given timing constraints. This is due to the fact that circuit characteristics such as area, timing, or power consumption can only be estimated because detailed low-level information, e.g., on placement and routing, is not available during the optimization phase.

3.4.3 Optimization Algorithms

Usually, multi-level logic circuits are represented by Boolean networks. These networks are then manipulated to optimize the network structure in order to meet given constraints. Evidently, active area can be minimized by detecting logic parts common to several nodes of the network which are then extracted and implemented only once.[1] Therefore, one major task of multi-level logic optimization is to detect these common logic parts. In a first approach, this process can be viewed as generating a set of suitable candidate expressions for each node of the network (often called kerneling), and then detecting those candidates which are common to several nodes. These expressions are then

[1] To estimate and minimize the area used for the connections between these logic parts is a much harder problem.

represented by a new node in the network and are substituted by a new literal within the nodes they are derived from. Consequently, area can be saved as these logic parts are implemented only once in the network.

3.4.3.1 Kernels and Cokernels

An algebraic expression is said to be *cube-free*, if there is no literal which is contained in every cube of the expression. Thus, an algebraic expression is not cube-free, if it consists only of one single cube.

Example:	$(ab + c)$	is cube-free, because $\{a, b\} \cap \{c\} = \emptyset$
	$(ab + ac)$	is not cube-free, because $\{a, b\} \cap \{a, c\} = a$
	(abc)	is not cube-free

The set of *primary divisors* D of an expression f consists of expressions of the form:

$$D(f) = \{f/c | c \text{ is a cube}\}$$

The set of the *kernels* K of an expression f is defined as:

$$K(f) = \{g | g \in D(f) \wedge g \text{ is cube-free}\}$$

The cube c used to obtain the kernel $k = f/c$ is called *cokernel* of f with respect to k.

Example [BeHS90]:

$$\begin{aligned} y &= adf + aef + bdf + bef + cdf + cef + g \\ y &= (a + b + c)(d + e)f + g \end{aligned}$$

A possible primary divisor of y is $y/a = df + ef$. However, it is not a kernel since it is not cube-free (the literal f is contained in both cubes df and ef). Dividing y by the cube af provides a kernel of y, because in this case the algebraic division $y/af = d+e$ results in a cube-free expression. The expression y itself is also a kernel, because y is cube-free and y is a primary divisor of y with respect to the trivial cube "1."

All kernels and cokernels of a given function can be computed by applying the definition. For efficiency, in most cases only the kernels or cokernels of *level 0* are computed. A level-0 kernel contains no other kernel except itself. In Table 3.1 it is shown that the kernels $(a + b + c)$ and $(d + e)$ are level-0 kernels, since they contain no other kernel. The expression $(a + b + c)(d + e) = ad + ae + bd + be + cd + ce$ contains two kernels of level 0 and is therefore said to be a kernel of level 1.

Kernels (y)	Cokernels(y)	Level
$(a+b+c)$	df, ef	0
$(d+e)$	af, bf, cf	0
$(a+b+c)(d+e)$	f	1
$(a+b+c)(d+e)f+g$	1	2

Table 3.1: Kernels and cokernels of function y.

This recursive classification of kernels gives a natural partitioning of the set of all kernels: $K^0(f) \subset K^1(f) \subset ... \subset K^n(f) \subset K(f)$, where $K^0(f)$ is the set of all kernels of f with level 0, $K^1(f)$ is the set of all kernels of f with level 1 or level 0 and so on. For the given example, these sets are:

$$K^0(f) = \{(a+b+c),(d+e)\}$$

$$K^1(f) = \{(a+b+c)(d+e)\} \cup K^0(f)$$

$$K(f) = K^2(f) = \{(a+b+c)(d+e)f+g\} \cup K^1(f)$$

A recursive algorithm [BRSW87] calculating all level-0 kernels of a function f can be stated as depicted in Algorithm 3.1.

```
Level_0_Kernels (f, j) {
    K = ∅;
    for (i = j; i ≤ n; i++) {
        if (literal l_i is contained in more than one cube) {
            c = largest cube dividing f/{l_i} evenly (without rest);
            if(l_k ∉ c for all k < i){
                K = K ∪ Level_0_Kernels (f/({l_i} ∪ c), i + 1);
            }
        }
    }
    if (K = ∅)    K = f;
    return (K); }
```

Algorithm 3.1: Level-0 kernel algorithm.

First, all the literals of the function f are numbered from 0 to n (e.g., in their lexicographical order) and f is made cube-free by dividing by its largest cube factor. Then the function Level_0_kernels$(f, 0)$ is called, where the parameter j is set to the first literal (literal referenced by 0). The function systematically builds up cokernels $(\{l_i\} \cup c)$ and calculates the related kernel $(f/(\{l_i\} \cup c))$. These kernels are examined recursively until no more subkernels are contained, which means that a level-0 kernel has been found. In order to avoid the generation of cokernels already calculated, cokernels are examined to find out whether they contain already processed literals ($l_k \notin c$ for all $k < i$). This results in a very efficient strategy for searching all level-0 kernels of the function f.

Example: Application of the kernel algorithm to a more complicated function [BRSW87] leads to the result shown in Figure 3.15. The leaf nodes of this tree represent the level-0 kernels.

$$\begin{aligned} y &= abeg' + abfg + abge' + aceg' + acfg + acge' + deg' + dfg + dge' + \\ &\quad bh + bi + ch + ci \\ &= (a(b+c)+d)(eg' + g(f+e')) + (b+c)(h+i) \end{aligned}$$

with the lexicographical order $\{a, b, c, d, e, f, g, h, i, e', g'\}$

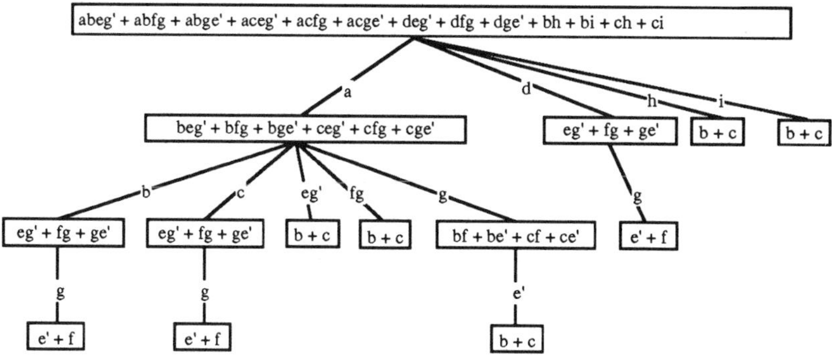

Figure 3.15: Recursion tree of the kernel algorithm.

3.4.3.2 Extraction of Common Subexpressions

Expressions common to more than one function can be grouped into two classes. First, there may be expressions consisting of multiple cubes, and second, *common subexpressions* may have only one single cube. Most algorithms for extraction of common subexpressions examine Boolean networks first for the existence of *multiple-cube expressions* and afterwards for existence of *single-cube expressions.*

Multiple-cube extraction. During this phase all nodes of the Boolean network are examined for some common subexpressions which are then substituted by a new variable and represented by a new node within the Boolean network. Two functions f and g have a *common multiple-cube divisor*, if a kernel $k_f \in K(f)$ and a kernel $k_g \in K(g)$ exist, such that $d = k_f \cap k_g$ has more than one cube. Therefore, the extraction of common subexpressions involves the following major steps:

1. Calculate the kernels of all functions
2. Calculate all intersections of kernels I(K)
3. Select the best candidates for substitution

Example: Consider the following equations:

$$\begin{aligned} f_0 &= aefhi + befhi + cefhi + defhi + k = efhi(a+b+c+d) + k \\ f_1 &= aeghi + beghi + deghi + m = eghi(a+b+d) + m \end{aligned}$$

First, all kernels of all functions are computed. The restriction to level-0 kernel leads to smaller sets of kernels, increasing the efficiency of finding intersecting kernels:

$$K^0(f_0) = \{(a+b+c+d)\}$$
$$K^0(f_1) = \{(a+b+d)\}$$

Next, the set $I(K)$ of all *intersecting kernels* is calculated. The set of all possible intersections of kernels represents the candidates which can be substituted. For the example, only one possible intersection is obtained:

$$I(K) = \{(a+b+d)\}$$

Now the gain for each of the candidates has to be estimated. One possible *cost function* is:

$$C(k) = (NF(k) - 1) * (L(k) - 1) - 1$$

where L(k) is the number of literals of the kernel k and NF(k) denotes how often the kernel k is contained in all functions of the Boolean network. It can be shown that this cost function gives the number of literals saved in the network by substituting the candidate if each node of the network is represented in its factored form. Only the best candidate is selected for substitution. However, to achieve an efficient algorithm usually several candidates are substituted at the same time.

In the example the cost function evaluates to "1", therefore, substituting this candidate truly saves one literal in the equation system (16 literals instead of 17). The resulting equations are:

$$\begin{aligned} f_0 &= efhiz_0 + cefhi + k = efhi(z_0 + c) + k \\ f_1 &= eghiz_0 + m \\ z_0 &= a + b + d \end{aligned}$$

The process of multiple cube extraction is repeated until no further common subexpressions are detected. In this example no common subexpressions remain after the first iteration. Now, all common subexpressions in nodes of the Boolean networks are single-cube expressions.

Single-cube extraction. For the single cube extraction process, first a new function G is computed which contains all product terms of the set of equations:

$$G = efhiz_0 + cefhi + k + eghiz_0 + m + a + b + d$$

Then the possible candidates for substitution are obtained by calculating the cokernels of the function G:

$$C^0(G) = \{efhi, ehiz_0\}$$

The best possible candidate is selected using the same cost function as for the multiple cube extraction. Substituting the two single cube candidates results in the following equations:

$$\begin{array}{ll} f_0 = z_0z_1 + cz_1 + k; & z_0 = a + b + d \\ f_1 = gz_2 + m; & z_1 = efhi \\ z_2 = ehiz_0 & \end{array}$$

The process of single cube extraction is also repeated until no further candidates are found. In this case, the next iteration provides the common single cube ehi (contained in z_1 and z_2), resulting for the whole extraction step in:

$$\begin{array}{ll} f_0 = z_0z_1 + cz_1 + k; & z_1 = fz_3 \\ f_1 = gz_2 + m; & z_2 = z_0z_3 \\ z_0 = a + b + d; & z_3 = ehi \end{array}$$

3.4.3.3 Elimination of Intermediate Variables

Due to the heuristic approach for extracting common subexpressions, some of the resulting intermediate expressions may no longer be profitable for the Boolean network (no saving of literals). For this reason the gain of all intermediate variables is recomputed and those variables that are no longer profitable are backsubstituted. The cost function provides the number of literals which are saved by introducing the intermediate variable:

$$value(y) = \{\sum_{g \in fanout(y)} N(g, y) - 1\} * \{L(y) - 1\} - 1$$

where L(y) is the number of literals of y and N(g, y) indicates how often y is used in the factored form g.

Considering the example the following gains are obtained:

$$\begin{array}{ll}
f_0 = z_0z_1 + cz_1 + k = z_1(z_0 + c) + k & \\
f_1 = gz_2 + m & \\
z_0 = a + b + d; & \Rightarrow value(z_0) = 1 \\
z_1 = fz_3; & \Rightarrow value(z_1) = -1 \\
z_2 = z_0z_3; & \Rightarrow value(z_2) = -1 \\
z_3 = ehi; & \Rightarrow value(z_3) = 1
\end{array}$$

In the example a cost value of -1 means that the intermediate variables z_1 and z_2 do not save literals. Hence, it is decided to backsubstitute them, resulting in:

$$\begin{aligned}
f_0 &= fz_0z_3 + cfz_3 + k \\
f_1 &= gz_0z_3 + l \\
z_0 &= a + b + d \\
z_3 &= ehi
\end{aligned}$$

3.4.3.4 Resubstitution of Intermediate Variables

During the extraction process simple heuristics are used for efficiency. Therefore, some common divisors may remain undetected. To find them, the *resubstitution* method checks whether an existing node itself is a divisor of other nodes. This is accomplished by checking whether the node i is an algebraic divisor of any other node $j \neq i$ (performing algebraic division) which means that the associated function of node i can be substituted into node j. Because algebraic techniques do not exploit all the Boolean properties of the network, one might also check whether a node is a Boolean factor of another node. Such a Boolean resubstitution method provides better results, but is much more expensive in terms of calculation times.

Combining the techniques discussed in this chapter leads to a simple but efficient synthesis method for multi-level logic consisting of the following steps [BRSW87, GBDH86, BeHS90]:

1. Starting point is a multi-level logic specification, e.g., a set of Boolean functions; each function is simplified to achieve an algebraic minimal expression. The simplification heuristic of Section 3.3.4 is sufficient for this task.

2. Extraction of multiple-cube expressions

3. Extraction of single-cube expressions

4. Elimination of intermediate variables

5. Resubstitution of intermediate variables

6. Repetition of steps 1–5 or 2–5 (faster) until there is no further improvement

7. Library mapping which is a technology-dependent step, shown in Chapter 4.

A special method where all intermediate variables are backsubstituted is called *collapsing*, with the effect that a multi-level logic is transformed into a two-level logic. This technique is useful when two-level minimization is to be applied to a multi-level logic specification. Afterwards, the optimization loop (steps 2–5) transforms the two-level logic back into a multi-level logic. Sometimes this strategy gives better results than minimizing each node of the initial multi-level logic.

3.4.3.5 Timing Optimization

Detailed information about the timing behavior is available only after technology mapping. At the technology independent level of logic optimization the timing behavior can be estimated only in a very rough way. A simple method is to use the fanout of nodes and the number of logic levels of the Boolean network (number of nodes between primary inputs and primary outputs) as a measure for the delay. During multi-level logic optimization, constraints for the fanout of nodes and for the number of logic levels can be considered in the cost function for the extraction and elimination process. Unfortunately, this method is not always sufficiently exact for practical use. Therefore, another approach, requiring a detailed timing analysis after technology mapping, is sometimes used. If some timing constraints are not met, resynthesis algorithms or local transformations [GBDH86] are applied to improve the timing.

3.5 Outlook

Optimization of sequential circuits. In general, circuits consist of sequential logic, containing combinational logic blocks separated by latches. Application of logic minimization techniques on each combinational logic block can not take advantage of logic dependencies across latch boundaries. Thus Malik et al. [MSBS91] propose to move latches to the boundaries of the circuit (called retiming), to apply logic minimization techniques to the new combinational logic partitions, and finally, to move the latches back (cf. Chapter 5). In another approach, a sequential circuit can be viewed as a set of interconnected finite state machines, which can be optimized using sequential don't care exploitation techniques as well as logic migration techniques, presented in [Deva88]. Finally, the relationship between state assignment of finite state machines and logic minimization across latch boundaries is discussed in [LiNe89a].

Wiring area. The multi-level logic optimization methods, described in this chapter, estimate the area of the circuit only very roughly, counting the number of literals. However, the number of literals reflects only the cell area of the final design, while the wiring area is not considered. Thus, in [ASSP90] an ordering of variables and output functions during the optimization process is proposed. The cost function, weighting the candidate subexpressions for extraction, is extended to consider also how much this ordering is respected by the candidates. It is shown in [ASSP90] that this strategy results in a smaller overall area, though the number of literals and cells is not primarily minimized. This is due to the fact that, as feature-sizes becomes smaller in current and future submicron technologies, while the number of gates per chip goes up, routing area increasingly dominates active area.

Logic synthesis and testing. Logic synthesis and testing are strongly related. Basically, each non testable fault of a combinational block indicates redundancy in the logic. Removing these redundancies will avoid non testable faults. Exactly this is a task of logic optimization tools. The theory covering this topic and its application to multi-level logic minimization can be found in [Bart88] and [SWBS88]. More details concerning the relationship between testing and logic synthesis are also given in Chapter 10.

3.6 Problems for the Reader

1. Given the functions y_1, y_2, y_3, determine whether the functions are cube-free.

$$y_1 = abc + cd + f$$

$$y_2 = a'bc + ab'c' + d$$

$$y_3 = a'bc + a'cd + ef$$

2. Calculate all level-0 kernels of the function y by systematically building up the cokernels as formulated in the kernel algorithm of Figure 3.1.

$$y = a'e + ac + ad + bc + bd$$

3. Given the following function in form of a truth table:

a	b	c	y
0	0	0	0
0	0	1	1
0	1	0	0
0	1	1	1
1	0	0	1
1	0	1	1
1	1	0	0
1	1	1	1

   a) Is the function a completely specified or an incompletely specified Boolean function?

   b) Transform the truth table representation of the function into a sum of products form.

   c) Apply the simplification heuristic from Section 3.3.4 to the function resulting from step b.

   d) Show that the simplification heuristic provides a prime and irredundant result for the given function. Use the geometrical representation of the result to prove primeness and irredundancy.

4. The *factorization* of Boolean functions can be performed with the same method as described for the extraction of common (multiple-cube) expressions. The only difference is that the methods are applied to one single function only, and thus the kernel intersection step is skipped. A quite simple factorization method can be stated as:

(a) Calculation of all level-0 kernels of the given function

(b) Selection of the most profitable kernel is quite simple. Select the kernel with most literals.

(c) Substitution of the selected kernel, by introducing a new intermediate variable

(d) Repeating steps a)–c) until the function contains no further kernels

(e) Rewriting of the obtained function by including the expressions represented by the intermediate variables in brackets.

Apply the above described scheme for the factorization of the following function:

$$y = adf + aef + bdf + bef + cdf + cef + g$$

Chapter 4

Technology Mapping

Anton Langmaier

4.1 Introduction

In its conventional meaning *technology mapping* refers to a design step during logic-level synthesis [Detj87].

Definition 4.1 (Technology mapping) *Given a technology independent multi-level logic structure, a cell library, in which each cell implements a Boolean function and is characterized by technology data (e.g., timing information), and a set of design constraints (e.g., timing constraints, fan-out constraints), technology mapping is the process of transforming the multi-level logic structure into a netlist of library cells which represents the given multi-level logic structure and fulfills the design constraints.*

Library mapping, *technology binding* and *module binding* are synonyms often used to describe the same process.

During the mapping process among many feasible solutions, one that minimizes an objective function (area, path-delay) is sought. In general, optimum technology mapping is an NP-complete problem, which can be solved approximately using heuristic methods (cf. Section 4.3).

According to the definition above, technology mapping selects one of all possible implementations of the technology independent structural description using a given target technology. Thus, technology data is added to the description.

Therefore, a generalized definition of technology mapping can be formulated:

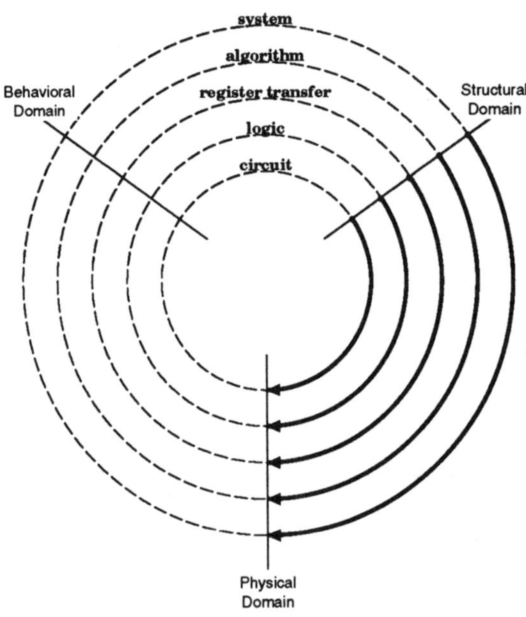

Figure 4.1: Technology mapping at several levels of abstraction.

Definition 4.2 (Technology mapping) *Given a technology independent structural description, a target technology, and a set of design constraints, technology mapping is the process of implementing the structural description in the physical domain at the same level of abstraction, where all design constraints are fulfilled. (Figure 4.1).*

During the transition from the structural domain to the physical domain (generation) no structural information is lost; only technology data is added. Therefore, the structural design representation can easily be extracted from the physical representation.

Depending on the level of abstraction and the target technology, technology mapping can be classified as follows:

Usually, for the *algorithmic level*, technology mapping is not applied in IC design. At this level, tools for developing microprograms which map a technology independent algorithmic structure to a specific architecture can be imagined, for which the target technology is a microprogrammed processor.

For the *RT-level*, technology mapping is a widely used approach in high-level synthesis. Target technologies can be macro cells, i.e., each functional block is realized by one macro cell or a netlist of cells from a standard-cell or a gate-array library.

For the *logic level*, standard-cells or gate-array cells can be considered to be the predominant target technology, to which the conventional technology mapping from Definition 4.1 can be applied. Other targets for mapping a design at the logic level are programmable devices (PLA, FPGA, etc.). In these cases, the existing circuit structure often has to be partitioned into subcircuits.

Usually, RT-level and logic-level technology mapping are applied in available synthesis systems. They are described in more detail in Sections 4.3 and 4.4.

4.2 Abstraction of Technology

Neglecting technology data in early design steps is a well-known strategy both for manual and automated design. In this section the reasons for and the difficulties resulting from this restriction are discussed. The design stage most suitable for fixing the target technology and performing the technology mapping is discussed, too.

4.2.1 Advantages and Disadvantages of Technology Abstraction

The following advantages are evident:

- A good technology independent design facilitates the choice of the best target technology.
- The application of technology independent design steps decreases the difficulties of switching from one technology to another during the design process.
- Tools for automated design that are implemented in a technology independent way are more flexible.

Since neglecting details implies ignoring aspects which could be of importance later on, there are also some disadvantages of technology abstraction:

- It is not possible to use optimization criteria that rely on technology data.
- It is not possible to verify a complete circuit description without referring to technological data.

These disadvantages may be circumvented without generating the complete physical design data. One method is the estimation of technology dependent data during circuit optimization or verification, while using a technology independent circuit structure. This approach is called *library-look-ahead* [Duzy89]. It leads to a technology independent description of the circuit structure which is optimized according to a specific technology. In addition, this method includes possibilities to change the target technology by

- mapping the circuit structure, which has been optimized according to the old technology, to the new technology without any changes. This procedure minimizes the design effort, but the synthesis results are not optimal.
- performing those optimization steps that estimate technology data again before mapping to the new technology. This leads to better synthesis results, but requires more design effort. The use of modern design tools, however, adds a high degree of practicality to this approach.

4.2.2 When to Select the Target Technology

Whereas in many cases the target technology is fixed, in other cases the target technology is a design parameter, which can be selected from a set of supported technologies during the design process.

A strong base for a decision on the right technology is ensured if the target technology is fixed after some technology independent design steps have been performed. The most useful target technology can be chosen if the designer experiments with various technologies. Each technology independent design step before branching into variants will reduce the design costs. Another reason for a late selection of a target technology is the ease of later redesign.

An early choice of technology, on the other hand, might lead to better results, since technology data can be taken into account in all subsequent design steps, either because technology mapping is done earlier (cf. Section 4.2.3) or because library-look-ahead strategies are used.

To conclude, the choice of the target technology should be done as late as possible, but prior to any transformation, that depends on technology data.

4.2.3 When to Perform Technology Mapping

Another important question is when technology mapping should be executed. Similar arguments as in Section 4.2.2 can be presented:

Late technology mapping minimizes the effort of redesign and allows for a simpler formulation of all earlier design steps. Early mapping, on the other hand, allows for direct access to technology data in the following optimization steps. As technology data can also be estimated using library-look-ahead strategies, technology mapping is often performed as the last synthesis step before continuing the IC design using a conventional back-end CAD system. However, sometimes the IC design process starts with a structural specification and technology mapping as its first step.

These two possibilities lead to two different design philosophies:

- Technology mapping as the last design step in synthesis incorporates the philosophy that synthesis tools have to be adaptable to any given back-end CAD system, which often have a specific and technology dependent entry format. The main advantage of this approach is the universality, but it asks for high implementation and adaptation efforts.

- Technology mapping as the first design step of the back-end CAD system supports the idea of a standard input format which allows for technology independent circuit specification. This philosophy is connected to the "Meet in the Middle Strategy" [DeMa87]. The main advantage here is that the standard interface is independent from technological details and the back-end CAD system. In practice, however, many CAD systems do not support a standard input format, which fulfills these requirements. Thus, this approach is restricted to a small class of back-end CAD systems. In reality, therefore, the adaptation effort is shifted from the synthesis tool to the back-end CAD system, which is required to support the standard output format of the synthesis tool.

4.3 Logic-Level Technology Mapping

The logic-level synthesis algorithms discussed in Chapter 3 construct an optimized technology independent structural description. Such a description can be given as a two-level representation or as a multi-level representation. A two-level representation can be used as input for a PLA (PLD) generator or has to be partitioned into several fractions which can be implemented as PLAs (PLDs). These techniques are not covered in this book. The interested reader is referred to the literature [ErDe91, FrRC90, FrRV91, GrRo91, Karp91a, Karp91b].

In this section, methods are presented that map a technology independent multi-level logic structure to elements of a specific semi-custom cell library. More precisely, one solution for the following problem is explained:

Given a description of a circuit in a directed acyclic graph representation, or, in other words, as a Boolean network N (cf. Section 3.2.1 e) and a representation of each cell c_i of the library as a Boolean function f_i[1] and some technology

[1] In general, the Boolean function representing a library cell can have multiple outputs

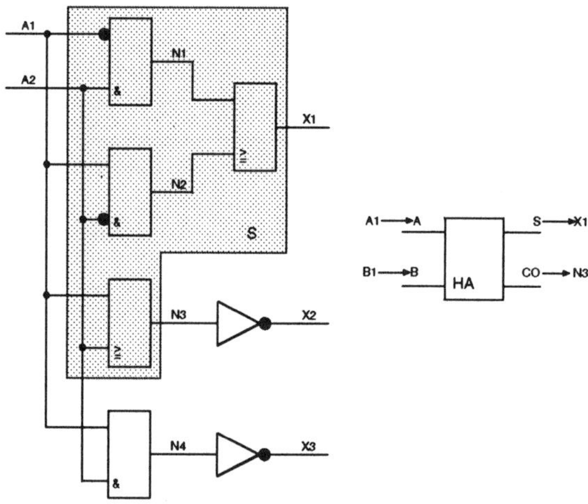

Figure 4.2: Subgraph S matches the library cell HA.

data t_i (e.g., cell area), find the optimum implementation with respect to some cost function in the physical domain for this circuit.

Definition 4.3 (Network matching) *A connected subgraph S of the given Boolean network N matches a library cell c_i, if there is an isomorphism between the input variables of the subgraph and the input variables of the library cell c_i and an isomorphism between the output variables of the subgraph and the output variables of the library cell c_i, such that the Boolean function g of the subgraph S and the function f_i of the library cell c_i are identical after substituting the variables (Figure 4.2).*

Definition 4.4 (Network covering) *A Boolean network N is covered by a given library if there is a partitioning of the given Boolean network N for which each partition is matched by one library cell (Figure 4.3).*

The problem of finding an optimum covering (e.g., optimum with regard to the overall cell area) of a Boolean network with a given set of Boolean functions is very similar to the problem of optimum code generation for arithmetic expressions, which has been extensively studied in the context of software compiler techniques [AhSU86]. It has been shown that this problem is in general NP complete. As a consequence, heuristic procedures searching for a "reasonably good" solution have to be applied. Common approaches include local

(e.g., a fulladder cell). In practice, however, only single value Boolean functions can be handled by most technology mappers.

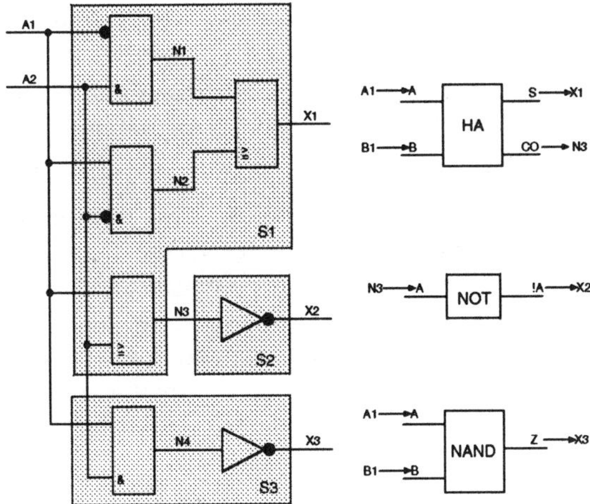

Figure 4.3: Subgraph S is covered by a library containing HA, INV and NAND.

methods, randomized methods, and tree covering methods which are explained in the following sections.

4.3.1 Local Methods

Local methods are intended to find a solution that is optimum only in a small environment of some edges of the Boolean network. This is a typical application of rule-based expert systems, using local transformations. The first example of a tool following this strategy is described in [GBDH86].

4.3.2 Randomized Methods

For randomized methods, a parameterized set of local methods has to be implemented. The mapping procedure is performed via multiple mappings, each one using randomly selected parameters. Randomization alleviates the negative effects of algorithmic dependencies on processing sequence and initial circuit configuration. The number of mappings can be defined by the user. Thus, the designer is able to trade CPU runtime for better results. A system using randomized methods is described in [LiBK88].

4.3.3 Tree Covering Methods

Tree covering methods have initially been developed for code generation. They also comprise the most commonly used methods for library mapping [Keut87,

Detj87]. The main idea is to partition a given Boolean network into a forest of trees and to cover each tree separately. On a tree, the covering problem can be solved in linear time with regard to both the size of the given Boolean network and the size of the target library. The key to this approach is the use of dynamic programming.

Partitioning. The simplest way to partition a Boolean network into a forest of trees is to take each node with a fan-out larger than 1 as the root of a new tree. After partitioning, the mapping is done separately for each subtree. By applying this method, any Boolean network can be broken into a forest of disjoint trees. The optimality of the mapping, however, is restricted to the scope of the tree. No optimization across tree boundaries can be achieved. Therefore, the overall mapping is only suboptimal on the whole network.

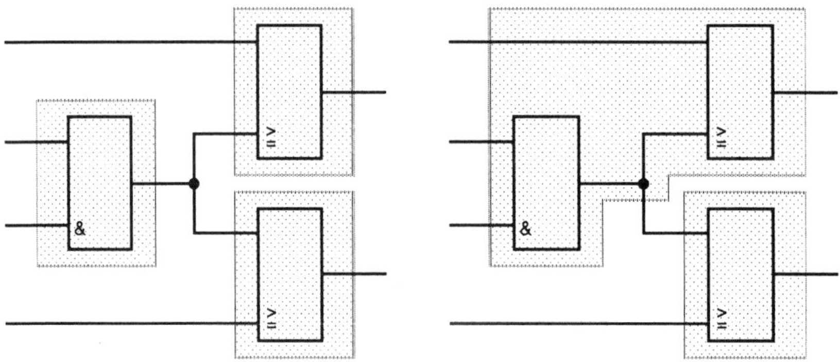

Figure 4.4: Breaking a Boolean network into a forest of trees.

Slightly better optimization results can be achieved if a less restrictive strategy is used to break the network into a forest of trees. Instead of assigning each multiple fan-out node to be the root of a tree, only primary outputs of the network are used as roots. Partitioning boundaries are inserted at all but one edge connected to multiple fan-out nodes. In this way, technology mapping can be applied across multiple fan-out nodes. The disadvantage of this approach is that redundancies are generated if a multiple fan-out node becomes an internal node of a mapped subgraph (see Figure 4.5). It can be used, however, to reduce the number of logic stages on critical paths [Detj87].

Tree covering. Canonical tree representations are commonly used both for the Boolean tree and the library cells, when covering a Boolean tree with a library of logic gates. Thus, testing for equivalence of Boolean functions can be performed by pattern matching methods. Moreover, the use of canonical representations consisting only of small gates (e.g., 2-input NAND gates and

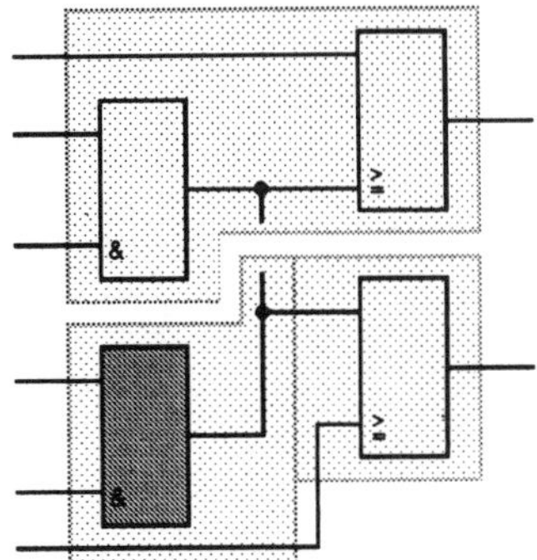

Figure 4.5: Inserting additional logic during tree construction to improve the timing.

inverters) has the advantage that the mapping is not restricted to the boundaries of the technology independent gates. Since such representations are not unique, all non-isomorphic and irredundant patterns have to be used for each library cell. At a first glance, this may produce too many patterns, but for important cases (e.g., 2-input NAND gates and inverters) a detailed analysis shows that even for fairly large libraries the pattern set is manageable in size [Detj87].

The dynamic programming algorithm can be described as a procedure with three phases:

Matching phase:
For each node of the tree a list of all possible matchings at this node is generated.

Optimization phase:
This is a recursive bottom-up procedure, which determines the optimum matching of each subtree. The recursion stops at the leaves (the primary inputs) of the tree, which have a trivial optimum matching with zero costs.
In the nontrivial recursion case, each possible matching (determined by the matching phase) is combined with the optimum matchings at all input nodes of the matched cell (determined by a former incarnation of the optimization phase) and the overall costs of the matching cell and the corresponding subtree costs are computed. Finally, the optimum matching at the root node is determined as the matching with minimum overall cost.

Accumulation phase (mapping phase):
This recursive top-down procedure accumulates the optimization results from the optimization phase to produce a cover of the whole tree. After the optimum matching of the root node is associated with the corresponding cell, the mapping procedure is recursively applied to all subtrees at all input nodes of the mapped cell. The optimum matchings of all internal nodes of a mapped subgraph, which are also computed in the previous phase, are not used by the accumulation phase, but they are necessary to complete the optimization phase.

Different canonical tree representations lead to different variants of tree covering algorithms. Particularly, representations for which only a small number of patterns per library cell exist are promising. The application of the conventional tree covering methods, as described above, to these representations, however, decreases the optimization potential. Thus, improved algorithms that split tree nodes during the mapping procedure (mapping "between cell boundaries") and that look faster and more directly for an isomorphism between parts of the initial tree and the standard cells of the library have to be applied [Berg88].
Tree covering methods are—as a solution for the technology mapping problem—suboptimal, because only the problem of covering subtrees of the overall graph is solved exactly. The combination of optimum covered subtrees, however, does not guarantee a global optimum. Moreover, there is no way of mapping to standard cells that have no tree representations (e.g., XOR, MUX, half adder). For these reasons, extensions of pure tree covering methods have been proposed. One possibility is the use of local methods to preselect standard cells without tree representation, before applying tree covering methods. Another approach is the use of more general tree representations (e.g., BDDs [Brya86]) which can help to increase the number of cells that can be represented. XOR and MUX gates can then be handled by tree covering methods.

4.4 Register-Transfer Level Technology Mapping

High-level synthesis usually ends with a structural description on a higher level of abstraction than the logic level, for example a structural RT-level description. This higher abstraction level allows hierarchical design representation, signal bundles and parameterized functional modules such as arithmetic components, decoders, registers and bus drivers.

Invoking technology mapping at this design stage allows for taking into account knowledge about these functional units. Thus, it helps to avoid exhaustive methods for regenerating this information at a later design stage, e.g., logic-level synthesis. It is easier, for instance, to generate an adder structure using some back-end technology than to detect adders in Boolean equations. On the other hand, a complete transition from the structural to the physical domain at this design state complicates the application of technology independent logic-level synthesis tools. A common compromise is to map the functional modules to an RT-level back-end library consisting of complex library gates and some technology independent logic gates. Then, logic-level synthesis and logic-level technology mapping can be applied to the technology independent combinational subcircuits.

The most common RT-level technology mapping methods are module generators and rule-based methods.

4.4.1 Module Generators

One possibility for RT-level technology mapping is the use of macrocell layout generators from the back-end CAD system. In this case, each functional block is realized by one corresponding macrocell. The only task remaining for the technology mapper is to adapt some interfaces:

- The parameter space covered by the generator tool(s) is usually not identical with the parameter space of the functional modules. Therefore, for each parameter tuple of a functional block a corresponding parameter tuple has to be found which can be used by the generator tool.
  For example, a functional module

  $$arith(function = addition, width = 16, Carry\_in = T, Carry\_out = T)$$

  can be realized by an adder/subtractor generator with the parameters:

  $$version = +, dimension = 16, 2\_complement = ON$$

- The ports of the functional block have to be bound to the corresponding ports of the generated macrocell.

- A one-to-one realization using one macrocell is not always possible using the available generator tools. In some cases, however, after introduction of additional logic gates these generators can be used with changed parameters.

  For instance, the introduction of some inverters allows for the realization of a functional module

  $$arith(function = subtraction, 2\_complement = T, width = 16)$$

using an adder generator.
Rule-based methods such as those described in the next section can be used for this purpose.

4.4.2 Rule-Based Methods

If a functional block can not be realized using a module generator, or if a strict semi-custom design style is required, a semi-custom library has to be selected as the target technology. Rule-based methods are usually applied to map RT-level structures to elements of semi-custom libraries.

A rule-based approach is a programming style which is characterized by three elements: first, the *data*, on which several manipulations can be performed and which usually are different for each application; second, the *knowledge base*, which consists of several rules given by a condition part and an action part. The condition part specifies all conditions that have to be satisfied to execute the action part, a set of legal data manipulations. Third, the *rule interpreter*, which is the kernel of a rule-based system, performs the task of finding and invoking that sequence of applicable rules that transforms the given data into a final state in an optimal way with regard to a given optimization criterion and several optimization constraints.

In a rule-based system used for technology mapping of a RT-level structure to elements of a semi-custom library data, knowledge base and rule interpreter can be understood as follows:

Data. Initially, the RT-level structure is represented as a network of functional blocks. During the application of rules this representation becomes a mixed network consisting both of functional blocks and library cells. Usually, it can be hierarchical, which means that each node of the graph can be either a leaf node (a functional block or a library cell) or a hierarchical network itself. The back-end library is a set of cells $\{c_i\}$, for which each cell is described by a functional specification f_i and some technology data t_i. The functional specification f_i can be given either as a Boolean function or as a cell classification, which consists of a cell class and several parameters. For example, a 4-bit adder can be specified by the cell classification:

$$arith(function = addition, width = 4, Carry\_in = T, Carry\_out = T)$$

Knowledge base. The condition parts of the rules are a predicate logic expression composed from the following primitive conditions:

a) **Module condition.** This condition checks the membership of a network node against a set of functional blocks, e.g., module m is an adder with $width > 4$

b) **Library condition.** The existence of a specific library cell is tested by a rule of this class, e.g.,

$$\exists c \in L : cell\_class(c) = arith,$$
$$parameter(c, function) = +,$$
$$parameter(c, width) >= 4$$

c) **Optimization condition.** This class of conditions describes certain optimization goals, (e.g., $optimization\_goal = min\_area$). Rules of this class are not used by every rule-based system because sophisticated rule interpreters automatically search for the best rule with regard to the optimization goal.

The expressions formed using these primitives can become complex logic expressions with bound variables, for example:

$$\text{(Module } m \text{ is adder with } width > 1) \wedge$$
$$(\exists w : (0 < w < width(m)) \wedge$$
$$(\exists c \in L : cell\_class(c) = arith,$$
$$parameter(c, function) = +,$$
$$parameter(c, width) = w))$$

The rules used for technology mapping can be divided into two rule classes: *mapping rules* and *expansion rules*. The action part of an expansion rule describes the substitution of one complex functional block by a network of smaller and simpler functional blocks. An example for this is the substitution of a 16-bit adder by a ripple carry network of four 4-bit adders. The substitution of one block by a network of functional blocks can be defined via a submodule specification and a connection specification. These describe the functional blocks that have to be inserted, the interconnection signals between the inserted functional block and the interconnections between the environmental network and the inserted blocks. Further rules need to be applied recursively to functional blocks that have been inserted this way.

The action part of a mapping rule describes the substitution of one functional block by one library cell. Such a rule stops the recursive process of rule application.

Rule interpreter. In general, there are many possible orders in which to apply the rules, and at any given state there may be many rules to choose from. Therefore, as examples two canonical methods are described and the resulting solutions, which comprise a good trade-off between these two extreme cases, are presented.

The first and simplest way to interpret the rules is to apply exactly one rule per functional module. This achieved either by ensuring that in each situation only one rule is applicable or by applying only the most plausible from several applicable rules. As this approach uses no backtracking, a simple but fast algorithm results. Only a few solutions are examined; therefore, the results are unlikely to be optimal. Improvements can be achieved only by a very complicated and hard-to-maintain rule base, which covers every possible special case and in which the basic advantages of the rule-based approach are lost.

The other extreme case is full backtracking. This asks for an exhaustive search in the solution space of the technology mapping problem. For this approach, a simple rule-based system can be used to find an optimum solution. However, due to the exponential complexity of this algorithm this approach is unpractical for all but the most trivial examples.

Compromises between these two canonical cases are achieved by *tree pruning* strategies similar to those used in applications of artificial intelligence [Wins84]. In this approach, the number of possibilities to be tested is reduced by considering only a few rules for each situation. Another possibility to limit the complexity is the introduction of a bound for the *look ahead depth*, the length of rule sequences that are tested before a final decision is made and the first rule of the sequence is definitely applied. In particular, a look ahead depth of 1 leads to the first canonical method without any backtracking and a look ahead depth of ∞ asks for full backtracking.

4.5 Outlook

At the logic level, the mapping of sequential circuits is of interest. In [MLSB89] a canonical representation of sequential circuits is described, which allows for the application of graph-covering algorithms. Further developments can be expected in methods that take timing information and estimated routing area into account.

RT-level technology mapping methods can be improved by a stronger interaction of technology mapping and related tools. Especially, a good interaction between the mapping tool and the synthesis tool allows for flexibility in choosing the stage in the design process at which the target technology is fixed and the mapping procedure is executed. Thus, different circuit parts can be treated separately according to requirements of the designer. Also, stronger interaction between logic-level synthesis tools and RT-level technology mapping is necessary to make use of the advantages of both design steps.

Mapping an abstract technology independent circuit to a specific target architecture also must fulfill testability requirements. Thus, design for testability (DFT) methods are of importance in the context of technology mapping. This topic is presented in Chapter 10.

As the starting point for synthesis moves to higher levels of abstraction (see Chapter 7) the mapping to technology dependent circuit representations is accomplished in earlier design stages. Consequently, formalisms to derive the structural technology independent design representation from technology dependent descriptions have to be introduced at the algorithmic level. This way, it becomes possible to implement a formally specified system as a composition of algorithmically described system components, using different target technologies to take advantage of the specific technological properties.

4.6 Problems for the Reader

1. Given a semi-custom library

$$L = \{\mathrm{IV}, \mathrm{AN2}, \mathrm{AN3}\ , \mathrm{AN4}, \mathrm{OR2}, \mathrm{OR3}, \mathrm{OR4}, \mathrm{AOR}, \mathrm{OAN}\}$$

with the following Boolean functions and (area) cost factors,

c_i	f_i	a_i
IV	$z = \neg a$	1
AN2	$z = a \wedge b$	2
AN3	$z = a \wedge b \wedge c$	3
AN4	$z = a \wedge b \wedge c \wedge d$	4
OR2	$z = a \vee b$	2
OR3	$z = a \vee b \vee c$	3
OR4	$z = a \vee b \vee c \vee d$	4
AOR	$z = (a \wedge b) \vee c$	3
OAN	$z = (a \vee b) \wedge c$	3

a) determine all multi-level logic representations of these cells that fulfill the following conditions:

(*i*) They consist only of 2-input NOR gates and inverters.

(*ii*) For all paths P_I consisting of inverters only the following holds:

$$1 \leq length(P_I) \leq 2$$

(*iii*) For all paths P_N consisting of NOR gates only the following holds:

$$length(P_N) \leq 1 \ .$$

(i.e., no NOR node is a direct successor of another NOR node.)

Use inverter pairs to realize (*iii*).

b) Suppose the result of logic-level synthesis and of partitioning the resulting Boolean network into a forest of trees leads to the following Boolean expression:

$$Z = \neg a \vee b \vee (c \wedge (d \vee e)) \ .$$

Represent this expression, using a multi-level logic structure which fulfills (*i*),(*ii*),(*iii*) and

(*iv*) Next to each primary input and output is an inverter node.

c) Execute the tree covering algorithm to map to the library $L^* = L \cup \{W\}$, where W is is a dummy library cell for a wire element with the Boolean function $f_W \ : \ z = a$ and the (area) cost factor $a_W = 0$.

Matching phase:
Determine the set of possible matchings at each tree node.

Optimization phase:
Compute the overall costs for each possible matching. Start at the leave nodes and continue by taking the optimum costs for the subtrees not covered by the matching library cells.

Accumulation phase (mapping phase):
Start at the root node to accumulate the optimization results.

Chapter 5

Register-Transfer Level Synthesis

Harald Vollmer, Norbert Wehn

5.1 Introduction

The result of high-level synthesis is typically an initial description at the RT-level of a data path and a controller. RT-level descriptions mostly characterize a system definition in terms of registers, multiplexors, and operations, as is explained in Chapter 1. These descriptions have already incorporated a notion of a specific architecture and a clocking scheme. Hence, an initial assignment of operations to clocking cycles has been made. The coarse timing frame has been fixed by the scheduling (cf. Chapter 6), but at this level of abstraction further optimization steps may change these initial choices slightly. At this design stage, the physical design characteristics of the various blocks are known in more detail than at the algorithmic level, thus permitting optimization steps which are based on more realistic design models than those that are available for high-level synthesis. RT-level descriptions, on the other hand, form the input to logic-level synthesis, for which the system is specified by blocks of combinational logic and storage elements. In the following, techniques for data path synthesis and controller synthesis are presented.

5.2 Data Path Synthesis

5.2.1 Optimization Strategies

In this chapter, some simple examples are explained to motivate RT-level synthesis and to give an impression of possible optimization steps at this level of abstraction.

Given the following RT-level description:

$$\begin{array}{ll} \text{control-step 1:} & r_1 \leftarrow c_1(v_i, ..) \\ \text{control-step 2:} & r_2 \leftarrow c_2(r_1, v_j, ..), \end{array}$$

where r_1, r_2 are registers, c_1, c_2 combinational blocks, v_i, v_j variables. The maximum clock frequency feasible for this RT-level sequence is determined by

$$f_{max} = \frac{1}{max\{delay(c_1), delay(c_2)\}}$$

In the following, it is assumed that block c_1 determines the frequency, i.e., $delay(c_1) > delay(c_2)$. There are various possibilities to increase the clock frequency without changing the global structure of the circuit. First, one can substitute a faster block for the slowest combinational block, e.g., if c_1 represents a ripple-carry adder, it can be substituted by a carry-look-ahead adder. A second possibility is to restructure the circuit and to change to positions of the registers. Assume that c_1 can be restructured, e.g., by partitioning:

$$c_1^{new} = g(f_1(v_i, ..), f_2(v_i, ..))$$

It is evident that f_1 and f_2 have a smaller complexity than c_1, thus their delays can be assumed to be smaller than the delay of c_1. With this assumption, the RT-level description can be resynthesized:

$$\begin{array}{ll} \text{control-step 1:} & r_1 \leftarrow f_1(v_i, ..);\ r_2 \leftarrow f_2(v_i, ..) \\ \text{control-step 2:} & r_3 \leftarrow c_2(g(r_1, r_2), v_j, ..) \end{array}$$

This solution is faster than the original one under the assumption that $delay(g) + delay(c_2) < delay(c_1)$. Hence, at the cost of one additional register the speed of the circuit is increased.

The modifications explained in the previous example are based on area/time trade-offs, i.e., additional area must be paid to increase the speed. In most cases, a ripple-carry adder consumes less area than a carry-look-ahead adder. For the second solution, an additional register and the composition function g must be provided.

In the next example it is shown how the property of mutually exclusive transfers can be used to optimize hardware. Assume that the following description is to be executed in one cycle:

```
if F=0 then B := A+C
       else B := C-1;
```

A straightforward implementation requires one adder and one decrementer. However, if one takes into account that, due to their data dependency on the same variable F, the two operations can not be executed simultaneously, one functional unit is sufficient. This functional unit has to perform two different operations: add and decrement. In a further optimization step the property of two's complement representation can be used. A decrement of 1 corresponds to an addition of a variable with all bits equal to one. Thus, only one adder is needed. Furthermore, the commutativity of addition can be used to reduce the number of multiplexor inputs. Only one multiplexor is needed at one input of the adder for A and the constant "11..1", respectively, if the operands, e.g., of the second addition, are swapped. The optimized final hardware structure is shown in Figure 5.1.

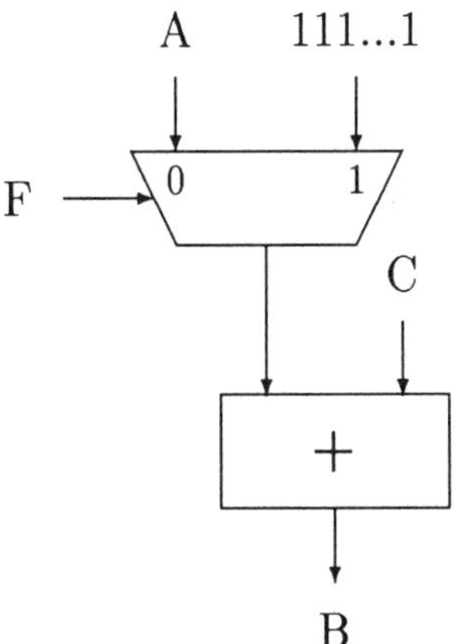

Figure 5.1: Optimized hardware structure.

In the third example, it is demonstrated how the controller complexity can be reduced by shifting part of the hardware into the data path.

Given is the RT-level description of Figure 5.2, where A, B and C represent 8-bit registers. The controller of this RT-level sequence contains at least 4 states (see Figure 5.3). In state 1, a value is read from the input port and written into register A. In state 2, which directly follows state 1, a value from the port is written into register B. The next state to be executed depends on the values of A and B: if $A[7] = 0$ and $A > B$, state 3 is performed, which transfers the

```
read_port(A);
read_port(B);
if A[7]=0 then
  begin
    if A>B then C:=A
             else C:=B
  end
else
  begin
    if A>B then C:=B
             else C:=A
  end;
```

Figure 5.2: RT-level description of Mealy/Moore automaton trade-off.

value of A to C. The same state is invoked if $A[7] = 1$ and $A \leq B$. In all other cases, state 4 is called which transfers the contents of B to C. Thus, a Mealy type of automaton emerges, since the controller outputs depend on values generated in the data path. In detail, the following is observed for the data path. In front of the register representing C, a 2-input multiplexor with A and B as inputs is found. This multiplexor is controlled by the controller states 3 and 4. Furthermore, the data path contains a comparator to perform the operation $A > B$. The most significant bit of register A can be used to perform the test for $A[7] = 0$. The two signals are fed into the controller. But, instead of providing two extra controller states for the loading of C depending on the values of these two signals, an EXOR combination is used directly in the data path. The output of this gate controls the multiplexor. Thus, the controller is simplified and the two states 3 and 4 are substituted by a single state. This state only generates the signal to load register C, as shown in Figure 5.3. Obviously no data path signal is left which is used by the modified controller. The controller is simplified to a Moore automaton and can even be reduced to a simple counter [MiSa91a].

In the previous example, the Mealy/Moore automaton trade-off allows for the reduction of the number of states at the cost of some additional hardware in the data path. This simplification is possible through a shift from a totally control flow oriented synthesis to a more data flow oriented synthesis. In addition, this trade-off also minimizes the interaction between the controller and the data path.

In the following, the various optimization techniques of the previous examples are considered in more detail.

5.2.2 Retiming and Resynthesis

Retiming is a transformation of sequential circuits at the RT-level, whereby

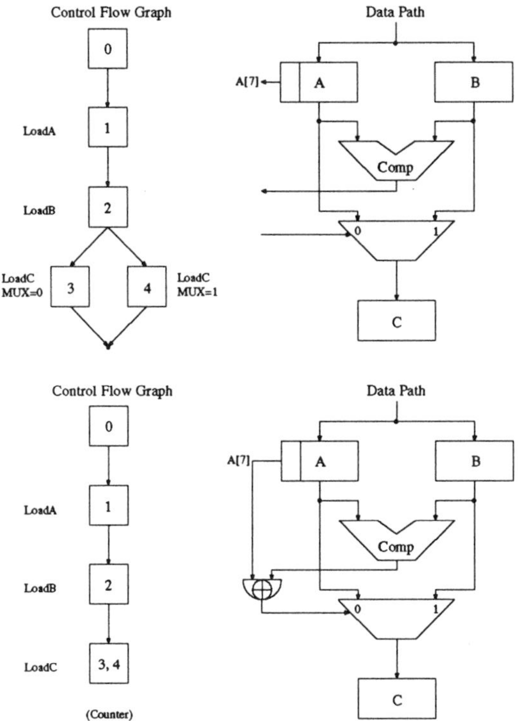

Figure 5.3: Mealy/Moore automaton trade-off.

registers are moved across combinational blocks in such a way as to minimize the clock cycle, or to minimize the number of registers. The function of the circuit is not changed by this register relocation process.

In order to motivate retiming, the design of a digital correlator at the RT-level is presented. A bit stream IS, denoted $s_0, s_1, \ldots$, is used as input to the correlator. A substream of length $k+1$ of IS is compared with a fixed-length pattern $C = c_0, c_1, ..c_k$ in each cycle. Thus, the correlator produces as output o_i in cycle i the number of matches between C and the $k+1$ bits $s_j, ..s_{j+k}$ of IS:

$$o_i = \sum_{j=0}^{k} \delta(s_{i-j}, c_j)$$

$\delta(x, y)$ is the comparison function which results in a 1 if $x = y$, 0 otherwise. In Figure 5.4 a design of a simple correlator for $k = 3$ is shown. The correlator contains two types of functional units: adders and comparators. The boxes represent registers. However, this design has a poor performance, since the critical path contains one comparison unit and 3 adders. If one assumes that the adder has a delay of $10ns$ and the comparison unit a delay of $5ns$, a total delay of $35ns$ results. This path determines the clock frequency. By moving the registers in an appropriate way between the functional blocks, a modified correlator is obtained which is shown in Figure 5.4. In this circuit, the critical path contains only 2 comparison units and one adder, and therefore the clock cycle is reduced to $20ns$. It is left to the reader to prove that both circuits are equivalent.

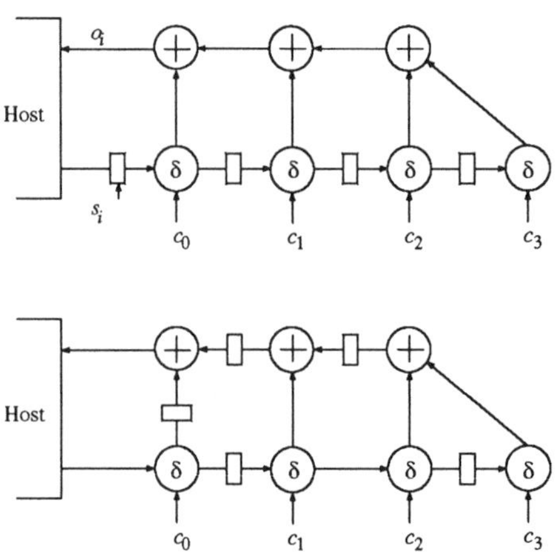

Figure 5.4: Correlator and its retimed version.

To consider the retiming in a more formal way, a sequential circuit is modeled by a directed acyclic, vertex- and edge-weighted graph $G = (V, E, d, w)$. The vertices V of G model the functional elements, i.e., the combinational blocks or I/O pins. Each vertex representing a functional element is weighted with the propagation delay $d(v)$ of the corresponding block. A directed edge $e \in E$ models the interconnections between two elements and is weighted with a register count $w(e)$ which represents the number of registers along the connection.

Retiming algorithms at the RT-level were proposed first by Leiserson et al. [LeSa81, LeRS83]. Here, a *lag* $l(v)$ of a vertex v is introduced representing the number of registers to be shifted from the output to the input of the corresponding functional block v in the retimed circuit. A retiming can then be viewed as an assignment of a lag to each vertex $v \in V$.

Definition 5.1 (Legal Retiming) *A retiming of a synchronous circuit, i.e., the assignment of an integer $l(v)$ to each vertex of G, is legal if:*

1. $l(v) = 0 \quad \forall v \in I/O$ *pin*

2. $w(e) + l(v) - l(u) \geq 0 \quad \forall e(u, v) \in E$

The edge weights (register counts) of the retimed circuit are $w_r(e) = w(e) + l(v) - l(u)$. A legal retiming has been proven [LeSa81] to generate a circuit that is functionally equivalent to the original circuit. Moreover, retiming is, in a sense, the most general method for changing the register counts within a circuit, without disturbing the functionality of the circuit.

Starting from these observations, some optimization techniques that are based on retiming are explained. Consider the following:

Given a circuit graph $G(V, E, d, w)$. Find a legal retiming $L(V)$ of G such that the clock period $\phi(G_r)$ of the retimed circuit G_r is as small as possible.

In [LeRS83] the conditions under which retiming produces a circuit whose clock period is not larger than a given constant are shown. To solve the problem the following theorem is introduced.

Theorem 5.1 (Legal Retiming) *Let $G(V, E, d, w)$ be a synchronous circuit, c an arbitrary positive real number, and L be a retiming. L is a legal retiming of G such that $\phi(G_r) \leq c$ if and only if*

1. $l(v) = 0 \quad \forall v \in I/O$ *pin*

2. $w(e) + l(v) - l(u) \geq 0 \quad \forall e(u, v) \in E$

3. $W(u, v) + l(v) - l(u) \geq 1 \; \forall u, v \in V$ *such that* $D(u, v) > c$. $W(u, v)$ *is the minimum number of registers on any path from vertex u to vertex v in G. $D(u, v)$ is the maximum propagation delay on any path from u to v with $W(u, v)$ registers.*

The constraints on the unknown variables $l(v)$ in the theorem are linear inequalities. Thus, the problem is similar to an Integer Linear Programming (ILP) problem, with the difference that a cost function, which needs to be minimized, does not exist. To solve these linear inequalities the Bellman-Ford algorithm [Leng90] is used. An artificial node v_0 is introduced, which is connected to each node of G with an edge of weight 0. The Bellman-Ford algorithm solves the single source shortest-paths problem on G, i.e., the problem to determine the least-weight path from v_0 to any $v \in G$. If the graph contains a cycle of negative weight, the least-weight paths are not defined. In this case, the inequalities cannot be satisfied. If the graph contains no negative cycles, the inequalities can be solved by interpreting the least-path values as solution values for the variables of the inequalities. Of course, this assignment is not the only one. But, it is only of interest whether any assignment for a given clock period c exists at all. The Bellman-Ford algorithm has the complexity $O(|V||E|)$. Thus, the complete optimization algorithm proceeds as follows:

1. Compute $W(u,v)$ and $D(u,v)$ for all paths in G from u to v. This can be done efficiently with the Floyd-Warshall method.

2. Sort the elements in the range of $D(u,v)$ in ascending order.

3. Binary search has to be performed among the elements $D(u,v)$ for the minimum clock period. To test if a clock period c is feasible, apply the Bellmann-Ford algorithm.

4. For the minimum clock period from Step 3, use the values $l(v)$ found by the Bellmann-Ford algorithm as the optimum retiming.

The algorithm runs in $O(|V|^3 log|V|)$ time. For an algorithm with better asymptotic performance the reader is referred to [LeRS83].

In [MSBS91] retiming is extended by introducing the concept of negative registers for edges that are connected to I/O vertices, i.e., the weights of these edges can be negative. A negative edge weight $w(e)$ on an I/O edge is equivalent to borrowing $w(e)$ registers from the environment. The registers are returned in a subsequent retiming step by forcing the corresponding edge weights to this negative value in the retiming procedure.

The idea of shifting all registers to I/O edges temporarily permits a global optimization of the sequential circuit by standard optimization techniques developed for pure combinational circuits. Before this optimization method is explained, the term of peripheral retiming is introduced.

Definition 5.2 (Peripheral Retiming) *A* <u>*peripheral retiming*</u> *is a retiming such that*

1. $l(v) = 0 \quad \forall v \in I/O$ *pin*

2. $w(e) + l(v) - l(u) = 0 \quad \forall e(u,v) \in E$ *with* $u \notin I/O$ *pin and* $v \notin I/O$ *pin.*

Thus, a peripheral retiming moves all registers to the peripheral edges, leaving only combinational logic between them. In [MSBS91] it is shown that the application of a peripheral retiming and a subsequent legal retiming results in a circuit equivalent to the original circuit. It is, however, obvious that not all circuits permit a peripheral retiming. The following theorem specifies the conditions for the existence of a peripheral retiming:

Theorem 5.2 (Peripheral Retiming) *A sequential circuit has a peripheral retiming if and only if*

1. *In case of more than one path between an input i and an output j, all paths have the same weight*

2. $\forall_{1 \leq i \leq |input|} \exists \alpha_i \in Z, \forall_{1 \leq j \leq |output|} \exists \beta_j \in Z$ *such that for the paths between the input i and the output j*

$$\alpha_i + \beta_j = \sum_{path(i,j)} w(e)$$

α_i represents the registers which must be placed after input vertex i, and β_j represents the number of the registers placed before output j of the peripherally retimed circuit.

The original circuit represents a synchronous circuit, i.e., no edge weights are negative. Peripheral retiming preserves the synchronous property, since retiming does not change the path weights. Thus, in the peripherally retimed circuit $\forall i, j$: $\alpha_i + \beta_j \geq 0$. However, the optimization procedures which are applied to the interior pure combinational part of the peripherally retimed circuit can create new paths from input to output nodes. Although the new paths represent pseudo-dependencies, since the output node of the new created path in fact does not depend functionally on the corresponding input, structural paths can be produced. Thus, for such a path $\alpha_i + \beta_j \geq 0$ cannot be guaranteed. Since the retiming algorithm does not take care of functionality and considers only path weights, no legal retiming can be found if $\alpha_i + \beta_j < 0$ for a pseudo-dependent path. Thus, the resynthesis, i.e., the optimization procedure, must not produce pseudo-dependencies that result in negative path weights. But this will not significantly reduce the usefulness of the retiming and resynthesis approach.

Based on these considerations, the extended retiming algorithm can be formulated:

1. Check Condition 1 of Theorem 5.2 for each input/output path. If there is no violation, proceed with Step 2, otherwise assume fail.

2. For all inputs i and outputs j evaluate α_i, β_j according to Condition 2 of Theorem 5.2. If successful, proceed with Step 3, otherwise assume fail.

3. Place α_i registers behind each input i, and β_j registers before each output j. Remove all other registers inside the circuit.

4. Resynthesize the interior combinational circuit by standard optimization techniques for combinational logic.

5. Check whether the new circuit represents a synchronous circuit, i.e., whether for all paths between any input/output node $\sum_{path} w(e) \geq 0$. If no violation occurs, proceed with Step 6, otherwise reject the resynthesis and continue with Step 4.

6. According to a cost criterion find a legal retiming for the circuit, e.g., minimize the clock cycle as described before.

In [MSBS91] the application of this technique to FSMs, the most relevant and important class of sequential circuits, is described.

The synthesis approach presented above is based on a separation of the combinational logic from the registers, and applies circuit transformations to the combinational block only. A different strategy is presented in [Mich91], in which the circuit equations and register positions are optimized concurrently by a set of algorithms based on logic transformations. In this approach, a model for synchronous logic-level synthesis is presented that combines retiming with combinational logic-level synthesis. Logic transformations are presented that can be used to optimize the circuit area under cycle-time constraints, or to optimize the cycle time under area constraints. These transformations are based on extensions of common network restructuring operations, such as elimination, resubstitution, extraction, and decomposition considering registers, too.

To summarize, retiming can improve the performance of sequential circuits substantially. The algorithms are well understood and efficient.

5.2.3 Resource Allocation and Assignment

The basic algorithms for binding/allocation, i.e., the mapping of operations and variables to basic hardware modules, registers, and interconnections, dependent on the schedule, are presented in Chapter 6. This section puts emphasis on the assignment of operator *implementations* to abstract hardware modules, called *operator selection* in the following.

Usually, the binding process in high-level synthesis assigns the operations of the algorithmic description to abstract functional modules, e.g., if an addition occurs in the algorithmic description, binding determines which functional adder performs this specific addition. Thus, for the considerations below it is assumed that the operations of the algorithmic description have already been scheduled and assigned to abstract functional modules. The task of operator selection is defined as the selection of an appropriate operator implementation from a library. A small example is presented to motivate this task. Assuming that the abstract functional module represents an unpipelined adder, operator selection determines the optimum hardware realization of this adder with regard to a specific objective function, which takes into account both area and speed. If the library contains, for example, a ripple-carry adder, a carry-look-ahead adder and a carry-select adder, one has to choose one of the three adders. Obviously, the choice depends on various criteria. If the abstract functional module has to perform additions that are on the critical path, a fast adder, e.g., a carry-select adder, has to be selected, at the cost of area. On the other hand, if no time critical operations are performed on the corresponding module, the adder with the smallest area is selected, e.g., the ripple-carry adder.

In the following, the operator selection problem is formulated as a linear programming problem LP, which can be solved with modest computational resources [Fogg89]. Since LP techniques are widely known, only a short summary of these optimization problems is given.

Definition 5.3 (LP Problem) *Given a real $m \times n$ constraint matrix A, a real $m-$vector b, and a real $n-$vector c, the LP problem is to find in the set of all $n-$vectors x which satisfy the linear constraints $A * x \leq b$, $x \geq 0$ the vector x_{min} that minimizes $c^T * x$.*

The linear expression $c^T * x$ represents the cost or objective function. The constraints given by A and b are linear. The configuration space described by the constraints represents a polytope P, i.e., if $s, t \in P$, all convex combinations $\alpha * s + (1 - \alpha) * t$, $0 < \alpha < 1$ are also $\in P$. It can be shown that the minimum cost on P with respect to c is attained in a corner of P. The most common algorithm to solve LP problems is the Simplex algorithm, which performs a greedy search on the skeleton of the polytope. An LP can also be defined by substituting $\leq$ with $\geq$, or $=$ in some of the inequalities of $A * x \leq b$. Using more modern algorithms than Simplex, linear programming can be done in polynomial time [Karm84].

To map the operator selection problem onto an LP problem some notations are introduced. Given is a set of abstract functional modules Ω of a non-branching data path. The functional modules are located between registers. Furthermore, a given clock period T is specified. Ω and the registers result from the high-level synthesis task; T can be a user constraint. One has to find an appropriate operator implementation i^* for each $o^* \in \Omega$ such that the overall area is minimized. Normally, for each i^* a limited set of different implementations exists in the library denoted $i^{\{*\}}$. The elements $i^* \in i^{\{*\}}$ differ mainly in their delay $\delta(i^*)$ and area $a(i^*)$ values. Thus, the problem is to find an implementation i^* for each $o^* \in \Omega$ that minimizes $\sum_{o^* \in \Omega} a(i^*)$. A constraint is introduced to satisfy the timing constraint, i.e., the propagation delay of all modules that are located on a path between any two registers must be less than T:

$$\sum_{i^* \in path} \delta(i^*) \leq T$$

In order to complete the LP, one has to find a mathematical representation of an operator implementation set $i^{\{*\}}$ for each o^* respectively. According to the LP formulation, an operator assignment is legal if it is located inside a convex hull P. Each $i^* \in i^{\{*\}}$ of the library is characterized by its two values $(a(i^*), \delta(i^*))$. If these pairs for each alternative implementation are arranged in a two dimensional area/time space, the corresponding convex hull $P_{i\{*\}}$ can be constructed in a preprocessing phase. $P_{i\{*\}}$ is described by a set of linear inequalities of the form:

$$\alpha^n_{i\{*\}} * a_{o^*} + \beta^n_{i\{*\}} * \delta_{o^*} \leq \gamma^n_{i\{*\}} \text{ or } \alpha^n_{i\{*\}} * a_{o^*} + \beta^n_{i\{*\}} * \delta_{o^*} \geq \gamma^n_{i\{*\}}$$

for $n = 1 \ldots N$. The a_{o^*} and δ_{o^*} represent the variables of the LP problem. The number N of linear inequalities depends on the complexity of the convex hull for the operator implementation set $i^{\{*\}}$.

Thus, the LP formulation can be completed. The constraints are given by:

$$\forall o^* \in \Omega : \qquad \alpha^n_{i\{*\}} * a_{o^*} + \beta^n_{i\{*\}} * \delta_{o^*} \geq (\leq) \gamma^n_{i\{*\}} \qquad \sum_{o^* \in path} \delta_{o^*} \leq T$$

The goal is to minimize:

$$\sum_{o^* \in \Omega} a_{o^*}$$

The variables a_{o^*}, δ_{o^*} of the solved LP problem yield the delay and area for the ideal operator implementation. If this procedure is repeated for all clock periods T (if more than one is available), the optimum, piecewise linear tradeoff between area and speed is generated. Unfortunately, the ideal LP solutions may not be realizable, because they use operator implementations which are linear interpolations between realizable operators, i.e., no $i^* \in i^{\{*\}}$ exists with $a(i^*) = a_{o^*}$ and $\delta(i^*) = \delta_{o^*}$. To find an exact solution of the real problem, one has to solve an integer LP (ILP) which is NP-hard. Thus, these operators are mapped in a postprocessing phase onto discrete realizable implementations by searching for the nearest discrete implementation. Although the discrete LP solutions are not guaranteed to be optimum, the ideal LP solutions provide a lower bound.

5.3 Controller Synthesis

5.3.1 Tasks of Controller Synthesis

As tasks in and research on controller synthesis span from the algorithmic level to the register-transfer level, it can not be considered a pure subtask of synthesis at RT-level. The task of the controller itself is to control the data path, namely loading of registers, switching of multiplexors or selecting the operation code for multi-functional units. It is the implementation of the scheduling task in hardware, whereby an initial schedule can be changed by retiming. Since a controller is specified by states and state transitions, it is called a Finite State Machine (FSM).

Finite state systems. A sequential system has the capability to capture the influence of past inputs $x(-\infty, t)$ on the present outputs $y(t)$. To determine the output y at time t, in most systems it is not necessary to "memorize" the complete input function $x(-\infty, t)$. The values of the input function can be grouped into classes in such a way that all time functions that have the same effect on the output at time t are in the same class.

The *state*, specified by an auxiliary variable S, represents the class. It summarizes the effect of past inputs on present and future outputs. Consequently, the output at time t depends on the input at time t and the current state. The *output function* is

$$y(t) = f_{output}(S(t), x(t))$$

To continue to reflect the influence of the input time function, the state has to change. The new state depends on the present state and on the input. The *state transition function* is

$$S(t + \delta t) = f_{nextstate}(S(t), x(t))$$

States $S(t)$ and $S(t + \delta t)$ are called the *present* and *next* states. A *state transition graph* is a directed graph that represents the transition and output functions. A node is assigned to each state and an arc to each transition. For example, in Figure 5.5 it is shown that for a present state S_i and an input x the next state is S_j and the output y. The *canonical implementation* of a sequential system is based directly on its state description. It consists of a *state register* to store the state, and a *combinational network* to implement the transition and output functions. Such a *finite state machine* is shown in Figure 10.16. In practice, a lot of variations exist for controller implementations but they can be derived from this basic finite state machine model.

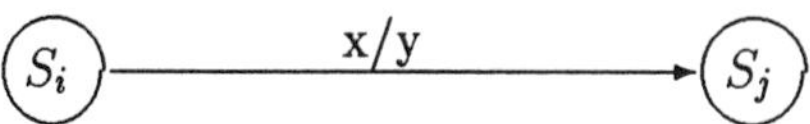

Figure 5.5: State diagram notation.

Generation of the state graph. One task of high-level synthesis is to extract the control flow from a specification at algorithmic level and to transform it into state transition graphs or microcode. For example, in [Camp91a] the state transition graph is constructed during the scheduling phase, taking into account resource constraints. There are other approaches addressing concurrency, e.g., [CBBG87] or [BKKR86], which map behavior into a modified Petri-net model, which is partitioned into strongly connected state machines. Both support concurrency and external synchronization, but the state explosion problem needs to be addressed.

Controller architectures. In some automated synthesis systems, the Read Only Memory (ROM) based microprogrammed controller style is applied [SuLP88, Goos87, Raba88]. Each microprogram command corresponds to one or several words in the ROM. Each bit controls a specific part of the data path in a certain clock cycle defined by a sequencing hardware within the controller. Heuristic approaches for controller optimizations, as for example minimizing the width of the microcode, are performed [SuLP88]. Another common synthesis method is to map the state transition graph to PLA-based FSMs or single/multiple FSMs realized by multi-level logic for the combinational part and D-flip-flops for the storage part. States can be encoded with minimal code length or, if one flip-flop is assigned to each state ("one-hot encoding") [Bray88, KuDe89], the logic for calculating the next-state and output signals is simpler and faster, at the expense of significantly more register area. Complex controller architectures have been proposed for specific target applications: For larger controller sizes with relatively low branching degree in the state transition graph a counter can be used as the storage part which reduces the combinational logic in the FSM [AmBa89]. For Digital Signal Processing (DSP) applications, a complex pipelined controller architecture has been proposed, which reduces the number of "no-operations," when jumps in the microprogram occur [Meer88].

Finite state machine optimizations for area and performance. The approaches follow several interacting objectives. Targeting just one objective does not provide the optimum solution. One objective is the reduction of states and transitions in the state graph; but it sometimes misses the minimal number of product terms or literals after logic minimization. Minimizing literals can lead to an increase in wiring area. Due to floorplanning constraints and/or performance considerations, a single large minimized FSM often is not suitable. Recent work in this area, therefore, concentrates on the decomposition of a FSM into several interacting FSMs. An important domain of controller synthesis is optimum state assignment and much work has been done in this area. Some methods are described in the following sections.

5.3.2 State Assignment for Controllers

For a given controller architecture and state transition table description, the cost of the logic required to realize a synchronous sequential network is strongly dependent on the way in which flip-flop states (state variables) are assigned to the states in the state table. If the number of states is small, it may be feasible to try all possible assignments, evaluate the cost of realization for each assignment and choose the assignment with the lowest cost. Suppose that three symbolic states a, b, c and two flip-flops, $Q1$ and $Q2$, are given; then, 24 possible state assignments can be distinguished as shown in Table 5.1.

	1	2	3	4	5	6	7	...	19	20	21	22	23	24
a	00	00	00	00	00	00	01	...	11	11	11	11	11	11
b	01	01	10	10	11	11	00	...	00	00	01	01	10	10
c	10	11	01	11	01	10	10	...	01	10	00	10	00	01

Table 5.1: State assignments for 3 states.

Permuting of columns does not affect the cost of a realization, because it is equivalent to relabeling flip-flop variables. Complementing of columns may require adding an inverter in the case of D-flip-flops, but the network generally can be designed to eliminate the inverter. (The exception is a PLA realization for which complementing columns can lead to a decrease in area.) Two state assignments are therefore called *equivalent* if one can be derived from the other by permuting and complementing columns; otherwise, they are called *distinct*. Let z be the number of states and t the minimal number of bits needed to encode the states ($t = \lceil log_2 z \rceil$); then, N distinct state assignments are possible:

$$N(z) = \frac{(2^t - 1)!}{(2^t - z)! \times t!}$$

The impressive growth of N, even for small values of z, is shown in Table 5.2. This can be compared with the moderate growth of the number of different code columns which indicates that a column-oriented state assignment strategy is much more efficient in terms of computer runtime.

Heuristic methods are used to solve the state assignment problem.

5.3.2.1 Objectives and General Assignment Strategy

Objectives. State assignment was studied intensively in the sixties and seventies. One objective was the simplification of the next state and output equations of controllers. With the advent of new logic minimization techniques for integrated digital circuits these approaches were reinvestigated.

Number of states (z)	Number of distinct state assignments N(z)	Number of different code columns
5	140	15
8	840	35
9	10 810 800	255
16	5×10^{10}	6435

Table 5.2: Number of different state codes.

Current state assignment methods follow one of two objectives:

1. Improvement of the testability [DMNS90], see also Section 10.4.4, and concurrent testing capabilities of controllers [LeSa89b]. Concurrent checking means that the controller is tested concurrently to its normal operation. In [LeSa89b] sequences of state codes are compacted producing the same signature at certain points of the state graph.

2. Minimization of the area after logic minimization

The majority of tools follow the second objective and some corresponding methods will be described in the sequel.

Looking closer at the second objective, the following cost functions are generally considered:

1. Minimization of the number of product terms after two-level logic minimization for PLA realizations [Demi86a, SaDS87]

2. Minimization of literals in the factored form, after multi-level logic minimization for multi-level implementations [DMNS88b], [LiNe89b], [SaPo89]

The approaches using the first cost function can be subdivided into those for single and for double PLA implementations. In the latter case, the next state functions and output functions can be handled separately by the assignment tool.

General state assignment strategy. Although not all state assignment methods follow the scheme given below precisely, three major tasks can be distinguished:

1. Identification of sets of states in the state table that, if encoded with minimal Hamming distance (number of differing code bits), lead to a simplification of the corresponding next-state and output equations after logic minimization. These sets (groups) of states, which are usually intersecting, are called *adjacency groups*. Every group represents a *coding constraint* for the binary values of the states included.

2. The groups and their intersections are analyzed with respect to the degree of potential minimizations during the subsequent logic minimization. Results are data structures like matrices, priority lists or weighted graphs, which reflect these relations and potential gains in the cost of the final logic.

3. Coding constraints and calculated gains control the encoding heuristics, which try to satisfy as much constraints as possible. This task is called *embedding*, relating to the fact that states are assigned to vertices in a Boolean n-cube.

5.3.3 State Assignment for Single PLA Controllers

The method described has been integrated in the tool KISS [DeBr85] and follows the approach taken in [DeBr85],[Demi86b],[Demi86a]. More recently, a tool called NOVA [ViSa89] has been developed, which incorporates new encoding methods.

The single FSM from Section 10.16 represents the architecture of the controller. The logic part is represented by the PLA that calculates the next-state and output functions. Under the assumption that no topological compaction techniques, such as folding, splitting or partitioning, are applied, the PLA area is proportional to the number of rows (product terms) times the number of columns. Both row and column cardinality depend on the state assignment. If n_b is the state encoding length (i.e., n_b state variables), $3 * n_b$ columns, namely n_b output-columns and $2 * n_b$ input-columns for the non-inverted and inverted present-state input are needed.

The state assignment problem can be defined as follows:
Problem: *Find the assignment of minimum code length among all possible assignments that minimize the number of rows of the PLA.*

As described previously, the first task of state assignment is to detect adjacency groups in the state transition graph. Such groups are, for instance, present states which have edges to the same next state with associated identical or compatible (FSM-)inputs and identical or compatible (FSM-)outputs. This case is called *join*-situation. An example is shown in Table 5.3.

If states $S1$ - $S4$ and $S5$ - $S6$ are encoded with close codes (small Hamming distance), for example, $S1$ = (111**00**), $S2$ = (111**01**), $S3$ = (111**10**), and $S4$ = (111**11**), a minimized state table (see Table 5.4) is obtained.

Let t be the number of transitions (edges in the state graph) for the join-situation. Then, $t - 1$ product terms can be saved. The (present-) state set $S1, S2, \ldots St$ forms an adjacency group. In the example above, $S1$ - $S4$ are embedded in a 2-cube which covers 4 encodings. Because several product terms

In	Present State	Next State	Out
01	S1	S1	010
01	S2	S1	010
−1	S3	S1	010
0−	S4	S1	010
01	S5	S6	111
10	S5	S7	111
10	S6	S7	111
⋮	⋮	⋮	⋮

Table 5.3: State table with state codes in symbolic notation.

Inp	Present State	Next State	Out
01	$\{S1, S2, S3, S4\} = (111--)$	S1	010
01	S5	S6	111
10	$\{S5, S6\} = (0000-)$	S7	111
⋮	⋮	⋮	⋮

Table 5.4: Minimized state table.

(cubes) collapse into one product term, the term *cube collapsing* is sometimes used for this situation.

Suppose only the three states $S1 \ldots S3$ of the example above are contained in the adjacency group. If the code (111**11**) can be used for any other state, cube collapsing is achieved.

It is possible to derive coding constraints from a join-situation with edges associated to incompatible inputs [Demi86a]. Such constraints are more difficult to extract and are not described here.

5.3.3.1 Generation of Coding Constraints

The method used to generate the adjacency groups is called *symbolic minimization*, which is a logic minimization performed with a symbolic (code independent) representation of the state table. Each symbolic string (state) takes a different logic value corresponding to a multiple-valued logic representation. The task of symbolic minimization is to find a minimum number of symbolic implicands (rows in Table 5.4). This is equivalent to finding a minimum sum of product representation independent of the encoding of the symbolic strings. In order to use a binary-valued logic minimizer for that purpose, a suitable encoding of the states must be provided, i.e., multiple-valued variables must be represented by two-valued variables. This is performed by "one-hot" encoding of each symbolic variable [SuCh72]: A p-valued logical variable is represented by a string of p binary symbols. Value r is represented by a 1 in the r-th position, all others being 0. For example, if a FSM has four states, state 2 is encoded (0100). The cube $C = (0110)$ for the multiple-valued state variables represents the condition "State 2 $\vee$ State 3." More details on multiple-valued logic minimization are given in [BHMS84].

With "one-hot" encoding of all states the Boolean minimizer tries to find a minimum multiple-valued input cover. The result of this step, applied to the example of Table 5.3, is shown in Table 5.5.

Input	Present state	Next state	Output
01	11110.......0	10.............0	010
01	000010.....0	0000010.....0	111
11	0000110...0	00000010...0	111
⋮	⋮	⋮	⋮

Table 5.5: Result of symbolic minimization.

Each row of the present state column defines an adjacency group of states (indicated by 1), which are mapped by some input into the same next state and which assert the same output. For example, the first row of the present state column indicates that this condition is true for state 1,2,3,4 and, therefore, the adjacency group (1,2,3,4) is defined. The sum of Hamming distances between the final encodings of each state contained in an adjacency group has to be minimum, i.e., the states are embedded in a minimum Boolean subcube containing the encodings of the states assigned to that group. This subcube must not intersect with encodings of states not belonging to that group. Therefore, a constraint encoding problem can be stated:
Problem: *Given a set of state groups, find an encoding such that each Boolean subcube of a group does not intersect the code assigned to any state not contained in the group.*
The number of product terms given by symbolic minimization is an *upper limit* for the number of product terms after final logic minimization has been performed.

5.3.3.2 Constrained State Encoding

The coding constraints are contained in the *present state column*, after symbolic minimization (cf. Table 5.5). Let n_s be the number of states, n_l be the number of adjacency groups and n_b be the code length. The present state column after symbolic minimization can be interpreted as a *constraint matrix* $A \in \{0,1,*\}^{n_l \times n_s}$ representing n_l adjacency groups.

Redundant rows in A can be eliminated, namely rows with no 0 entry, rows with only one 1, identical rows, and rows that are equivalent to the intersection of two or more rows of A.

The encoding strategy explained in the following is called *column encoding.* It iteratively computes one encoding bit in each step for all states. In contrast to this method, *row encoding* refers to encoding of one state in each step. Column encoding is more run time efficient, because only $n_b = log_2 n_s$ encoding steps are needed. This technique is based on a *state code matrix* $S \in \{0,1\}^{n_s \times n_b}$ representing n_s states with encoding length n_b.

The encoding problem is to find a state code matrix S with minimal number of columns that satisfy the constraint relation given by the constraint matrix A.

The encoding method is explained with the following example. Given a redundancy-free constraint matrix A with non-identical columns:

$$A = \begin{pmatrix} 1 & 0 & 1 & 0 & 0 & 1 \\ 0 & 0 & 1 & 1 & 0 & 0 \\ 0 & 1 & 0 & 0 & 1 & 0 \\ 0 & 0 & 0 & 1 & 1 & 0 \\ 0 & 1 & 0 & 0 & 0 & 1 \end{pmatrix}$$

Suppose, the first row of A is taken as the first column of S: $S^T = [101001]$.

This encoding fully satisfies the first coding constraint of A, because the state codes of states 1,3,6 are included in a (0-) subcube, which does not contain the code of states 2,4,5. The second constraint in A, given by its second row, is not fulfilled, because states 3,4 are included in a (1-) subcube, which contains all the other state codes (0). However, the code column partially satisfies some other coding constraints. For instance, looking at the third row of A, the adjacency group (2,5) is coded with 0 and states 1,3,6 are not included in this subcube. The only state, which must still be excluded by further encoding bits, is state 4, because it is encoded with 0. The next coding column of S has to be chosen such that it *maximally satisfies* all remaining constraints. Thus the encoding length is kept low. For that purpose, one needs to keep track of the remaining coding constraints.

Further notations are introduced in the following to allow for a formal treatment of the problem. The matrices which are considered below have pseudo-

Boolean entries from the set $\{0, 1, *, \emptyset\}$ where $*$ represents the don't care condition and $\emptyset$ represents the empty value. The conjunction $\cap$ and disjunction $\cup$ on pseudo-Boolean variables are defined as shown in Table 5.6.

$\cap$	0	1	$*$	$\emptyset$
0	0	$\emptyset$	0	$\emptyset$
1	$\emptyset$	1	1	$\emptyset$
$*$	0	1	$*$	$\emptyset$
$\emptyset$	$\emptyset$	$\emptyset$	$\emptyset$	$\emptyset$

$\cup$	0	1	$*$	$\emptyset$
0	0	$*$	$*$	0
1	$*$	1	$*$	1
$*$	$*$	$*$	$*$	$*$
$\emptyset$	0	1	$*$	$\emptyset$

Table 5.6: Conjunction and disjunction of pseudo-boolean variables.

The *face matrix* $F \in \{0, 1, *, \emptyset\}^{n_l \times n_b}$ is the matrix whose rows are the group faces (subcubes of the encodings of the states included in the groups). n_b is the code length.

In this representation the empty group corresponds to the empty face represented by n_b $\emptyset$ entries.

To simplify the notation in the algorithms below, the *selection* symbol "$\bullet$" is introduced. Let $a \in \{0, 1\}$ and $b \in \{0, 1, *, \emptyset\}$. The *selection* of b according to a is

$$a \bullet b = \begin{cases} b & \text{if } a = 1 \\ \emptyset & \text{if } a = 0 \end{cases}$$

Selection can be extended to two-dimensional arrays and is similar to matrix multiplication.

Let $A \in \{0, 1\}^{p \times q}$ and $B \in \{0, 1, *, \emptyset\}^{q \times r}$. Then $C = A \bullet B = \{c_{ij}\}^{p \times r}$ where $c_{ij} = \cup_{k=1}^{q} a_{ik} \bullet b_{kj}$.

With this notation, the face matrix F can be obtained by performing the selection of S according to a constraint matrix A:

$$F = A \bullet S$$

Continuing with the example, after choosing the first column of S, the matrix F is computed:

$$S = \begin{pmatrix} 1 \\ 0 \\ 1 \\ 0 \\ 0 \\ 1 \end{pmatrix} \qquad F = \begin{pmatrix} 1 \\ * \\ 0 \\ 0 \\ * \end{pmatrix}$$

F is used, then, to compute the remaining coding constraints, i.e., to reduce the amount of coding constraints in A. The meanings of the variables in F are:

$f_{ij} = 1$: The j-th bit position of all states belonging to the group i (row i in A) is encoded with 1, but not necessarily the j-th bit position of all others with 0. In that case, the corresponding coding constraint is only partially satisfied by the code column j.

$f_{ij} = *$: The j-th bit position of some of the states belonging to the group is encoded with 1. The j-th bit position of the others belonging to the group is encoded with 0. Therefore, the corresponding coding constraint can not be satisfied by the code column j.

$f_{ij} = 0$: The j-th bit position of all states belonging to the group i is coded with 0, but not necessarily the j-th bit position of all others with 1. In that case, the corresponding coding constraint is only partially satisfied by the code column j.

The states already satisfying the non-intersection with the group face (subcube formed by the adjacency-group) are marked with "$*$" in A. Thus, the reduced constraint matrix A is as follows:

$$A = \begin{pmatrix} 1 & * & 1 & * & * & 1 \\ 0 & 0 & 1 & 1 & 0 & 0 \\ * & 1 & * & 0 & 1 & * \\ * & 0 & * & 1 & 1 & * \\ 0 & 1 & 0 & 0 & 0 & 1 \end{pmatrix}$$

The algorithm of the constraint reduction is described in Algorithm 5.1.

```
BEGIN
        F := A • S;
        FOR i := 1 TO n_l DO
            FOR j :=1 TO n_s DO
                IF (a_ij = 0 and f.i ∩ s.j = ∅ )
                THEN a_ij := *
END.
```

Algorithm 5.1: Constraint reduction.

The first row of A does not contain any 0 and does not need to be considered any more. The task is to find the next code column which maximally satisfies the remaining coding constraints. The reduced constraint matrix may consist of compatible rows which are satisfied by one code column. Two rows in A are compatible if either all pairs of entries contain equal values (1 and 1 or 0 and 0) or if all pairs contain inverse values. In both cases, the "*" may be interpreted

a1	0	1	*	0	1	*	*
a2	0	1	*	*	*	0	1

a1	0	1	*	0	1	*	*
a2	1	0	*	*	*	0	1

Table 5.7: Compatible pairs of entries in constraint-matrix A.

as 1 or as 0. In Table 5.7 two examples of compatible rows in the constraint matrix A are shown.

For each pair of compatible rows there exists a 0/1 mapping of the "$*$" places of one of the two rows which forms a code column satisfying the coding constraints given by the two rows. In the example, rows 4,5 and rows 2,3 of A are compatible and a possible code column for the latter pair is $S^T = [001100]$. To find the code column that maximally satisfies the constraints, the *maximal compatibility classes* must be computed. Methods (e.g., graph coloring or clique partitioning, see Section 6.5.4.2) are available to perform this task. With the new code column given above new matrices are computed:

$$S = \begin{pmatrix} 1 & 0 \\ 0 & 0 \\ 1 & 1 \\ 0 & 1 \\ 0 & 0 \\ 1 & 0 \end{pmatrix} \quad F = \begin{pmatrix} 1 & * \\ * & 1 \\ 0 & 0 \\ 0 & * \\ 0 & * \\ * & 0 \end{pmatrix} \quad A = \begin{pmatrix} 1 & * & 1 & * & * & 1 \\ * & * & 1 & 1 & * & * \\ * & 1 & * & * & 1 & * \\ * & 0 & * & 1 & 1 & * \\ 0 & 1 & * & * & 0 & 1 \end{pmatrix}$$

The last two rows are compatible and a possible new column is $S = [01*001]$. If "$*$" in S is mapped to 1, the second row of F gets a second "$*$." This subcube, then, covers four state encodings, although two states are contained in the corresponding group. Therefore, the "$*$" is mapped to 0, the subcube in the second row of F being of minimal size. After reducing A, the matrices are as follows:

$$S = \begin{pmatrix} 1 & 0 & 0 \\ 0 & 0 & 1 \\ 1 & 1 & 0 \\ 0 & 1 & 0 \\ 0 & 0 & 0 \\ 1 & 0 & 1 \end{pmatrix} \quad F = \begin{pmatrix} 1 & * & * \\ * & 1 & 0 \\ 0 & 0 & * \\ 0 & * & 0 \\ * & 0 & 1 \end{pmatrix} \quad A = \begin{pmatrix} 1 & * & 1 & * & * & 1 \\ * & * & 1 & 1 & * & * \\ * & 1 & * & * & 1 & * \\ * & * & * & 1 & 1 & * \\ * & 1 & * & * & * & 1 \end{pmatrix}$$

All coding constraints are satisfied, because A does not contain 0 anymore.

In case of identical columns in the initial constraint matrix A, identical state codes (rows in S) are produced. In this case, A has to be extended by additional rows before encoding starts. Every additional row may contain one 1 in different places, thereby obtaining A with non-identical columns.

In conclusion the column encoding algorithm is outlined below:

Step 1: Choose a column code $\{0,1\}^{n_s}$ that maximally satisfies the coding constraints. Insert the column into the state matrix S.

Step 2: Compute the face matrix F and reduce the constraint matrix A.

Step 3: If S satisfies the coding constraints, stop, otherwise go to step 1.

The method described can be extended so that next state encoding is taken into account [Demi86a].

5.3.3.3 Column Encoding with Minimal Encoding Length

The column encoding method described above does not necessarily produce the minimum code length $n_b = ld(n_s)$.
Example: Let $n_s = 8, n_b = 3, A = [11000000]$. The first code column chosen is $S^T = [11000000]$ or $S^T = [00111111]$.

With this initial encoding, only 6 states can be encoded with $n_b = 3$. To encode 8 states with different encodings, a fourth bit is necessary. Therefore, heuristics for column encoding have been proposed that try to keep the encoding length at minimum. The idea used in [RiNe90] is, first, to partition the coding constraints into disjoint sets and to encode according to these partitions, thereby obtaining a more evenly distributed 0/1 encoding. In subsequent steps, the encoding procedure described in the previous section is used. The process stops when the minimal code length is reached. Even then, it may occur that not all states are encoded differently. Therefore, the encodings of identical states are dropped and the respective states are eliminated from the coding constraints. Subsequently, new free encodings are computed. As a final step, a significant improvement of the result after logic minimization is obtained by encoding that (present-) state "0...0" for which most output variables are 0. All other states are re-encoded accordingly.

5.3.4 State Assignment Targeting Multi-Level Logic

Due to area and/or performance reasons, large controllers often cannot be synthesized as a single PLA. Therefore, multi-level logic implementations are generally used. Specific state assignment techniques have been developed, since those used for optimal PLA implementations are inadequate for multi-level implementations [DMNS88b].

In the following, a method is described, which minimizes the number of literals in the factored form after multi-level logic optimization [DMNS88b, LiNe89b, SaDP90]. However, one has to be aware of the fact that minimizing

Input	Present State	Next State	Output
any states	S1	any states	11111
	S2		01000
	S3		11101
	S4		00010
	S5		11110
	S6		00001

Table 5.8: State table for present-state encoding.

literals globally can increase the wiring area after technology mapping. Often, wiring area is the predominant part of the whole controller area if a standard cell or gate array based design is used. Therefore, a primary target is *cube collapsing*, as described in Section 5.3.3. Cube collapsing naturally leads to smaller gate and wiring area. The assignment tool in ASYL [SaDP90] gives primary attention to these situations.

5.3.4.1 Generation of Coding Constraints

The method described in the sequel [LiNe89b] tries to maximize the number and size of common single cubes. The assignment tool JEDI aims at solving a general symbolic encoding problem for which the assignment of states is seen as a subproblem. The main ideas of the technique are given in the following.

Common cube optimization is accomplished by assigning *minimal distance codes* to symbolic (state-)variables. Two basic encoding procedures are considered:

- Encoding of the present state field, taking into account next state- and output field (*present state encoding*)
- Encoding of the next state field, taking into account present state- and input field (*next state encoding*)

Coding constraints for present state encoding. The encodings of present states that assert similar outputs are chosen to have more common literals in their intersection. Thereby the size and number of common cubes in the output function increases. An example is shown in Table 5.8.

States S1, S3, and S5 are supposed to get close codes which yield a large common cube in many output functions. As a coding constraint, an *input assignment matrix* M_I for the present states is constructed. It is a $n \times n$ symmetric matrix, with each entry m_{xy} corresponding to a cost relationship between a pair of symbolic (present-)state values $(s_x, s_y) \in S$ if S is the set of states. The value of m_{xy} is the relative gain that can be achieved if minimal distant codes are assigned to the pair of corresponding state values. To calculate m_{xy} two auxiliary variables are introduced:

The proximity $P(c_i, c_j)$ between two cubes c_i and c_j is the number of non-empty literals in their intersection.

O_n is the subset of output combinations (1-entries in the output field) whose input part is the symbolic state variable s_n. o_{nj} is the j-th element of that subset.

For a pair of symbolic state variables $(s_x, s_y) \in S$ the entry m_{xy} of the matrix is calculated as follows:

$$m_{xy} = \sum_{a=1}^{|O_x|} \sum_{b=1}^{|O_y|} P(o_{xa}, o_{yb})$$

The value m_{xy} is used to predict statically the effect of logic optimization by comparing how far apart the cubes in O_x are from those in O_y.

The next state field has not been considered so far. Therefore, present states leading to the same next state can be taken into account by an appropriate increase of the corresponding entries m_{xy}.

Coding constraints for next state encoding. This approach is similar to the method described above. A further auxiliary variable is introduced:

I_n is the subset of input cubes, whose output part is the symbolic state variable s_n. i_{nj} is the j-th element of that subset.

The entry m_{xy} in the *output assignment matrix* for the next states is calculated as follows:

$$m_{xy} = \sum_{a=1}^{|I_x|} \sum_{b=1}^{|I_y|} P(i_{xa}, i_{yb})$$

The state values are to be encoded in such a way that the input cubes with many common literals result in the same encoded ON-sets of the next state functions.

The present state field has not been considered. Therefore, next states with edges from a common present state can be taken into account by an appropriate increase of the corresponding entries m_{xy}.

For the subsequent encoding process both, present state- and next state coding constraints can be considered concurrently by constructing a new assignment matrix which is the sum of the input and output assignment matrix.

For state encoding the interested reader is referred to [LiNe89b].

5.4 Outlook

One of the major topics for RT-level synthesis of the future concerns a better link of RT-level synthesis to high-level synthesis, and to the physical design process. Several of the RT-level approaches are still isolated in the design cycle. For example, in high-level synthesis no efficient techniques are known for the estimation of the final size of the resulting controller. Data path related timing optimizations are well elaborated. However, if the controller is part of the critical path, no global technique is known to resolve this problem. Another important topic is RT-level synthesis combined with testing aspects. The interested reader is referred to Chapter 10.4.4.

Retiming. Several problems of retiming are not yet solved satisfactorily. Specifically, there exists a strong need for advanced retiming techniques which support an extended propagation delay model with explicit fan-out dependency and which satisfy both upper and lower bounds on propagation delays, thus, permitting the processing of circuits with critical clock skews. In addition, an appropriate set of logic transformations for synchronous sequential circuits similar to that of pure combinational circuits is missing.

Input/output encoding of controllers. A current research issue is the optimal encoding of symbolic inputs and outputs of controllers. For this case, the encoding problem is similar to the state encoding problem. The final goal is to encode inputs, outputs and states simultaneously such that optimal area and/or performance are achieved.

Decomposition of controllers. The decomposition of larger controllers into several smaller, interacting controllers is a further important research topic. It is done for several reasons: The total area of the decomposed controller can be less than the area of the original controller, and even if this is not the case, a better floorplanning may be achieved after decomposition. The delay of the critical path can be reduced, since the decomposed machines are smaller and possibly run in parallel. Finally, heuristics for state assignment and Boolean minimization perform better for smaller machines. One approach [MuGe91],[GeMu91] applies the algebraic structure theory of [HaSt66] in order to find parallel or serial decompositions of the original machine. Floorplanning constraints are considered in [GrMu89]. Other approaches are shown in [AsDN91].

5.5 Problems for the Reader

1. Prove that the correlator circuits of Figure 5.4 are equivalent.

2. Consider the correlator example of Figure 5.4. Assume that the delay of the adder is $10ns$ and the delay of the comparison unit $5ns$. Find a retiming of the correlator such that the circuit is running with a frequency of 60 MHZ. How many registers are necessary?

3. Given the state table from Table 5.9. Find a state assignment for a single PLA controller using the column encoding method described in Section 5.3.3.2. Why are rows in the constraint matrix A redundant if they contain no "0" or just one "1"?

In	Present State	Next State	Out
01	S2	S1	010
01	S4	S1	010
−1	S5	S1	010
00	S1	S2	001
11	S2	S3	111
11	S3	S3	111
10	S4	S1	011
10	S5	S1	011
01	S3	S4	110
−1	S4	S4	110
01	S5	S4	110
00	S1	S5	101
00	S3	S5	101

Table 5.9: State table with state codes in symbolic notation.

Chapter 6

High-Level Synthesis

Sabine März

6.1 Introduction

Synthesis takes a specification of the functionality of a digital system and a set of constraints and finds a structure that implements the intended behavior and satisfies the constraints. At the algorithmic level, the specification takes the form of an algorithm. This implies that basic implementation decisions have already been made—thus the term algorithmic "description", introduced in Chapter 1, is used rather than the term "specification." But, of course, in comparison to the RT-level, many implementation details are left open at the algorithmic level and have to be filled in by automated synthesis.

6.1.1 The Starting Point

An algorithm is a precise method for the computational solution of a problem. In the design hierarchy, the term "algorithmic description" refers to behavior specified in terms of operations and computation sequences on certain inputs to produce certain outputs. The basic elements of an algorithmic description correspond to the basic elements of programming or hardware description languages. They comprise arithmetic and logic operations, variables or values, and control constructs like if/case, loops, and procedure calls. Regardless whether procedural languages, functional languages, or even graphical input are used, an algorithmic description captures the intended data and control flow in order to specify the behavior of a digital system.

Most synthesis systems use a textual form of an algorithmic description as input. Only a few systems accept graphical input of the data and control flow [HaEl89]. Graphical interfaces to the data and control flow representation may, however, be very useful for modifications once an initial algorithmic description has been entered in a textual form. Many systems take (subsets of) procedural programming languages such as PASCAL as input [Stok91, PeKL88]. The actual choice of the language is usually determined by the application domain and the environment the synthesis system has to fit in. Examples of hardware description languages used are ISPS [Barb81], SILAGE [Hilf85, Raba88, PoRa89], ELLA [WhNe90], and VHDL [VHDL87, Thom90, Duzy89, Camp91b, BhLe90] (cf. Chapter 2). Sometimes only a subset of the language is supported by the synthesis system.

```
use  work.syn_pack.all ;

 ENTITY diffeq IS
 PORT ( inport   : IN  integer;
        outport  : OUT integer;
        sysclock : IN  bit);
 END diffeq;

 ARCHITECTURE hls_synthesis OF diffeq IS
 BEGIN
   PROCESS
     VARIABLE a, dx, x, u, y: integer;
     VARIABLE x1, y1: integer;
   BEGIN
     cycles(sysclock,1);
     a := inport;
     cycles(sysclock,1);
     dx := inport;
     cycles(sysclock,1);
     y := inport;
     cycles(sysclock,1);
     x := inport;
     cycles(sysclock,1);
     u := inport;
     LOOP
       cycles(sysclock,7);
       x1 := x + dx;
       y1 := y + (u * dx);
       u := u - 5 * x * (u * dx) - 3 * y * dx;
       x := x1; y := y1;
       EXIT WHEN NOT (x1 < a);
     END LOOP ;
     outport <= y;
   END PROCESS ;
 END hls_synthesis;
```

Figure 6.1: VHDL code for Differential Equation example.

As one example of a HDL-description VHDL code[1] is shown in Figure 6.1.

[1]Using the CALLAS VHDL subset. For more details refer to Chapter 2.

The code describes the solution of the differential equation[2]

$$\frac{d^2y}{dx^2} + 5\frac{dy}{dx}x + 3y = 0$$

The corresponding graph representation of the data and control flow is depicted in Figure 6.6.

Depending on the application domain, specific assumptions are made or additional features are introduced into algorithmic descriptions. An algorithmic description in the area of digital signal processing (DSP) is often assumed to be embedded in an implicit infinite outer time loop such that inputs are sampled and corresponding outputs are produced repeatedly [Hilf85]. For interface circuits and for circuits that have mostly controller functionality, means to specify I/O protocols or temporal relations between sets of inputs and outputs [DeKu88, GlUm90, Umbr90] are needed in order to give a sufficiently precise specification of the intended behavior.

Finally, ***pragmas***, known from some conventional programming languages, can be used in algorithmic descriptions as well to provide hints that guide the synthesis of the ***structural implementation***.

6.1.2 The Target

The result of high-level synthesis is a description of a digital synchronous system in the structural domain at the register-transfer level. The system consists of a data part which manipulates the input data to provide the desired output, and a control part which controls the sequence and the type of data manipulations. Data part and control part communicate via condition flags and control signals. Multiprocessor architectures with hierarchical control [Raba88, KrRo89, WaTh89] and systems with distributed control synthesized from communicating automata [WoTL91] have been studied within high-level synthesis. Most of the target architectures, however, consist of a single data part and a centralized controller. Typical implementations at the register-transfer level can be characterized as follows:

- The data part consists of a set of functional units such as adders, comparators, arithmetic and logic units (ALU), multipliers, etc., a suitable amount of storage, such as registers, latches, register files, or memory, as well as interconnect hardware, such as busses, multiplexors, and nets.
- The controller is specified in terms of a symbolic state-transition table that is given to subsequent controller synthesis (cf. Chapter 5) for the final implementation.

---

[2] The Differential Equation example was proposed for illustration of synthesis approaches by [Paul88].

- The control nets enable or address storage, switch multiplexors or bus drivers, and provide opcodes for multi-functional units.
- The condition nets feed results from test expressions for data dependent branching to the next-state and output logic of the controller.

A synchronous system is characterized by processing in discrete *time steps*, independent of whether a single-, a two-, or a four-phase clocking scheme is chosen in the final realization. The simplest realization of a synchronous system is achieved by a circuit where edge-triggered registers are controlled by a single system clock. Here a time step corresponds to a clock cycle. As long as non-pipelined designs are considered, a time step also corresponds to a *control step*. A control step is the transition from one controller state to the next.

Data part architectures differ in the type of arithmetic and logic units, in interconnection schemes, and the type of storage hardware, as well as in clocking schemes. The most popular data part architectures fall into two categories basically differing in their interconnection scheme—the *multiplexed* and the *bidirectional bus* data part architecture. The following assumptions are made for the multiplexed data part architecture:

- Arithmetic and logic operations are mapped to combinational functional units, with the exception only of internal storage of pipelined functional units.
- Storage is provided by distributed registers.
- Interconnect is established by multiplexors and nets.
- A single system clock controls all registers in the circuit with the same edge.

Examples for the multiplexed architecture are found in the HAL system [PaKn89a] in which unidirectional busses may substitute multiplexors, in the S(P)LICER system [PaGa87] where either two-level multiplexing or unidirectional busses are used, and in the LYRA/ARYL [HCLH90] tools where interconnect is implemented via multiplexor trees. In Figure 6.10 a register-transfer level structure for the Differential Equation example using the multiplexed architecture is shown.

For the bidirectional bus architecture the following assumptions are made:

- Functional units have storage capabilities. Either inputs and outputs are latched, or the inputs or the outputs are associated with a register.
- Instead of distributed registers, register files (usually single port) are often supported.

- Interconnect hardware comprises bidirectional busses that are understood as resources, multiplexors, drivers, and nets.
- A two-phase clocking scheme may be needed.

Examples for this type of architecture are found in the SPAID [HaEl89] and in the CADDY/CALLAS [KrRo89] system. The SPAID system assumes a two-phase clocking scheme to provide separate read and write cycles. In the CADDY/CALLAS system a two-phase clocking scheme for latch designs and a single clock for register designs are supported. In Figure 6.39 the bidirectional bus architecture is shown using the Differential Equation example again.

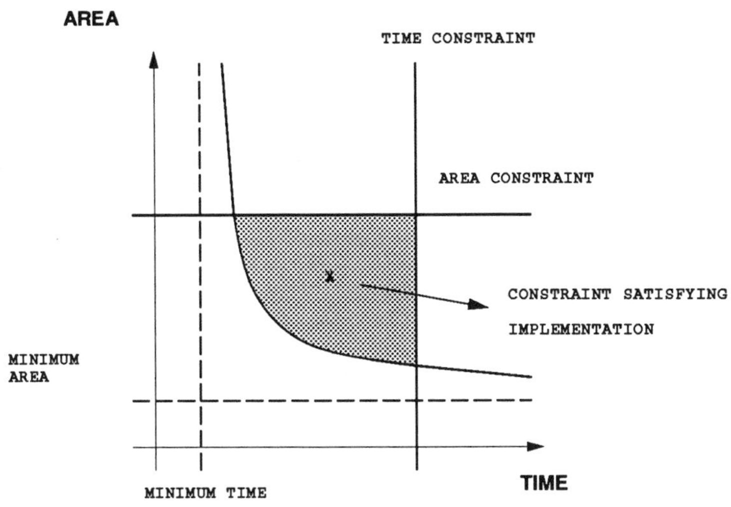

Figure 6.2: Design space and constraint satisfying implementation.

6.1.3 Design Space and Constraints

Mapping behavior to structure is not unique. Unlike von Neumann computers in which operations specified in the source program have to be executed sequentially, hardware synthesized to realize a single algorithm may range from sequential up to completely parallel solutions. The basic trade-off that has to be faced within high-level synthesis is the serial/parallel trade-off modeling the range between area-efficient but slow and area-intensive but fast designs.

Structural implementations that satisfy the specified behavior are considered points in the *design space.* In principal, the design space is spanned by all physical parameters—area, performance (delay or throughput), power consumption, clock cycle time, etc.—that are relevant for meeting hardware constraints. The basic area/time trade-off is illustrated in Figure 6.2. The points that correspond to *valid* implementations reside in the grey-shaded region which has as lower bounds the minimum feasible area and the minimum feasible time and as upper bounds the constraints on area and time.

As in manual design, the problem in high-level synthesis is to find an appropriate register-transfer level structure that allows for a final implementation which satisfies the physical constraints. Whether these constraints are met, however, is known only after the whole design process has been completed, that is after subsequent logic-level and RT-level synthesis (see Chapter 3 and 5), technology mapping (Chapter 4), and physical design. Consequently, metrics for physical parameters have to be abstracted in order to guide high-level synthesis tools which, on the other hand, have to interact with estimation tools in order to find an implementation that satisfies the constraints.

6.1.4 The Classical High-Level Synthesis Tasks

There are several tasks to be performed in the automatic transformation of an algorithmic description to an appropriate register-transfer level structure. An overview of the classical high-level synthesis scheme is given in Figure 6.3.

The first step is to derive an internal graph-based representation equivalent to the algorithmic description for both the data flow and the control flow. The data flow originates from the operations and their data dependencies and the control flow from the control constructs of the algorithmic description. The derivation of the internal representation usually is combined with code optimizations borrowed from conventional compiler techniques in order to remove inefficiencies from the algorithmic description. An example of such a code optimization is the elimination of the common subexpression $(u \times dx)$ in the Differential Equation example. Processing of the algorithmic description, derivation of the internal representation, and compiler optimizations are often viewed as *front end* tasks of synthesis. The internal representation of the data and control flow provides the essential starting point for the synthesis of the register-transfer level structure.

Before transforming data and control flow to structure, graph transformations, often called *behavioral transformations*, can be applied. Assuming that behavior is defined by the specified data and control flow, every graph transformation that changes the data or the control flow modifies the behavior of the digital system. An example of changing the data flow is the application of the associative law to arithmetic expressions. A corresponding example of modifying the control flow is the unrolling of loops with a known, constant iter-

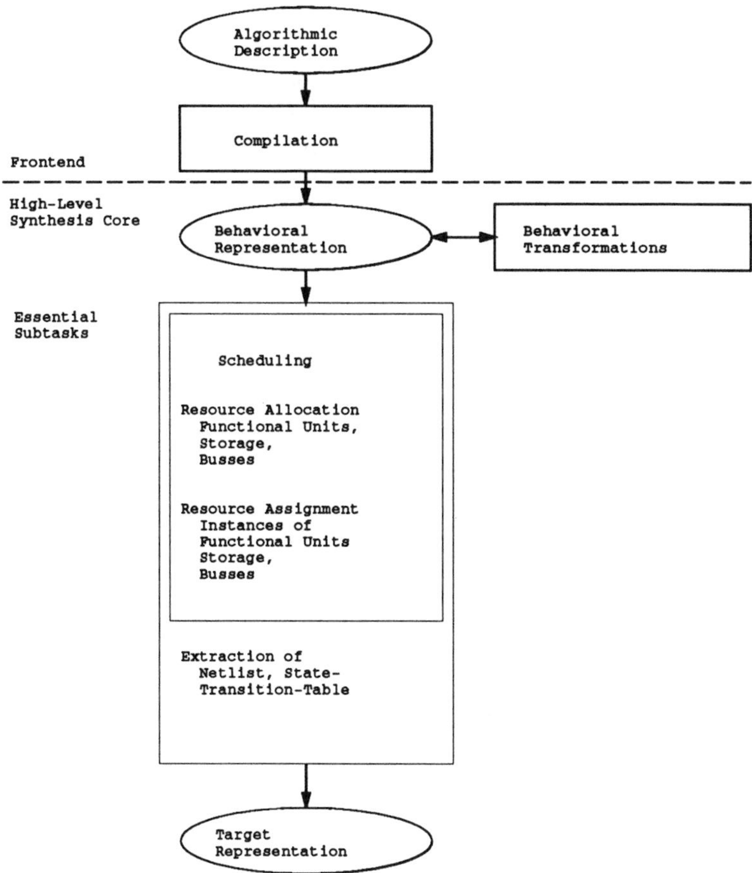

Figure 6.3: The classical high-level synthesis scheme.

ation count. From a conservative point of view, it is not recommended to apply graph transformations that do not preserve the specified data and control flow. However, a fixed data and control flow may overspecify the intended behavior or, in other words, a certain algorithmic description may be only one of a whole class of algorithmic descriptions that satisfy the intended interaction of the digital system with the external world. Graph transformations that modify the data or the control flow are therefore useful to investigate variants resulting from *one* generic algorithmic description. In order to control the modifications of the data and control flow, behavioral transformations are usually applied on user's demand.

Creating the register-transfer level structure means mapping the data and control flow in two dimensions— time and hardware. Two different mappings of the Differential Equation example are shown in Figure 6.4. Time steps, or control steps are indicated by rows; hardware components (here functional units only) correspond to columns. The first mapping represents a design

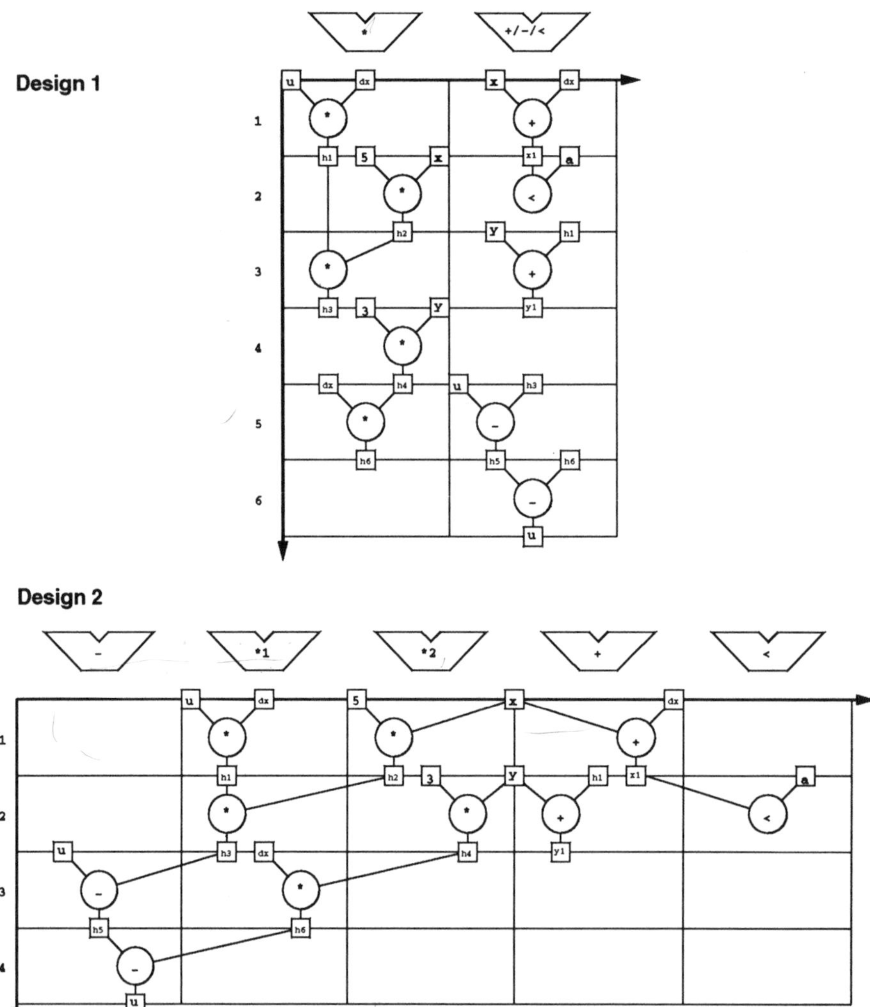

Figure 6.4: Two mappings of the Differential Equation example.

in which one multiplier and one ALU are available for the execution of the operations; the number of time steps needed is 6. The second design in which two multipliers, one adder, one subtractor, and one comparator are available, needs only 4 time steps. Finding an appropriate solution with respect to the serial/parallel trade-off is essential in synthesizing the register-transfer level structure. The more functional units are allocated, the more operations can be scheduled in parallel. The second design variant shows another important degree of freedom in synthesizing the register-transfer level structure. Since two multipliers are available, each multiply operation has to be assigned to a specific multiplier. Besides the functional units, other resources have to be considered. Data transferred between operations in different time steps need to be stored. A suitable number of storage elements such as registers has to be allocated and the values to be stored have to be assigned to specific instances of storage elements. For the transfer of data between storage hardware and functional units, interconnect hardware has to be provided. If busses are required, they have to be allocated and data transfers have to be assigned to them. Finally, drivers and/or multiplexors have to be extracted.

Transforming the data and control flow into a register-transfer level structure can now be introduced systematically. Synthesis of the register-transfer level structure consists of the essential subtasks: *scheduling*, *resource allocation*, and *resource assignment*.

- *Scheduling* is the assignment of operations to time steps subject to certain constraints and minimizing some objective function.
- *Resource Allocation*[3] is the determination of the type and number of resources required.
  - Number and types of functional units
  - Number and type of storage elements
  - Number and type of busses
- *Resource Assignment* is the assignment to resource instances.
  - Operations to functional unit instances
  - Values to be stored to instances of storage elements
  - Data transfers to bus instances

The strong interrelation of allocated resources and schedule leads to two basic scheduling problems. If the number of resources—most likely the functional units—is fixed and the goal is to minimize the number of time steps

[3] Often, the determination of the number (*allocation*) and the determination of the types (*selection*) are distinguished.

needed, ***resource constrained scheduling*** has to be solved in order to provide the mapping to time steps. If the number of time steps is specified and the goal is to minimize the amount of resources, ***time constrained scheduling*** has to be performed in order to provide the mapping to time steps.

Since scheduling, resource allocation, and resource assignment are interrelated, the synthesis subtasks are neither independent nor do they have to be performed in a certain sequence.[4] Some of the tasks involved are NP-hard and some of them are NP-complete. Hardly any scheduling problem can be reduced to a problem that is solvable in polynomial time and the most general resource allocation problems have been proven to be NP-complete. Thus, solving all subtasks at once optimally is in principle intractable. Several different strategies to break the interrelation of the subtasks and heuristics that obtain near-optimal results for separate subtasks have been developed.

After scheduling, resource allocation, and resource assignment, all the necessary information is available to build the register-transfer level structure. The data flow has been mapped to functional units, storage, and interconnect hardware and is finally realized as the data part. The control flow implied by the control constructs in the algorithmic description has been refined by scheduling. After the data part has been generated, condition and control signals can be encoded and a final controller specification in terms of a symbolic state-transition table can be derived. At this stage the high-level synthesis task is finished—the netlist and the state-transition table represent the desired register-transfer level implementation.

6.1.5 Practicality of High-Level Synthesis

Understanding high-level synthesis from the classical point of view is not sufficient. Making high-level synthesis practical means putting emphasis on meeting hardware constraints, on handling interface issues, and on considering testability.

Hardware constraints are physical constraints on area and delay that are only known after placement and routing are finished. Even if detailed information from the component library on area and delay is available, a netlist at the register-transfer level does not supply sufficient physical information. Moreover, controller logic and combinational logic blocks are not yet completely synthesized when high-level synthesis is finished. Therefore, a top-down design strategy with the abstractions made from physical parameters may fail completely. Overall synthesis strategies need to incorporate means to overcome this problem. Among them are estimation routines in general, especially an

[4] Of course, some precedence relations exist—an operation can not be assigned to a functional unit without having allocated this functional unit.

early floorplanning step, a transparent link to low-level synthesis to feed back results of or to feed through constraints to low-level synthesis, additional synthesis routines such as partitioning into tasks for logic-level synthesis, retiming capabilities at register-transfer level, and finally, analysis tools, especially for timing analysis.

Regardless of whether a complete circuit or only a subcircuit is to be synthesized, there is always the hardware environment which the synthesized circuit has to fit into. The environment imposes structural and timing constraints on the circuit to be synthesized. Furthermore, even though high-level synthesis usually deals with synchronous designs, behavior often is of an asynchronous nature at the interface. At least, a synthesized circuit has to react to asynchronous events. However, both *interface synthesis* [Bori88, Nest87] and asynchronous circuit synthesis are beyond the scope of this chapter.

A crucial point is to define an overall synthesis strategy that matches the application domain. Circuits that mainly have controller functionality (*control-dominated* circuits) typically do not require many arithmetic operations, nor do many arithmetic operations have to be executed in parallel. Algorithmic descriptions of control-dominated ASICs tend to make extensive use of procedural control constructs. Both considerations imply the use of scheduling approaches that are capable of meeting complex I/O protocols rather than meeting resource constraints. Control-dominated ASICs usually contain a considerable amount of combinational logic blocks, and storage usually consists of distributed registers. Consequently, logic-level synthesis, retiming capabilities, and register-transfer level optimizations have to be an integral parts of the corresponding overall synthesis strategy.

In contrast to control-dominated ASICs, digital signal processing applications are characterized by intensive arithmetic computation. Many arithmetic operations—including multiplication—have to be executed in parallel in order to meet throughput constraints. Scheduling approaches have to meet resource constraints and must support pipelining. In addition, a large amount of array data has to be stored, resulting in a major part of the DSP chip being occupied by memory. Consequently, *memory management*, i.e., allocating memory, assigning data to background or foreground memory and to locations within memories, as well as providing address calculation units, becomes an essential synthesis task of the overall synthesis strategy [NGCD91].

6.2 Internal Representation

The starting point for synthesizing the register-transfer structure is the intended behavior of the circuit to be synthesized. Since data flow and control flow are given only implicitly in the algorithmic description and a textual representation is—for computer programs—inconvenient to work with directly, a suitable internal data structure is needed, which represents data flow and control flow explicitly and possibly separated.

In order to introduce typical elements for the internal representation of behavior, the *data flow model* and the *control flow model* of computation and the corresponding implications for high-level synthesis have to be considered in more detail.

The control flow model of computation originates from the conventional von Neumann computer, in which, controlled by a program counter, instructions are selected for sequential execution. Accordingly, imperative programming languages have been designed that allow for an explicit specification of the flow-of-control: sequence of statements, conditional branches and loops. For code optimizations, conventional compilers derive a *control flow graph*, a directed graph $CFG(V, E)$ in which vertices correspond to *basic blocks* and edges describe the flow-of-control. A basic block is a block of sequential statements that is entered and left as the result of a control statement and that is not able to halt or to change the flow-of-control.[5] The evolution of computer architectures towards highly parallel machines has inspired new models of computation suited to the parallel nature of hardware—the *data flow models* [Denn84].

Data flow models describe computations in terms of locally controlled events. A *data flow graph* is a directed graph $DFG(V, E)$ in which the vertices correspond to *actors* representing operations; the directed edges, called *links*, represent paths over which data values are conveyed from one actor to another, i.e., they represent data dependencies between operations. In the following, the terms "token" and "firing" are adopted from Petri-net [Pete77] terminology. The state of a computation in progress is shown by placing tokens on the edges of the data flow graph. A computation can be regarded as a succession of snapshots between which tokens are placed and removed by the firing of actors. The data flow graph specifies a partial order of operations due to data dependencies. In order to represent not only data flow but also control flow, special actors and firing rules for conditional branches and loops are added.

Since high-level synthesis aims at hardware architectures ranging from processor-like architectures to highly parallel architectures, both the data flow and the control flow model of computation are useful in principle. Starting from sequential descriptions, the control flow in basic blocks is relaxed by deriving a data flow graph and leaving the decision on sequence to scheduling. The control flow due to control constructs, however, is usually kept and can be rediscovered in the state-transition table. Starting from applicative descriptions the natural internal representation is a data flow graph including the special actors and

[5]For more details about conventional compilers and flow graphs the reader is referred to [AhSU86].

firing rules for conditional branches and loops. Conditional branches and loops are again rediscoverable in the state-transition table, whereas sequence or flow-of-control for the remaining parts of the data flow graph are introduced by scheduling.

Depending on the synthesis goals and implementational details, control flow and data flow may be differently emphasized, and may be kept in separate graphs or combined in a single graph. The *Value Trace* (VT) [Snow78], for example, is basically a control flow graph whose vertices correspond to basic blocks, containing the data flow graph of a straight-line sequence of statements. A similar internal representation is found in the control/data flow graph (CDFG) of the HAL system [Paul88]. In the synthesis system HIS [Camp91b] both the data flow and the control flow are derived and represented in separate graphs. The control flow graph—used as the main data structure for path-based scheduling—is interlinked with disconnected data flow graph pieces. Separate graphs for the control flow and the data flow are also used in the CADDY/CALLAS system [März89]. The data flow graph graph provides the basis for scheduling. The control flow graph is used during scheduling for the recursive processing of nested loops and the detection of mutual exclusion in conditional branches. Examples of the use of a single graph with data flow semantics as internal representation are found in the internal representation proposed in [Sche91] and also in the internal representation of the EINDHOVEN SILICON COMPILER [Stok91].

The internal representation of the structural implementation consists of a netlist which is the result of mapping to hardware, and a state-transition table which is the result of mapping to time or to control steps. Many systems, for example the CALLAS system, have transformational character. During synthesis the behavioral representation is successively transformed into a structural representation. Other systems, for example CATHEDRAL II/2ND [Raba88] and HIS, annotate the internal behavioral representation during synthesis and finish by building the data structures for the structural representation separately.

Internal representations outlined so far are not sufficient to serve as a design representation for synthesis-based design automation. The integration of the behavioral and the structural domain for incremental synthesis strategies and the coexistence of manual and automatic design steps are required. Many other aspects such as the representation of design constraints including area, delay or throughput, or clocking frequency, the representation of temporal behavior, and the representation of asynchronous behavior or concurrency need to be covered. Considerable effort has been spent in defining design representations that account for—at least some of—these aspects (see, e.g., [AmBW89]).

6.2.1 Internal Representation of the Algorithmic Description

The internal representation presented in this section is similar to the internal representation of the EINDHOVEN SILICON COMPILER [Stok91]. A single graph $DFG(V_o, V_c, E, E_c, E_s)$ with data flow semantics representing both data flow and control flow is used.

The vertices represent actors of two different classes. The first class V_o corresponds to operations. The actor fires by removing a token from each of its inputs and placing a token on each of its outputs. The second class V_c corresponds to actors that have been introduced in order to represent the control flow. These include branch and merge vertices for conditional branches and loop entry and loop exit vertices for loops. The edges in E are called *data flow edges* and represent links between two operations $o_i, o_j \in V_o$. Links connecting an actor $o_i \in V_o$ and an actor $o_j \in V_c$ are called *control edges* and represented by E_c. Finally, links connecting two actors o_i, o_j either $\in V_o$ or $\in V_c$ do not reflect data values to be transferred but specify a sequence to be maintained. They are called *sequence edges.* Sequence edges represented by E_s are needed, for example, to specify an ordering of two I/O access operations that are not data dependent.

Data flow edges and control edges. Properties of data flow edges and control edges include the *data type* and an associated *data width.* The data type determines how the value has to be interpreted—examples are two's complement, signed, or unsigned fix point number, signed, unsigned, or binary coded integer, or Boolean. The data width specifies the number of bits needed for this interpretation.

Operation. The most obvious vertex type is an operation in its restricted sense. Operations may be arithmetic such as $-, +, *, /$, logic or relational such as $AND, OR, <, >$, or more complex functions or non-standard operations. The operation type—two's complement, signed or unsigned, fix point—is attached to the operation vertex definition, and for non-commutative operations the sequence of the inputs has to be defined as well.

Functional units are the corresponding structural elements for the operation vertices and therefore often called operators. A one-to-one mapping between the operation vertex type and the functional unit type, however, does not exist. Both an add as well as a subtract operation may be mapped onto an ALU, and both a pipelined multiplier or a parallel multiplier may be used to execute a multiply operation.

Read and write. The read and write vertices are needed to describe the access to external data. Read vertices have outputs only and write vertices have inputs only. As mentioned above, a sequence of read and write operations can be represented by sequence edges.

The structural equivalent for read and write vertices are ports. Again, a one-to-one mapping between read or write vertices and I/O ports or I/O devices does not exist. As an example, a bidirectional port is capable of executing both a read and a write access.

Constant. This vertex is used to represent the generation of a constant value. Consequently, there is no input to that vertex type.

A constant value can be implemented in hardware by specializing a functional unit for a constant input. An example of a direct structural equivalent to a constant value is a connection to power supply.

Select. This vertex type is used to select one value among two or more simultaneously calculated values according to a simultaneously calculated conditional value. Among the inputs, the select input and the data inputs are to be distinguished. The edge incident to the select input, however, has to be considered as data flow edge. A direct structural equivalent of the select vertex is a multiplexor.

Delay. This vertex type is used to represent a delay operation which is caused by feedbacks in the algorithmic description. The delay vertex corresponds to the z^{-m} symbol in Z-domain filter descriptions [Geni90]. The data input represents a value produced in state S_{n+m} and the output represents a value produced in state S_n.

The structural equivalent for a delay vertex is a storage element. A simple delay operation representing the z^{-m} symbol with $m = 1$ corresponds to a single register or a single location in a register file or memory. Delay operations with $m > 1$ correspond to shift registers or a similar implementation using other types of storage hardware.

The select and the delay vertices are not necessarily present in the set of vertices $v \in V_o$. The select vertex is needed only to distinguish between conditional assignments and conditional branching. A corresponding behavioral transformation from conditional branches to pure data flow using the select vertex has been proposed in [WaTh89]. Applicative description languages such as SILAGE have to provide means to capture recursion. The delay vertex is needed for handling feedback values in data flow descriptions containing recursion, a situation that, for example, occurs in recursive filters.

Branch/merge. A *branch* actor fires by removing a token from the data input $\in E$ and a token from the control input $\in E_c$ and placing a token on one of its outputs $\in E$. A *merge* actor fires by removing a token from the control input $\in E_c$ and a token from one data input $\in E$ and placing a token on its output $\in E$.

An **IF** statement or a **CASE** statement[6] occurring in the algorithmic description is represented by introducing a branch actor for each variable defined outside the conditional branch statement and used in one of the branches, and a merge actor for each variable defined in one of the branches and used outside the conditional branch statement. Both the branch and the merge actor have the same control edge as input. The internal representation for the sample **IF** statement

```
IF (x < a) THEN
    x := x + 1;
ELSE
    x := x - 1;
END;
```

is shown in Figure 6.5.

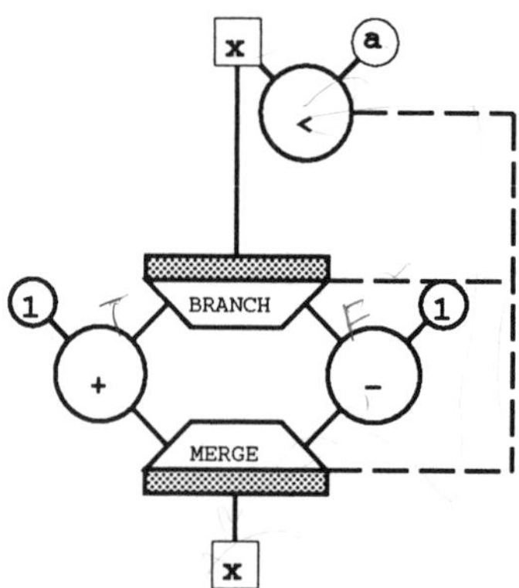

Figure 6.5: Representation of **IF** statement.

Loop entry/loop exit. A *loop entry vertex* fires by removing a token from one of its data inputs and a token from its control input and placing a token on its output. A loop entry vertex corresponds to a merge vertex. A *loop exit vertex* fires by removing the tokens from its data input and its control input and by placing a token on one of its outputs.

[6]For the **CASE** statement it is assumed that exactly one out of n branches is executed according to a conditional value.

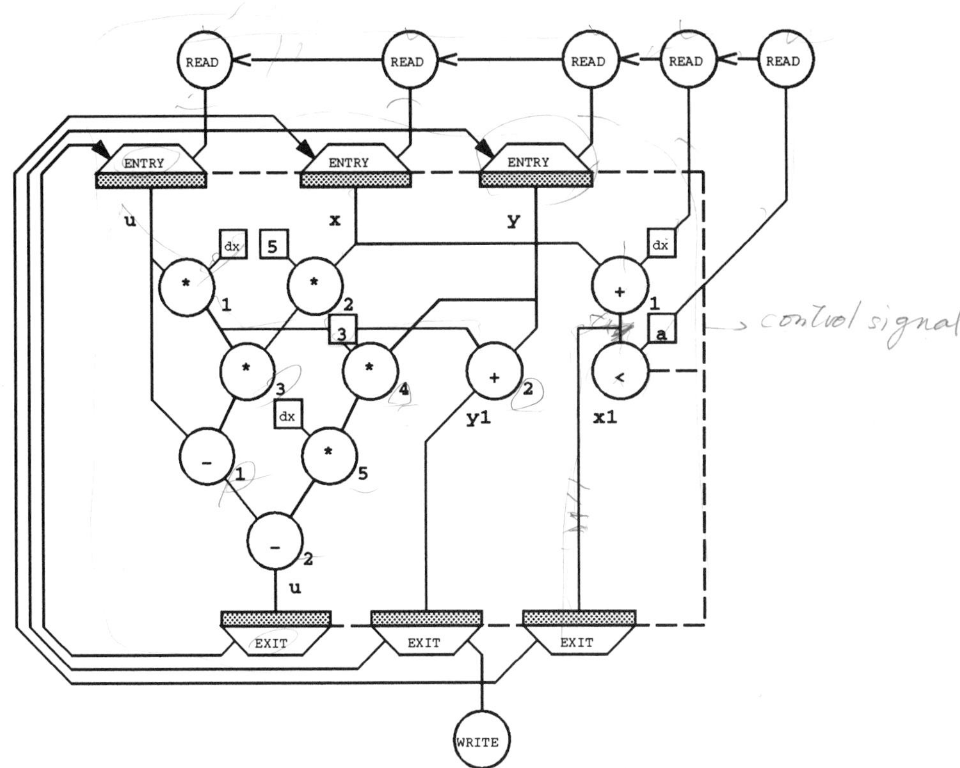

Figure 6.6: Representation of the Differential Equation example.

Whenever a loop statement occurs in the algorithmic description, a loop entry/loop exit vertex pair[7] has to be introduced for each variable defined in the loop and used in the next loop iteration. An example of a **REPEAT** loop scheme can be seen in the internal representation of the Differential Equation example shown in Figure 6.6. Again, the same control edge is incident to the loop entry and loop exit vertex. The test operations for a **REPEAT** and for a **WHILE** loop scheme, and the loop body operations of a **REPEAT** or a **FOR** loop scheme are encapsulated between the output of the loop entry vertices and the

[7]Within high-level synthesis it is usually assumed that a loop has only one exit condition.

input of the loop exit vertices. The while loop body operations are encapsulated between the output of the loop exit and the inputs of the loop entry vertices.

Vertices in V_c have equivalents in the state-transition table, a branch and a loop exit vertex correspond to a state which has more than one immediate successor states, and a merge and a loop entry vertex correspond to a state with more than one immediate predecessor states.

6.2.2 Internal Representation of the Synthesized Structure

The netlist $G(C, N, E)$ is a bipartite graph in which the vertices C correspond to instantiations of hardware components—functional units (adders, multipliers, ALUs), storage hardware (registers, register files, multiplexors), and the controller. The vertices N correspond to busses or nets, and the edges E correspond to terminals of components. An edge represents the incidence between a component and a net or a bus.

Descriptions of hardware components are kept in the component library. Physical information (area, delay, latency, number of stages, shape, fanin and fanout), the range for parameterization or configuration, the encoding of ALU opcodes or multiplexor select inputs, and the sequence of inputs for non-commutative operations are stored. Instantiations refer to the component type and the parameter settings such as the bit-width, the number of inputs, the number of words (in the case of read only memories (RAMs) or register files), the operand types (e.g., fix point, two's complement), and the configuration (in the case of ALUs).

In Figure 6.7 a small **REPEAT** loop example and the corresponding netlist are shown.

The *state-transition table* $FSM(S, X, Y, f, g)$ describes a finite state machine where $S = \{s_1, \ldots, s_N\}$ represents the set of the states, $Y = \{y_1, \ldots, y_K\}$ the control outputs, and $X = \{x_1, \ldots, x_M\}$ the condition inputs. Finite state machines are characterized by a transition function f

$$s(t_{n+1}) = f(s(t_n), X(t_n))$$

describing the next state $s(t_{n+1}) \in S$ as a function of the current state $s(t_n) \in S$ and the current condition values $x_1(t_n), \ldots, x_M(t_n)$, and an output function g:

$$\begin{aligned} \text{MOORE: } Y(t_n) &= g(s(t_n)) \\ \text{MEALY: } Y(t_n) &= g(s(t_n), X(t_n)) \end{aligned}$$

The output function describes the current output $y_1(t_n), \ldots, y_K(t_n)$ as a function of the current state only (MOORE) or as function of the current state and the current condition values (MEALY).

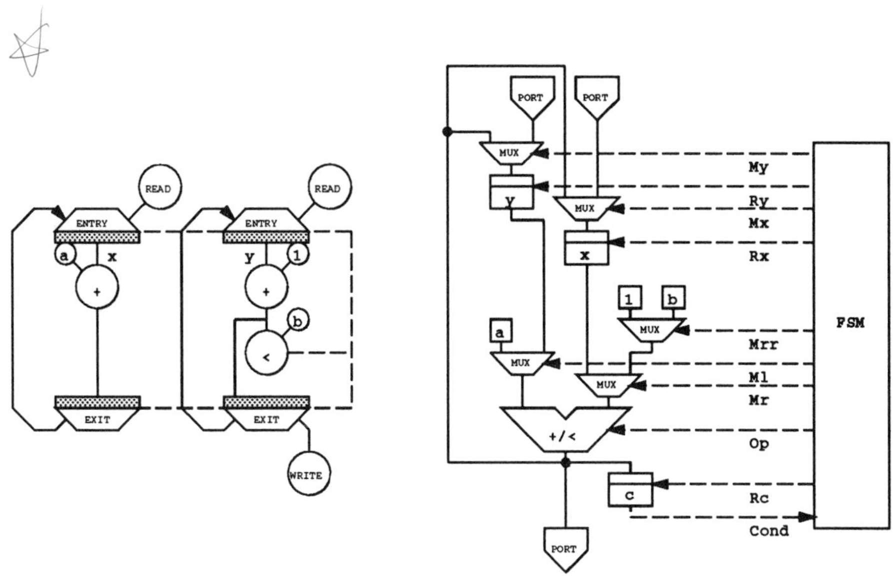

Figure 6.7: Small **REPEAT** loop example and the corresponding netlist.

A state-transition table with MEALY characteristics associated with the implementation of Figure 6.7 is listed in Table 6.1.

Cond	State	Next	Rc	My	Ry	Mx	Rx	Mrr	Ml	Mr	Op
x	...	S1	x	0	1	0	1	x	x	x	x
x	S1	S2	x	x	0	1	1	x	1	1	1
x	S2	S3	x	1	1	x	0	1	0	0	1
x	S3	S4	1	x	0	x	0	0	0	0	0
0	S4	S2	x	x	0	1	1	x	1	1	1
1	S4	...	x	x	0	x	x	x	x	x	x

Table 6.1: Part of the state-transition table for the small **REPEAT** loop example.

6.3 Synthesis of the Register-Transfer Level Structure

Once the internal representation of the data and control flow has been derived, the essential subtasks of scheduling, resource allocation, and resource assignment can be solved in order to collect the information needed to build the register-transfer structure.

Synthesizing an appropriate register-transfer level structure implies meeting hardware constraints such as area, clocking frequency, delay, and power consumption. Physical parameters, however, can only be estimated from the physical parameters of the hardware components in the library.

Chip area is abstracted to the cell area of the hardware components used. As more and more structural information is collected during synthesis steps, more precise estimation of, and control, over these parameters is possible. The contribution of interconnect—busses, drivers, multiplexors and wiring in a bus architecture, multiplexors and nets in the multiplexed architecture—to area is significant and has to be taken into account. The typical measure for area due to interconnect is the number of multiplexor inputs in the multiplexed architecture and the number of the busses, multiplexor inputs, and drivers in the bidirectional bus architecture. In order to meet an area constraint, the contribution from interconnect has to be kept low. During several synthesis phases interconnect is an optimization criterion as is demonstrated by the example synthesis process using the Differential Equation example in the following section.

Time is abstracted to the number of time steps needed. The remaining problem is to meet the cycle time(s). A first guess is based upon the hardware component delays and a rough estimate of the contributions due to storage and interconnect. In order to meet the cycle time constraint, the bus load and the depth of multiplexor trees have to be balanced.

The basic approach is to abstract the main contribution to area or delay in terms of hardware components, to iterate with alternate possibilities, and to optimize with respect to unconstrained resources in order to stay within the constraint. Depending on the application domain, however, some schemes may be more appropriate than others. The most obvious decision to be made is whether the time constraint(s) or the area constraint is more difficult to meet. Depending on that, one may choose an approach with resource constrained scheduling in order to be close to the critical area constraint. Meeting the time constraint is then the goal of iteration and optimization. Or one may choose an approach with time constrained scheduling in order to be close to the critical time constraint(s). In this case, meeting the area constraint is attempted by iteration and optimization. Depending on the approach taken, the strong interrelation of subtasks, especially scheduling, functional unit allocation and functional unit (type) assignment is broken. Synthesis approaches are classified accordingly.

Component	Delay *ns*
`ALU(+,-,<)`	40
`Adder`	35
`Subtractor`	35
`Comparator(<)`	10
`Parallel Multiplier`	80
`Register`	2
`2:1 Multiplexor`	2
`Tristate Driver`	2
`2-Stage Pipelined Multiplier`	90

Table 6.2: Component library.

6.3.1 Synthesis of the Differential Equation Example

In the following a case study is presented pushing the loop of the Differential Equation example through a sample synthesis approach. The starting point is the internal representation as shown in Figure 6.6.

First, a hardware component library (see Table 6.2) has to be provided. Multiplexed and bidirectional bus data part architectures can be supported by this library. For the case study, the multiplexed data part architecture with a single system clock is chosen. According to the delays of the hardware components, a clock cycle time of 50 *ns* is reasonable. Arithmetic and logic operations, except multiplication, can be performed within one clock cycle when adding a delay of 10 *ns* per clock cycle as a rough estimate for the contribution of interconnect delay and register delay. The multiply operation takes two clock cycles. Another reasonable clock cycle time is 100 *ns*. In this case, the multiply operation is executed in a single cycle when assigned to the parallel multiplier and two consecutive fast arithmetic operations can be chained, i.e., executed within one control step.

Most synthesis approaches rely on handling directed acyclic data flow graphs. For this purpose the loop is cut, i.e., only the loop body and the test expression are considered. For proper function one has to make sure that variables $x1, y1$ are assigned to variables x, y before the next iteration of the loop is started. It is assumed that the initial values of x, u, y are read from one read port and the result value of y is written out to a write port. Finally, it is assumed that the constants $3, 5$ and the parameters a, dx are held in registers.

The example synthesis approach consists of the following steps:

- Functional unit allocation
- Resource constrained scheduling
- Functional unit assignment
- Register allocation
- Register assignment
- Multiplexor extraction

Functional unit allocation comprises the decision how many and which hardware components need to be provided for the execution of the operations. For the Differential Equation example a functional unit set of two parallel multipliers and one ALU(+, − <) is chosen.

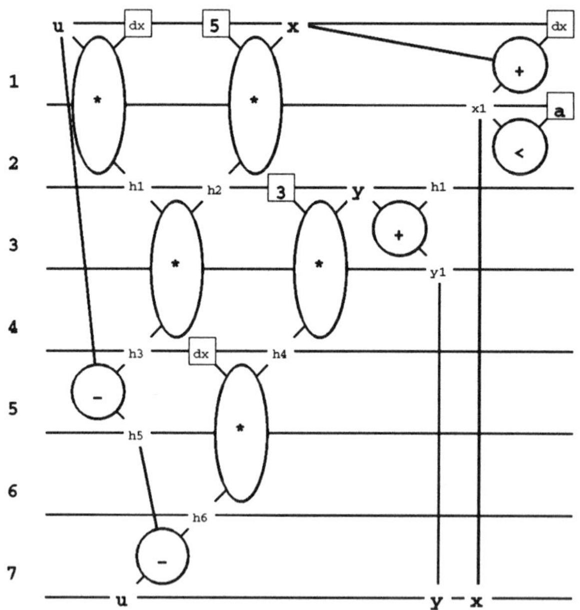

Figure 6.8: Schedule for 2 parallel multipliers and 1 ALU.

The functional unit set represents the resource constraints for the subsequent resource constrained scheduling. Considering the start vertices of the

Functional Unit	Multiplier $M1$	Multiplier $M2$
Operation	$1,3,5$	$2,4$

Table 6.3: Functional unit assignment.

data flow graph, one observes that three multiply operations and one add operation are ready to be scheduled. However, only two multipliers and one ALU are available. Hence, only two multiply operations can be assigned to the first control step. In the second control step the ALU is free again, and the compare operation that is now ready to be scheduled can be assigned. However, the two multipliers are still occupied, so the third multiply operation is delayed to the third control step. Proceeding from control step to control step, the schedule shown in Figure 6.8 is obtained. It needs seven control steps. An intelligent choice was made when choosing Multiply 1 and Multiply 2 among the three multiply operations ready to be scheduled in the first control step, since these two multiply operations lie on the critical path.

It must now be decided which operation is executed on which functional unit, i.e., functional unit assignment has to be performed. All add and subtract operations as well as the compare operation need the ALU, so for these operations, functional unit assignment is already decided. For the multiply operations, the first pair of operations that conflict in their active time is considered. Multiply 1 and Multiply 2 are assigned to Multiplier $M1$ and Multiplier $M2$ respectively. Since the assignment of operations to functional units influences interconnect area, it is advisable to control functional unit assignment with some measure for this influence. By assigning the multiply operations subsequently to the multipliers, one may exploit structural information obtained in intermediate situations. The source values of the two multiply operations are u, dx and $5, x$ and the destination values are the values $h1$ and $h2$ respectively. Inspecting the source and destination values of the next two multiply operations Multiply 3 and Multiply 4, there are only values $h1, h2$ in common. Multiply 3 takes both the output $h1$ of Multiplier $M1$ and the output $h2$ of Multiplier $M2$. Hence the choice of the multipliers for the multiply operations 3 and 4 is free; Multiply 3 is assigned to Mulplier M1, and Multiply 4 to Multiplier M2. The best assignment for Multiply 5 is Multiplier M1 since Multiply 1 and Multiply 5 have the source value dx in common. The resulting functional unit assignment is shown in Table 6.3.

The next steps are register allocation and register assignment. Both steps can be combined or performed separately. Inspecting the scheduled data flow graph one can see that, for example, value u is used in the first and the second control step by Multiply 1 and also by the subtract operation in Control Step 5. From the beginning of the first control step up to the fifth control step excluding the start time of the sixth control step, value u has to be available or in other words value u is *alive*. Since it cuts control step boundaries it has to be kept in a register. The time a value is alive is called its *lifetime*. The lifetime table for all values is shown in Figure 6.9. The largest number of values

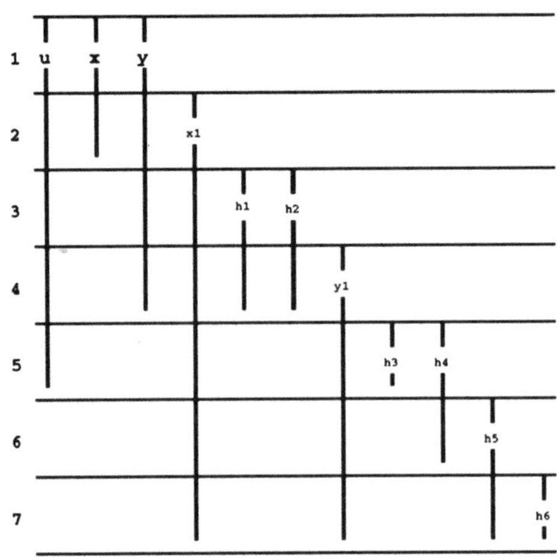

Figure 6.9: Lifetime table.

that have to be stored occurs when stepping from Control Step 3 to Control Step 4 where 6 values cut the control step boundary. Therefore 6 registers $R1, R2, \ldots, R6$ are allocated that have to be shared by 11 values. Two values can share a register if they have disjoint lifetimes. Again, the influence on interconnect area has to be taken into account. From the preceding functional unit assignment step, one can now exploit more precise structural information by inspecting the source and the destination hardware components a register has according to the values already assigned to it. In Table 6.4 a snapshot is shown where values $u, x, y, x1, h1, h2, y1$ are already assigned to registers. These have been assigned by proceeding from control step boundary to control step boundary. Inspecting the lifetime table one can see that the values $h3, h4$ have to be assigned to registers next. Due to the lifetimes of the values already assigned to registers, $h3$ and $h4$ can be assigned to $R3, R5, R6$. Comparing sources and destinations the best choice is to assign $h3$ to $R5$ and $h4$ to $R6$. The complete register assignment is shown in Table 6.5.

In order to complete the multiplexed data part architecture, multiplexors have to be extracted. As an example, values transferred to Multiplier $M1$

Register	$R1$	$R2$	$R3$	$R4$	$R5$	$R6$
Value	u	x $y1$	y	$x1$	$h1$	$h2$
Source	R-$Port$	$R4$ ALU R-$Port$	$R2$ R-$Port$	ALU	$M1$	$M2$
Destination	$M1$ ALU	$M2$ ALU $R3$	$M2$ ALU W-$Port$	ALU_l $R2$	$M1$ ALU	$M1$

Table 6.4: Intermediate result for register assignment.

Register	$R1$	$R2$	$R3$	$R4$	$R5$	$R6$
Value	u $h5$	x $y1$	y	$x1$	$h1$ $h3$ $h6$	$h2$ $h4$

Table 6.5: Final result for register assignment.

are considered. In Table 6.6 the control steps and the corresponding registers whose outputs are active at the left and the right input of the multiplier are listed. Counting the number of different registers active, one obtains the size of the multiplexor or more precisely the number of multiplexor inputs required at the corresponding input. In this case, three multiplexor inputs result for the left and two multiplexor inputs for the right input of Multiplier $M1$. There is still potential for optimization. Since the multiply operation is commutative inputs may be swapped in control steps 1 and 2 resulting in two multiplexor

Control Step	1	2	3	4	5	6	7
Register left input of $M1$	$R1$	$R1$	$R5$	$R5$	Rdx	Rdx	-
Register right input of $M1$	Rdx	Rdx	$R6$	$R6$	$R6$	$R6$	-

Table 6.6: Inputs to multiplier $M1$.

inputs for both inputs. Swapping inputs as described here is often referred to as *operand alignment*. Extraction of the multiplexors so far does not provide the mapping of the multiplexing structure to library components. For estimation or comparison of results two-to-one multiplexor trees are often used as a default mapping.

Most of the information needed for a register-transfer level implementation has been collected at this stage. There is still optimization potential at the interface between data part and controller with respect to the encoding of condition and control signals. For the purposes of this book, it is sufficient to assume that for each enable input of a register, for each select input of a multiplexor, and for each opcode select input of an ALU, control nets are provided and that control inputs for the multiplexors and the ALU opcodes are encoded straightforwardly. According to these assumptions the state-transition table is derived, and the netlist is generated. In Figure 6.10 the resulting netlist is shown, and in Table 6.7 the control outputs in Control Step 3 are depicted.

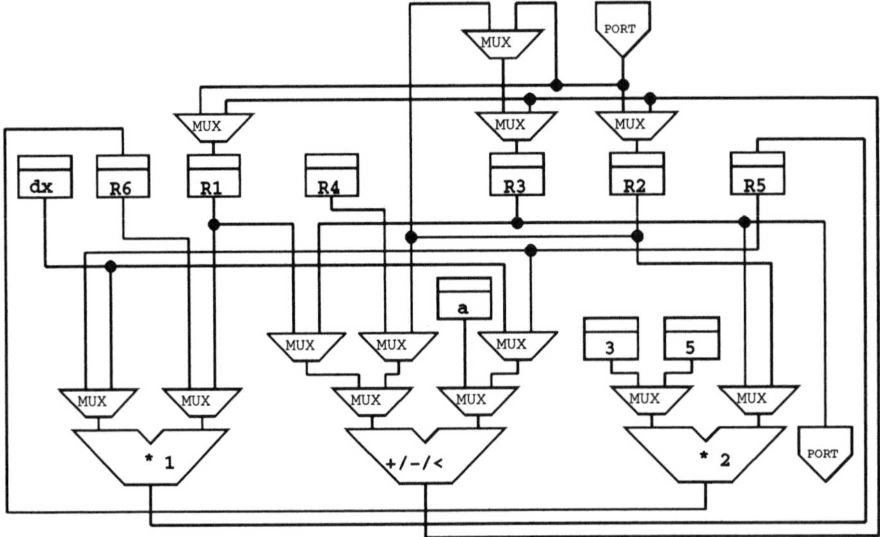

Figure 6.10: Multiplexed data part architecture of Differential Equation example.

6.3.2 Classification of Synthesis Approaches

In the eighties many synthesis approaches were developed. It is worthwhile to define criteria for classifying synthesis approaches.

$R1$	$R2$	$R3$	$R4$	$R5$	$R6$
0	1	0	0	0	0
ALU opcode	MUX_{M1_l}	MUX_{M1_r}	MUX_{M2_l}	MUX_{M2_r}	-
00	1	1	1	1	
$MUX_{ALU_{ll}}$	$MUX_{ALU_{lr}}$	MUX_{ALU_l}	$MUX_{ALU_{rr}}$	MUX_{ALU_r}	-
0	x	0	0	0	

Table 6.7: Control outputs in Control Step 3.

Referring to the basic serial/parallel trade-off, one may assume that a point in the design space is already well-defined by the schedule and the functional unit set satisfying the schedule. For this purpose, scheduling, functional unit allocation and functional unit assignment—at least the assignment of operations to functional unit types—have to be solved. How the close interrelation of these subtasks is broken, or whether it is broken at all, provides therefore a significant criterion for the classification of synthesis approaches. Independent schemes are characterized by completely separating the decision on the functional unit set from scheduling. Iterative schemes iterate the derivation of the required functional unit set with the derivation of a schedule. Simultaneous schemes derive the schedule and the functional unit set simultaneously.

Independent schemes. There are two basic independent schemes. Either scheduling is performed first and a resulting functional unit set is derived from the schedule, or functional unit allocation is performed by specifying resource constraints and a resource constrained scheduling follows. Starting with a fixed functional unit set is very popular. The CADDY program [GuKR91] is an example of this approach. Scheduling and assigning operations to functional unit types are performed simultaneously in CADDY.

An alternative approach to the second case—often referred to as partitioning approach—is to perform both functional unit allocation and functional unit assignment before scheduling. This approach has been implemented in the BOTTOM-UP DESIGNER (BUD) [McKo90], and a similar approach is described for the CALLAS system in [SGLM90].

Iterative schemes. An example of an iterative solution of scheduling and functional unit allocation is found in the HAL system [Paul88]. Scheduling and functional unit allocation are invoked repeatedly in order to obtain an acceptable solution for the schedule and the functional unit set. First, a reasonable number of the slowest functional unit types are allocated, then time constrained scheduling is performed which allows for refinement of the functional unit set as the third step. Finally, scheduling with respect to the refined functional unit set follows.

Simultaneous schemes. Scheduling, functional unit allocation and functional unit assignment can be solved simultaneously in several ways. All three tasks (and possibly others) can be performed in a constructive manner by either proceeding from control step to control step or by proceeding from operation to operation chosen according to a cost function. All three tasks can also be performed globally with respect to control steps and operations.

One representative of constructive approaches is found in the ADAM system [PaPM86]. In this system, first the critical paths are evaluated, based on an average delay of all functional units capable of executing a certain operation. Then a feasible clock cycle time is derived from the maximum average delay and a time constraint is set according to the corresponding minimum number of control steps. Finally, the MAHA tool [PaPM86] is invoked which schedules the critical path operations first, and concludes by scheduling the noncritical path operations. Those non-critical path operations that have the least freedom (the least number of available control steps) are scheduled first. Simultaneously with each operation-to-control step assignment decision, new functional units are allocated as needed and operations assigned to functional unit instances.

Another example of constructive approaches is found in the SAM tool [ClTh90]. According to a cost function, SAM proceeds from one operation-to-control step assignment and operation-to-functional unit assignment to the next considering in each step all not yet scheduled operations and all control steps.

A global approach to simultaneously solving scheduling, functional unit allocation and functional unit assignment is presented in [DeNe89]. Solving all three tasks simultaneously is modeled by placing the operations in a two-dimensional matrix. One dimension represents the control steps, the other the functional units. Placement of operations in the two dimensions is performed by simulated annealing.

6.4 Scheduling

Scheduling problems have applications in a broad variety of domains including computer science, economics, or management sciences. Within the field of operations research scheduling problems have been studied for decades. Scheduling problems are concerned with optimally executing a given set of tasks by utilizing several processors and other resources, subject to constraints such as precedence constraints between tasks or task deadlines. The goal is to minimize some objective function such as the total processing time or the cost of resources. Among the scheduling problems identified within operations research *precedence constrained scheduling* [GaJo79] (see Definition 6.1) fits into the context of high-level synthesis. Within precedence constrained scheduling the execution of tasks is subject to precedence constraints; this corresponds to the operation precedence constraints of the data flow graph. Although there are features of scheduling problems in high-level synthesis that are not covered by precedence constrained scheduling as given in Definition 6.1, it is worthwhile to introduce it for complexity considerations.

Definition 6.1 (Precedence constrained scheduling) *Given a set T of tasks of equal length 1, a partial order $\prec$ on T, a number of $m \in Z^+$ processors, and an overall deadline $D \in Z^+$. Precedence constrained scheduling is defined as the following problem: Is there a schedule $\sigma : T \rightarrow \{0, 1, \ldots, D\}$ such that*

$$|\{t \in T : \sigma(t) = s\}| \leq m \, \forall s \in \{0, 1, \ldots, D\}$$

and

$$t_i \prec t_j \Rightarrow \sigma(t_i) < \sigma(t_j) \text{ ?}$$

The precedence constrained scheduling problem can be related to high-level synthesis by identifying tasks with operations, task lengths with operation delays, processors with functional units, the overall deadline minus one with the number of control steps needed, and the partial order with the partial order defined by the data flow graph. Precedence constrained scheduling is computationally "inherently difficult"—it has been proven to be NP-complete [GaJo79]. Definition 6.1 formulates precedence constrained scheduling in such a way that just a yes-no answer is expected. Problems of this type are called *decision problems.* The two instances of the scheduling problem in high-level synthesis, however, correspond to *optimization* problems in which either the deadline or the processor cost is to be minimized. According to the definition given in [GaJo79] optimization problems involving an NP-complete decision problem are NP-hard.

Moreover, precedence constrained scheduling as given in Definition 6.1 is too restrictive. Assuming uniform processors that are capable of executing any task and uniform task lengths independent of the processor an operation is assigned to, is not a realistic model for applications in high-level synthesis. In the following section the basic terminology and assumptions for scheduling in high-level synthesis are introduced. These include the underlying data flow graph structure, the notion of time, and the assumptions made about operations and functional units.

6.4.1 Terminology and Assumptions

In Section 6.2 a graph representation $DFG(V_o, V_c, E, E_c, E_s)$ for the data and control flow was introduced as starting point for synthesizing the register-transfer structure. In general, data flow graphs are cyclic graphs. Cycles are caused by feedbacks in inner loops of the algorithmic description or by the implicit infinite outer loop that is assumed for DSP algorithms. Most of the scheduling algorithms are restricted to directed acyclic graphs. As already illustrated with the Differential Equation example in Section 6.3.1, a directed acyclic graph is cut out from a loop construct in the algorithmic description. Only the loop body including the test operations—encapsulated between loop entry/loop exit pairs associated with the loop in the algorithmic description—are given to the scheduler. Pure data flow descriptions are handled in a similar way. Delay operations are cut; the outputs of delay elements become inputs, the inputs of delay elements outputs of the directed acyclic graph. Accordingly, loop entry, loop exit, and delay vertices are not considered in most of the following sections. For the purposes of this chapter it is also sufficient to consider only data flow graphs without conditional branches (no branch and merge vertices) and without control flow resulting from sequence edges. As a starting point for scheduling, an acyclic data flow graph $DFG(V, E)$ without control flow elements remains. V is the set of operations V_o (without the delay operations). The edges E of the graph represent the precedence constraints. Acyclic data flow graph portions that have been cut out correspond to the basic blocks introduced above. Considering primarily basic blocks can be justified if applications where the length of the critical path in basic blocks is large in comparison to complete paths through the algorithmic description are the main target. Assumptions and extensions for scheduling when control flow is to be considered are treated separately in Section 6.4.5.

Scheduling has to provide the assignment of operations to time steps or control steps. Time step and control step (denoted s in the following) can be used as synonyms as long as non-pipelined designs are considered. For pipelined designs, however, one control step may contain operations of many different time steps associated with the computation of overlapping input samples. For convenience, the term "control step" is maintained until pipelining is presented. Assuming a single system clock, the time required for a computation is the product of the number of control steps needed and the cycle time t_{cycle}.

Since a one-to-one correspondence between operations and functional units does not exist, one must differentiate between operation types and functional unit types: o_t is an operation of type t and r_k is a functional unit of type k. R is the set of functional units, R_k the set of functional units of type k, $|K|$ is the number of functional unit types, and $|T|$ the number of operation types. In contrast to precedence constrained scheduling not all functional unit types

k are capable of executing all operations of type t. Relation $o_t \in r_k$ expresses that functional unit r_k is capable of executing operation o_t. The following hierarchy of assumptions regarding operation types and functional unit types is used throughout this section:

- *Uniform functional unit type*

  All types of operations can be executed on a single functional unit type such as a universal ALU:

$$o_t \in r_1, \forall t \in T$$

- *Disjoint operation type sets*

  There are disjoint sets of operation types that match a functional unit type. As an example one may consider operation types $+, -, <$ compatible with an ALU and a multiply operation compatible with a multiplier.

$$\{t \in T : o_t \in r_{k_1}\} \cap \{t \in T : o_t \in r_{k_2}\} = \emptyset \, \forall k_1 \neq k_2 \in K$$

- *Overlapping functionality*

  Operation type sets overlap. As an example consider add and subtract operations and a functional unit set containing adders and ALUs$(+, -)$. Obviously, the functionality of adders and the ALUs overlap. Even the mere existence of two types of multipliers (e.g., parallel and pipelined) causes overlapping operation type sets.

$$\{t \in T : o_t \in r_{k_1}\} \cap \{t \in T : o_t \in r_{k_2}\} \neq \emptyset \text{ for some } k_1 \neq k_2 \in K$$

Depending on the clock cycle time, only the fastest operations may complete within one cycle and slower operations take two or more clock cycles. This is called *multi-cycling*. Multi-cycling has been used for the multiply operation in the synthesis of the Differential Equation example (Section 6.3.1). If the clock cycle time is large enough, a chain of data dependent and fast operations may fit into one clock cycle. This is called *chaining*. In general, however, operation delays $\delta(o_t, r_k)$ (in *ns*) depend on the functional unit type k the operation is assigned to. This, of course, makes scheduling difficult. It is therefore often assumed that one has disjoint operation type sets which allows the association of operation delays with the operation type.

Accordingly, operations may be *fast* operations that can be chained with one or more other fast operations in one control step

$$\delta(o_{t_1}) + \delta(o_{t_2}) + \ldots + \delta(o_{t_n}) < t_{cycle}. \tag{6.1}$$

They may be *single-cycle* operations covering one control step

$$\delta(o_t) \leq t_{cycle}, \tag{6.2}$$

or *multi-cycle* operations that cover more than one control step

$$\delta(o_t) > t_{cycle}. \tag{6.3}$$

According to the terminology presented so far, the scheduling problem for high-level synthesis and the terms *schedule* and *schedule length* S can be introduced as given in Definition 6.2. For simplicity, disjoint operation type sets and single-cycle operations are assumed.

Definition 6.2 (Simple scheduling problem) *Given an acyclic $DFG(V, E)$ with single-cycle operations, $|K|$ different types of functional units, $|R_k|$ functional units of type k, disjoint operation type sets, and a schedule length S. The simple scheduling problem is defined as the following problem: Is there a schedule $\sigma : V \rightarrow \{1, \ldots, S\}$ such that*

$$|\{o_t \in V \wedge o_t \in r_k : \sigma(o_t) = s\}| \leq |R_k| \;\; \forall s \in \{1, \ldots, S\}, k \in \{1, \ldots, K\}$$

$$(o_i, o_j) \Rightarrow \sigma(o_i) < \sigma(o_j)?$$

If unlimited functional units are available ($|R_k| \rightarrow$ arbitrarily large, for $1 \leq k \leq K$), the minimum schedule length S corresponds to the critical path. Given a certain schedule length S and allowing for unlimited resources, operations can start execution in several different control steps that lie between the earliest $\sigma_{ASAP}(o)$ and the latest control step $\sigma_{ALAP}(o)$ feasible for operation o. The two values $\sigma_{ASAP}(o)$ and $\sigma_{ALAP}(o)$ depend on the precedence constraints. For calculating both the critical path and the ASAP and ALAP time of operations an algorithm is given in Section 6.4.3.

Definition 6.3 (ASAP and ALAP time) *Given an acyclic $DFG(V, E)$, a schedule length S and unlimited resources, the ASAP time $\sigma_{ASAP}(o)$ is the earliest time step feasible and the ALAP time $\sigma_{ALAP}(o)$ the latest time step feasible for operation o to start execution.*

6.4.2 The Two Basic Instances of the Scheduling Problem

In order to provide a formal reference, the two basic instances of the scheduling problem in high-level synthesis, resource constrained scheduling [HwHL90] and time constrained scheduling [LeHL89], are stated in terms of Integer Linear

Programming (ILP). Integer Linear Programming is the following optimization problem [PaSt82]. Given a cost function $c^T x$ and a constraints set of linear equations $A x = b$, A_{ij}, b_i integer,[8] find a parameter configuration x meeting the constraints such that the cost function is minimized and entries x_i are positive and integer.

The ILP formulation for resource constrained scheduling is presented in Figure 6.11. A set of functional units R with $|R_k|$ functional units of type k and $|K|$ different functional unit types are given. The objective is to minimize the schedule length S. The decision whether an operation o_i has been scheduled to control step s is represented by 0-1 integer variables $x_{i,s}$, such that $x_{i,s}$ equals 1 if o_i has been assigned to s, and $x_{i,s}$ equals 0 otherwise. The essential resource constraints are represented as set of constraints Relation 6.5. They have to be obeyed in each control step $1 \leq s \leq S$. Constraints 6.6 ensure that each operation o_i is scheduled. The control step $\sigma(o_i)$ operation o_i is assigned to is defined by the set of constraints Equation 6.7.[9] Inequalities Relation 6.8 express the precedence constraints, and finally, the set of constraints Relation 6.9 ensures that the control steps the end vertices are assigned to do not exceed the schedule length S. For convenience, additional integer variables $\sigma(o_i)$ have been introduced [HwLH91], which can be eliminated again if constraints Equation 6.7 are inserted into the sets of constraints Relation 6.8 and 6.9.

The typical objective in time constrained scheduling (see Figure 6.12) is to optimize the utilization of functional units, or—in other words—to distribute operations requiring the same functional unit type as much as possible over control steps. The cost function to be minimized is the resource cost where c_k is the cost of a functional unit r_k of type k. Constraints 6.11 ensure that no control step requires more than $|R_k|$ functional units of type k. Constraints 6.12 again reflect that all operations have to be scheduled, and Relation 6.13 represents the precedence constraints.[10]

The decision problem of Integer Linear Programming in general has been proven to be NP-complete [GaJo79]—again reflecting the complexity of the scheduling problem. However, since significant progress has been made in the past decade in the development of efficient ILP algorithms,[11] many scheduling approaches have been published that rely on solving the ILP problem di-

[8] In case of inequalities, slack variables can be introduced to yield equations.

[9] Constraint sets 6.6 and 6.7 exploit the fact that operations are scheduled between their ASAP and ALAP times. ASAP and ALAP times are calculated using the upper bound on the schedule length S derived from a preliminary list scheduling (see Section 6.4.3).

[10] Again, constraint sets 6.12 and 6.13 exploit the fact that operations are scheduled between their ASAP and ALAP times. ASAP and ALAP times are calculated using the specified schedule length S.

[11] For instance, *Combinatorial Optimization* by Papadimitriou and Steiglitz [PaSt82].

$$\min S \tag{6.4}$$

$$\sum_{o_i \in r_k} x_{i,s} - |R_k| \leq 0, \text{ for } 1 \leq s \leq S, \ 1 \leq k \leq |K| \tag{6.5}$$

$$\sum_{s=\sigma_{ASAP}(o_i)}^{\sigma_{ALAP}(o_i)} x_{i,s} = 1, \text{ for } 1 \leq i \leq |V| \tag{6.6}$$

$$\sum_{s=\sigma_{ASAP}(o_i)}^{\sigma_{ALAP}(o_i)} (s * x_{i,s}) - \sigma(o_i) = 0, \ \forall \, o_i \tag{6.7}$$

$$\sigma(o_i) - \sigma(o_j) \leq -1, \ \forall \, (o_i, o_j) \in E \tag{6.8}$$

$$\sigma(o_i) - S \leq 0, \ \forall \, o_i \in V_{end} \tag{6.9}$$

Figure 6.11: ILP formulation for resource constrained scheduling.

$$\min \sum_{k=1}^{K} c_k * |R_k| \tag{6.10}$$

$$\sum_{i=1}^{N} x_{i,s} - |R_k| \leq 0, \text{ with } o_i \in r_k \text{ and } 1 \leq s \leq S, \ 1 \leq k \leq |K| \tag{6.11}$$

$$\sum_{s=\sigma_{ASAP}(o_i)}^{\sigma_{ALAP}(o_i)} x_{i,s} = 1, \text{ for } 1 \leq i \leq |V| \tag{6.12}$$

$$\sum_{s=\sigma_{ASAP}(o_i)}^{\sigma_{ALAP}(o_i)} s * x_{i,s} - \sum_{s=\sigma_{ASAP}(o_j)}^{\sigma_{ALAP}(o_j)} s * x_{j,s} \leq -1, \ \forall \, (o_i, o_j) \in E \tag{6.13}$$

Figure 6.12: ILP formulation for time constrained scheduling.

rectly [HwHL90, LeHL89, HwHL91]. Rather than introducing the corresponding methods for ILP problems, heuristics that are often used are presented in the following sections. The basic assumptions that have been made for the ILP formulations presented here are too strict for a wide application to scheduling problems. Using heuristics, more realistic scheduling problems that include multi-cycling, chaining, and overlapping functionality become feasible.

6.4.3 Resource Constrained Scheduling

There is not much hope of finding an optimum polynomial-time algorithm that solves the scheduling problem including multi-cycling, chaining, and overlapping functionality. The typical heuristic for resource constrained scheduling is *list scheduling* which can be followed back to early publications in operations research (see, e.g., [Coff76]). The basic idea is to sort the operations in a priority list in order to provide a selection criterion if operations compete for resources. List scheduling is a constructive method proceeding from control step to control step. Those operations whose predecessor operations have been scheduled and completed earlier are candidates for being scheduled in the current control step. Candidate operations are called *ready* operations. If the number of ready operations exceeds the number of functional units available, the operations with the highest priority are selected for being scheduled.

6.4.3.1 Hu's Algorithm

It is well known that restricting NP-complete problems often leads to subproblems which are solvable in polynomial time. Aiming at optimum polynomial-time scheduling, however, restricts the solvable scheduling problems to an unacceptable small set. It is, however, worthwhile to introduce an important result for a subproblem of precedence constrained scheduling obtained by Hu [Hu61] to provide a simple outline for list scheduling.

This subproblem is derived by restricting precedence constrained scheduling as presented in Definition 6.1 to partial orders whose graph representation corresponds to a forest—a set of trees. Given a data flow graph $DFG(V, E)$ consisting of a set of trees with single-cycle operations and $|R_1|$ uniform functional units, find the minimum schedule length S. This problem is solved by Algorithm 6.1 (Hu's algorithm).

Hu's algorithm consists of two routines. The first routine labels the operations in order to provide the priority criterion and the second routine performs the essential scheduling step. From control step to control step, operations that are ready for being scheduled are collected. Since the number of ready operations $|V_{ready}|$ may exceed the number of functional units $|R_1|$ available, only ready operations with the highest label, or the highest priority, are selected—added to V_{sel}—to be assigned to the current control step. This is repeated until all operations are scheduled.

Hu's algorithm can be modified to be used as a heuristic for general acyclic data flow graphs; optimum results are then no longer guaranteed. However, the restrictions of Hu's algorithm to a uniform functional unit type and single-cycle operations are not acceptable for realistic applications. Furthermore, the priority function used is static and deals with single operations. Better results

A polynomial-time algorithm for finding a minimum schedule length S if DFG(V, E) is a forest. Without loss of generality, a single tree is assumed.

```
{ Labelling routine: Assign labels P(o) to operations }
Λ ← ∅;
label ← 1;
WHILE V\Λ ≠ ∅ DO
BEGIN V_end ← {o ∈ V\Λ : Succ(o, DFG) ⊆ Λ};
      P(o ∈ V_end) ← label;
      Λ ← Λ ∪ V_end;
      label ← label + 1;
END;

{ Scheduling routine: Assign operations to control steps σ(o) }
Σ ← ∅;
s ← 0;
WHILE Σ ≠ V DO
BEGIN
      s ← s + 1;
      V_ready ← {o ∈ V\Σ : Pred(o, DFG) ⊆ Σ};
      IF |V_ready| > |R_1| THEN
      BEGIN
         V_sel ← ∅;
         L_sorted ← Sort V_ready with decreasing P;
         WHILE L_sorted ≠ NIL ∧ |V_sel| < |R_1| DO
         BEGIN
               V_sel ← V_sel ∪ {HEAD(L_sorted)};
               L_sorted ← TAIL(L_sorted);
         END;
      END ELSE V_sel ← V_ready;
      σ(o ∈ V_sel) ← s;
      Σ ← Σ ∪ V_sel;
END;
writeln('Schedule Length S:', s);
```

Algorithm 6.1: Hu's algorithm for finding the minimum schedule length S.

can be expected when dynamic priority functions are used that depend on the schedule constructed so far and that can handle subsets of operations in the ready set.

6.4.3.2 ASAP and ALAP Scheduling

Allowing for arbitrary acyclic data flow graphs but unlimited resources one may use Hu's algorithm for ASAP and ALAP scheduling. In the scheduling routine of Algorithm 6.1 the **THEN** branch will never be entered since there are always functional units available. Thus, all ready operations are scheduled immediately or "as soon as possible." They are assigned to the earliest control step $\sigma_{ASAP}(o)$ possible. This is called ASAP scheduling. The resulting schedule length S_{ASAP} corresponds to the length of the critical path(s). By using the same procedure but starting from the end vertices of the data flow graph all operations are scheduled "as late as possible." They are assigned to the latest control step $\sigma_{ALAP}(o)$ possible. Accordingly, this is called ALAP scheduling. ASAP scheduling has been used in some of the early approaches [Paul88], but usually ASAP as well as ALAP scheduling are performed in order to calculate ASAP and ALAP times, the critical path(s), or to find an average distribution of operation types in a control step.

Both ASAP and ALAP scheduling as presented here are restricted, since refinements for multi-cycling and chaining are missing. Extensions necessary for these scheduling features are introduced in the next section.

6.4.3.3 List Scheduling

For the refined list scheduling heuristic given in Algorithm 6.2 disjoint operation type sets are assumed. The algorithm uses a dynamic priority function, and supports multi-cycling and chaining. Multi-cycling and chaining are, however, kept separate. An example of mixing both is a multi-cycle operation that is chained with a fast operation in the first or last control step covered by the multi-cycle operation. Keeping multi-cycling and chaining separate allows further simplification of the treatment of operation delays. They can be expressed in terms of cycles. The delay of operations is set to a value $\Delta(o) < 1$ for fast operations, to $\Delta(o) = 1$ for single-cycle operations, and to $\Delta(o) = \lceil \frac{\delta(o)}{t_{cycle}} \rceil$ for multi-cycle operations.

In order to keep track of whether a fast operation can be chained with a fast operation that is already scheduled in the current control step, absolute start times modulo the number of the current control step ($\tau(o)$) are stored along with the unscheduled operations.

The function $Resource\_Free\_For(o, s)$ returns **TRUE** if in control step s for operation o a functional unit of the correct type is available. The procedure $Reserve\_Resource(o, s)$ reserves a corresponding functional unit, and the function $Resource\_Free\_In(s)$ returns **TRUE** if there is any functional unit available in control step s. It is worthwhile to realize that even though fast operations may be chained within one control step, resources for fast operations are oc-

List scheduling with multi-cycling, chaining, disjoint operation type sets and a dynamic priority function $\mathcal{P}$.

```
s ← 0;
Σ ← ∅;
WHILE Σ ≠ V DO
BEGIN
    s ← s + 1;
    FOR all o ∈ V\Σ DO τ(o) ← 0;
    V_ready ← {o ∈ V\Σ : Pred(o, DFG) ⊆ Σ ∧
              ∀ o' ∈ Pred(o, DFG) : σ(o') + Δ(o') ≤ s};
    FOR all o ∈ V_ready DO calculate P;
    L_sorted ← Sort V_ready with decreasing P;
    WHILE L_sorted ≠ NIL ∧ Resource_Free_In(s) DO
    BEGIN
        op ← HEAD(L_sorted);
        L_sorted ← TAIL(L_sorted);
        IF Resource_Free_For(op, s) THEN
        BEGIN
          σ(op) ← s;
          Σ ← Σ ∪ {op};
          Reserve_Resource(op, s);
          FOR all o ∈ Succ(op, DFG) DO
              τ(o) ← max(τ(o), τ(op) + Δ(op));
          IF Δ(op) < 1 THEN
          BEGIN
            V_chain ← {o ∈ Succ(op, DFG) :
                        Pred(o, DFG) ⊆ Σ ∧ τ(o) + Δ(o) ≤ 1};
            FOR all o ∈ V_chain DO calculate P;
            L_sorted ← Sort SET(L_sorted) ∪ V_chain with decreasing P;
          END;
        END;
    END;
END;
writeln('Schedule Length S:', s);
```

Algorithm 6.2: List scheduling.

cupied for the complete control step. Furthermore, combinational functional units with a multi-cycle delay are occupied for the complete delay time, whereas pipelined functional units are free after the completion of the first stage. The use of pipelined functional units is often referred to as *structural pipelining*.

The conditions for ready operations have been modified in order to account for multi-cycling. In order to support chaining, the set of ready operations needs to be updated within the control step currently considered. For the ready operations, priorities are calculated in each control step.

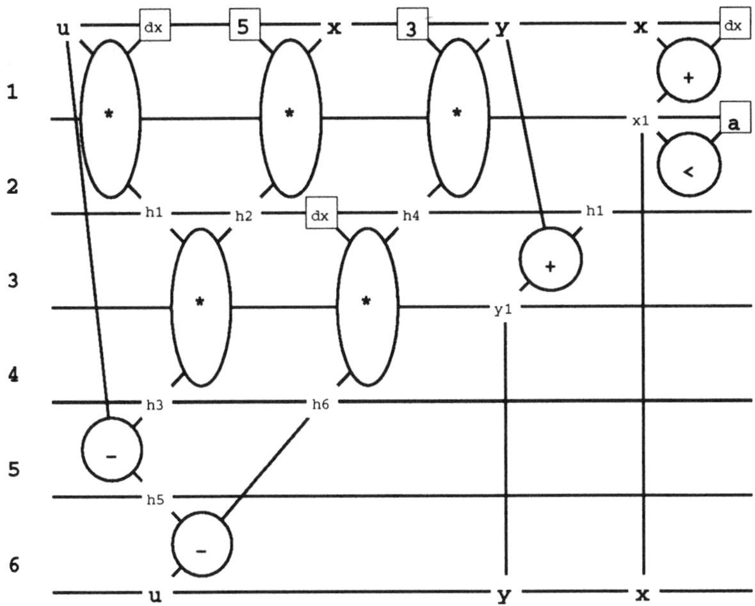

Figure 6.13: ASAP schedule for Differential Equation example.

Many different priority functions can be used. One of the simplest and best-known is *mobility* which is used as a representative for dynamic priority functions in this section. Mobility priority is used in the S(P)LICER system [PaGa87]. Mobility $\mathcal{M}$ is the number of control steps from the earliest up to the latest feasible control step in which an operation can start execution such that the critical path(s) are not lengthened. For the control step first processed this is simply the difference between the ALAP time and the ASAP time obtained from ASAP and ALAP scheduling

$$\mathcal{M}_o = \sigma_{ALAP}(o) - \sigma_{ASAP}(o).$$

In Figure 6.13 the ASAP schedule for the loop of the Differential Equation example is shown. The library presented in Section 6.3.1 was used to calculate this schedule. The ASAP schedule results in a minimum schedule length S_{ASAP}

of 6. Operations on the critical path obtain mobility $\mathcal{M} = 0$. The mobility of the $x + dx$ operation is 3, whereas the mobility of $u \times dx$ is 0. The higher the mobility, the lower is the need to schedule an operation in the current control step. Operations are therefore sorted according to increasing mobility. Since operations that have not been assigned to their earliest possible control step reduce in mobility, ASAP and ALAP times have to be updated for the rest of the graph for every control step. If mobility priority is used in Algorithm 6.2, application to the Differential Equation example with 2 parallel multipliers and 1 ALU as resources results in the schedule already shown in Figure 6.8. The ASAP schedule with schedule length $S = 6$ given in Figure 6.13 requires 3 multipliers, whereas the list schedule for 2 multipliers results in a schedule length of $S = 7$.

6.4.3.4 The CADDY Priority Function

In this section the *delay reduction look-ahead* priority function of the CADDY tool [CaRo89, KrRo89, KNRR88, GuKR91, KrRo90] is introduced.

The mobility priority gives priority to scheduling of critical path operations. This seems very reasonable; but there are cases where preferring critical path operations results in a bad schedule. A more refined priority function that may delay critical path operations often yields better results. This is the rationale behind the delay reduction look-ahead priority.

Another important feature of the CADDY list scheduler is that operation-to-functional unit type assignment is performed during scheduling in order to allow for overlapping functionality. The following two examples illustrate the importance of this feature. First, consider a functional unit set containing one pipelined multiplier, and one two-cycle parallel multiplier and a control step where only one multiply operation may be scheduled followed by two other multiply operations in the data flow graph. It is crucial to decide whether to reserve the pipelined or the parallel multiplier in the current control step, since the pipelined multiplier is available in the next control step again whereas the parallel multiplier is not. Second, consider as an example a data flow graph containing add and subtract operations and a functional unit set containing an ALU$(+, -)$ and an adder. The ALU matches with the add and the subtract operations, whereas the adder matches with the add operations only. Again the operation types for the adder and the ALU overlap, but are not identical. During scheduling, operations have to be assigned to functional unit types. The subtract operation may only be scheduled in the same control step with the add operation if the add operation is not assigned to the ALU and hence does not block the subtract operation. Extensions concerning overlapping functionality are considered in Section 6.4.3.6.

As mobility scheduling, delay reduction look-ahead list scheduling relies on repeatedly calculating the ASAP and ALAP schedules for the unscheduled part of the data flow graph. From the ASAP and ALAP schedules the number of control steps an operation may start execution in can be derived. This is called the *time frame* ΔT.

$$\Delta T(o) = \sigma_{ALAP}(o) - \sigma_{ASAP}(o) + 1 \tag{6.14}$$

Assuming that all assignments to feasible control steps have equal probability, fast and single-cycle operations o cover control step s with probability

$$p_o(s) = \begin{cases} \frac{1}{\Delta T(o)} & \text{if } \sigma_{ASAP}(o) \leq s \leq \sigma_{ALAP}(o) \\ 0 & \text{else.} \end{cases} \tag{6.15}$$

In order to account for multi-cycle operations as well, Expression 6.15 has to be extended to

$$p_o(s) = \begin{cases} \frac{1}{\Delta T(o)} \sum_{k=\sigma_{ASAP}(o)}^{\sigma_{ALAP}(o)} 1_{[k,k+\Delta(o)-1]}(s) & \text{if } \sigma_{ASAP}(o) \leq s \leq \sigma_{ALAP}(o) \\ 0 & \text{else} \end{cases} \tag{6.16}$$

with

$$1_{[i,j]}(s) = \begin{cases} 1 & \text{if } i \leq s \leq j \\ 0 & \text{else.} \end{cases}$$

The average value of the number of operations matching with functional unit type k covering control step s is

$$n_k(s) = \sum_{\{o \in V \setminus \Sigma : o \in r_k\}} p_o(s). \tag{6.17}$$

The average value is used to estimate how overloaded functional units are. The overload $\mathcal{C}(r_k, s)$ for functional units of type k in control step s is:

$$\Delta\mathcal{C}(r_k, s) = \mathcal{C}(r_k, s) - 1 = \frac{n_k(s)}{|R_k(s)|} - 1 \tag{6.18}$$

To account for multi-cycling, the number of type k functional units $|R_k(s)|$ available is a function of the control step s.

To give an example of overload a functional unit set containing 2 adders is assumed. In a situation where 3 add operations cover control step s with probabilities $p_{+_1}(s) = 1$, $p_{+_2}(s) = 0.5$, and $p_{+_3}(s) = 1$, respectively, the following load values result. For the average value of operations matching with the adder one obtains 2.5 in control step s, and for the overload of the adder resources $\Delta\mathcal{C}(r_{ADD}, s) = 0.25$.

Overload is used as a statistical measure to indicate how much the schedule has to be lengthened to satisfy the resource constraints. In the current control step s_{cur} all[12] possible subsets V_{sel} are generated from the candidate operations V_{ready} and judged according to the predicted schedule delay. For every V_{sel} the ASAP schedule, starting in control step $s_{cur}+1$, and the corresponding ALAP schedule are derived. From that the loads, or overloads, are calculated for control steps $s_{cur}+1$ up to S_{ASAP} according to Equation 6.18. The estimated schedule delay $\mathcal{D}(s, V_{sel})$ is assumed to be the sum over control steps $s_{cur}+1$ up to S_{ASAP} of the maximum overloads with respect to functional unit types:

$$\mathcal{D}(s, V_{sel}) = \sum_{s=s_{cur}+1}^{S_{ASAP}} Max_{k \in K}(Max(\Delta\mathcal{C}(r_k, s), 0)) \tag{6.19}$$

Taking into account that S_{ASAP} is a function of V_{sel} and s, results in the desired delay reduction look-ahead priority function:

$$\mathcal{P}(s, V_{sel}) = S_{ASAP}(s, V_{sel}) + \mathcal{D}(s, V_{sel}) \tag{6.20}$$

In contrast to the mobility priority which is dynamic but still "local" with respect to the operations, the delay reduction look-ahead priority is associated with the whole set V_{sel} of selected operations. It therefore provides a more global view of the effect of a chosen subset V_{sel} on the schedule length. To illustrate this feature, a "big and non-trivial" example which is a famous high-level synthesis benchmark, the 5th Order Digital Wave Filter, is introduced in the following section.

6.4.3.5 The 5th Order Digital Wave Filter

The 5th Order Digital Wave Filter (DWF) originally presented in [DeDN85] was chosen as a high-level synthesis benchmark in 1988. The starting point is the wave filter specification shown in Figure 6.14. The rectangular boxes containing a D are the delay operations that allow storage of different states in the circuit (see Section 6.2). The Digital Wave Filter contains 26 additions and 8 multiplications with filter constants. It is therefore an attractive benchmark for scheduling, resource allocation and resource assignment.

Under the assumption that an add operation needs one cycle and a multiply operation needs two cycles, the ASAP schedule yields 17 cycles as the minimum possible schedule length S_{ASAP}.[13] The critical paths are highlighted in Figure 6.15. A quick inspection shows that 4 adders and 4 parallel multipliers are needed under these assumptions. For the illustration of delaying

[12] Of course, this may explode, and heuristics for the choice have been introduced [RoKr91].

[13] Retiming allows to reduce S_{ASAP} to 16 cycles.

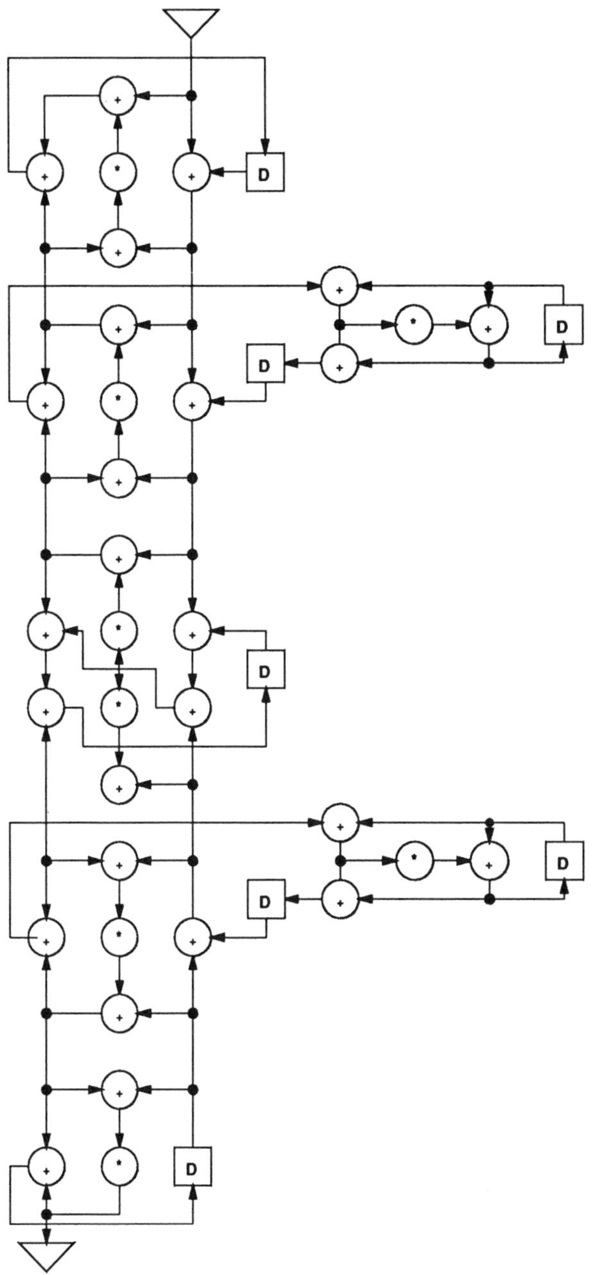

Figure 6.14: 5th Order Digital Wave Filter specification.

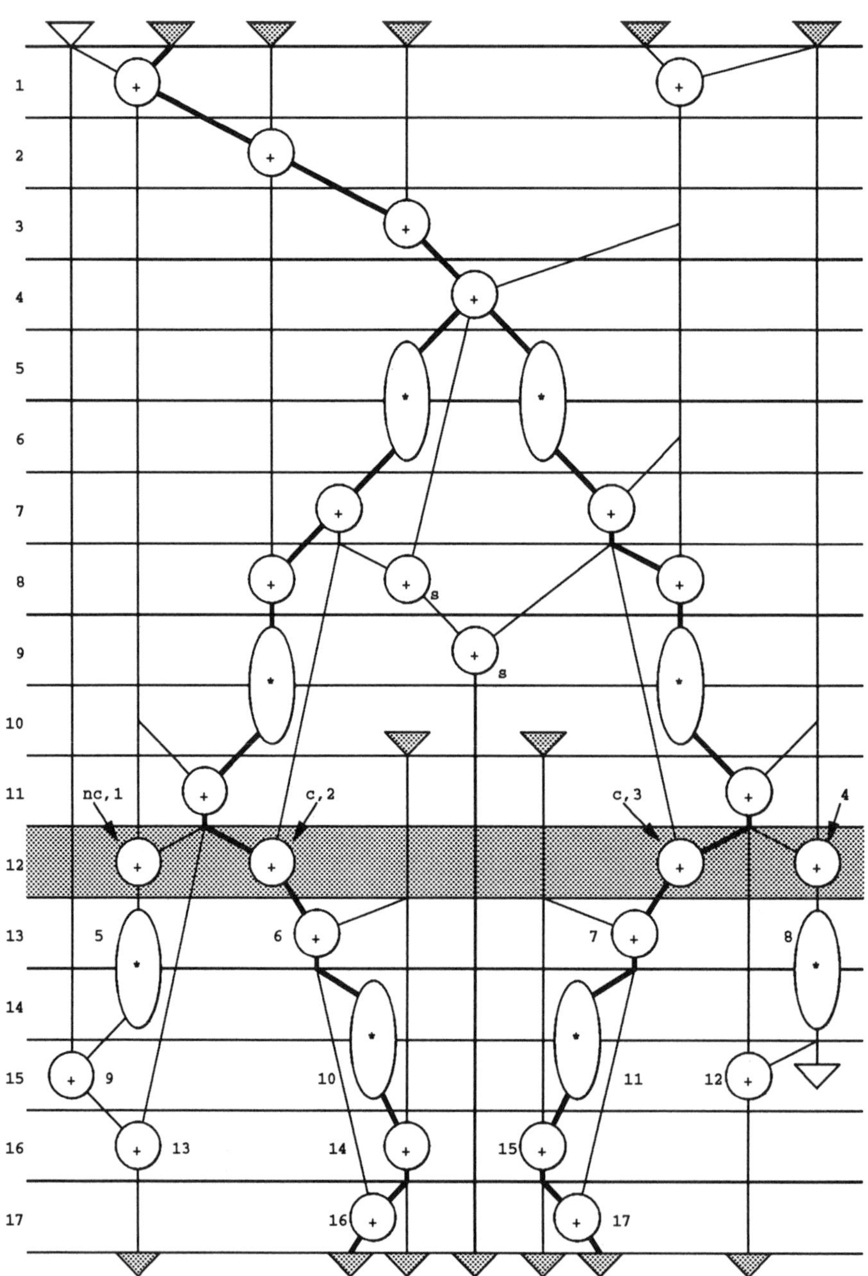

Figure 6.15: ASAP schedule of 5th Order DWF.

critical path operations the best resource constraints are 2 adders and 2 parallel multipliers. One may apply list scheduling up to the 12th control step "by hand," by shifting the two add operations marked with an s to Control Steps 9 and 10. In the 12th control step 4 add operations are ready to be scheduled, whereas only 2 adders are available. Two out of four add operations have to be chosen, resulting in 6 different V_{sel} sets. Mobility priority chooses the two add operations on the critical path marked with a c in Figure 6.15. This results in a final schedule length S of 19 cycles, whereas other operation sets containing only one or even no critical path operation result in $S = 18$.

In the following the delay reduction look-ahead priority (Equation 6.20) is checked on two V_1, V_2 sets, the first containing the two critical path operations and the second containing the left critical path and the non-critical path operation marked with nc. Using V_1 the ASAP schedule results in $S_{ASAP}(12, V_1) = 17$. ASAP and ALAP schedule of the unscheduled part of the data flow graph are shown in Figure 6.16. In Table 6.8 the resulting overloads $\Delta\mathcal{C}(r_{ADD}, s)$,

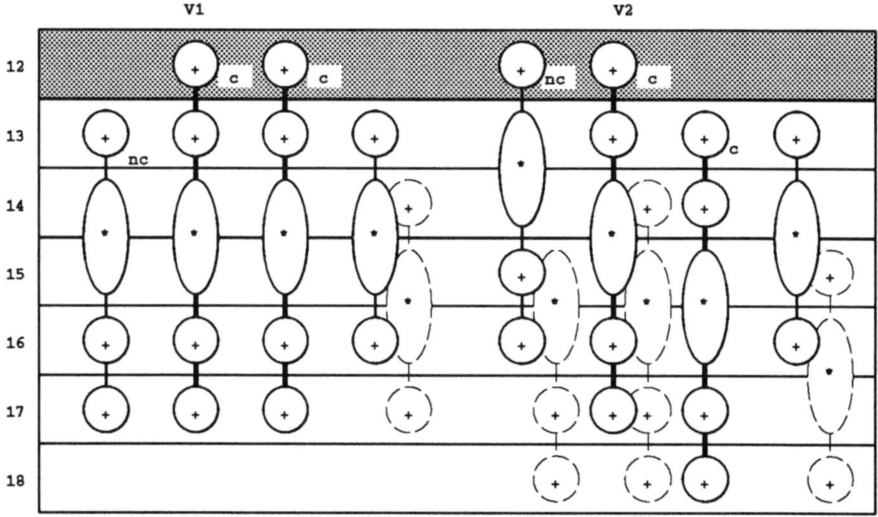

Figure 6.16: ASAP/ALAP schedules for the unscheduled part of the data flow graph.

$\Delta\mathcal{C}(r_{MULT}, s)$ are listed. Evaluating the priority function according to the preceding paragraph yields $\mathcal{P}(12, V_1) = 21$. In the second case the ASAP schedule results in $S_{ASAP}(12, V_2) = 18$, the corresponding overloads are listed in Figure 6.9, and the priority value becomes $\mathcal{P}(12, V_2) = 19.5$. Checking the other candidate subsets shows that V_2 yields the lowest predicted schedule length,

Control Step	$\Delta\mathcal{C}(r_{ADD}, s)$	$\Delta\mathcal{C}(r_{MULT}, s)$
13	0.75	0
14	0	0.75
15	0	1
16	0.75	0.25
17	0.75	0

Table 6.8: Overloads for V_1.

Control Step	$\Delta\mathcal{C}(r_{ADD}, s)$	$\Delta\mathcal{C}(r_{MULT}, s)$
13	0	0
14	0	0
15	0	0.6667
16	0	0.25
17	0.5	0
18	0.08333	0

Table 6.9: Overloads for V_2.

hence the add operations of V_2 are chosen for assignment to Control Step 12. Completing the remaining control steps finally yields the optimum schedule length of 18 cycles.

6.4.3.6 Extensions for Overlapping Functionality

The extensions needed for overlapping functionality are presented next. For that purpose the distinction between operation types and functional unit types has to be maintained which leads to the following modification. Instead of calculating the average value of operations matching the functional unit type k (see Equation 6.17), average values with respect to the operation type t

$$n_t(s) = \sum_{o_t \in V \setminus \Sigma} p_{o_t}(s)$$

are calculated from which the overloads of functional unit types are derived iteratively.

As an example consider add and subtract operations having average values of $n_+(s) = 1.5$ and $n_-(s) = 2$ in control step s. The functional unit set contains

2 ALUs(+, −) and 1 adder. Intuitively, the more constrained operation type is the subtract operation; therefore, it is assigned to functional unit types first. This yields

$$\mathcal{C}^{(0)}(r_{ALU}) = \frac{n_-}{|R_{ALU}|} = 1$$

$$\mathcal{C}^{(0)}(r_{ADD}) = 0$$

in the first iteration. In the next iteration—the last iteration for this example—the add operations are assigned to functional unit types. One can either assign the complete average value to the adder

$$\mathcal{C}(r_{ALU}) = \mathcal{C}^{(0)}(r_{ALU})$$

$$\mathcal{C}(r_{ADD}) = \frac{n_+ + \mathcal{C}^{(0)}|R_{ADD}|}{|R_{ADD}|},$$

or share the average value between the adder and the ALUs

$$\mathcal{C}(r_{ALU}) = \mathcal{C}(r_{ADD}) = \frac{n_+ + \mathcal{C}^{(0)}(r_{ALU})|R_{ALU}| + \mathcal{C}^{(0)}(r_{ADD})|R_{ADD}|}{|R_{ALU}| + |R_{ADD}|},$$

or assign the complete average value to the ALU

$$\mathcal{C}(r_{ALU}) = \frac{n_+ + \mathcal{C}^{(0)}(r_{ALU})|R_{ALU}|}{|R_{ALU}|}$$

$$\mathcal{C}(r_{ADD}) = \mathcal{C}^{(0)}(r_{ADD}).$$

In the first case, the ALUs are reserved for the subtract operations exclusively, yielding $\mathcal{C}(r_{ALU}) = 1$ and $\mathcal{C}(r_{ADD}) = 1.5$. In the second case, the loads are balanced: $\mathcal{C}(r_{ALU}) = \mathcal{C}(r_{ADD}) = 1.1\overline{6}$. The adder load is also reduced by using the ALUs. The third case, with $\mathcal{C}(r_{ALU}) = 1.75$ and $\mathcal{C}(r_{ADD}) = 0$ is definitely a bad choice. The second case is obviously the best choice, since it yields the least maximum overload of $0.1\overline{6}$. The best solution is achieved when "smearing" the current operation type over all those compatible functional unit types whose loads are smaller than the resulting load when adding the current operation type.

6.4.4 Time Constrained Scheduling

6.4.4.1 Time Constraints

The second basic instance of the scheduling problem is at least as important for high-level synthesis as resource constrained scheduling. Rather than iterating with different functional unit sets for finding a solution satisfying the time constraints, scheduling is performed subject to time constraints with the objective function to minimize the hardware to be allocated.

Two categories of time constraint patterns can be distinguished in general. The first category refers to typical digital signal processing applications where constraints on throughput or sampling rate have to be met. The corresponding time constraints usually are global with respect to a given algorithmic description. The second category refers to control-dominated applications where time constraints between operation pairs are spread over an algorithmic description or—in other words—over the data and control flow. As an example the reader may recall the CALLAS VHDL subset which allows specification of different time constraints between I/O operations in different program paths.

Time constraints may be exact or may specify upper bounds (maximum time constraints) or lower bounds (minimum time constraints). Scheduling with respect to minimum time constraints $T_{min} = \{\tau_{min}\}$ and maximum time constraints $T_{max} = \{\tau_{max}\}$ between pairs of operations whose delays are fixed corresponds to a compaction problem with upper and lower bounds on distances between objects, a problem well-known from physical design. For the HERCULES system a corresponding formulation has been outlined and extended to the so-called *relative scheduling* problem to handle operations with unbounded delay [KuDe90, KuFD91].

For the purpose of this chapter, however, it is sufficient to assume that a certain portion of the data flow graph, most likely a basic block, is subject to a single, global and exact time constraint τ_{exact}. This corresponds to specifying the schedule length S.

6.4.4.2 Force-Directed Scheduling

The most popular method for scheduling under a global and exact time constraint τ_{exact} is force-directed scheduling. This method was introduced in [Paul88, PaKn89a, PaKn89b] and has become a classical method; for variants of force-directed scheduling see [VAKL91, StBo88].

The intent of force-directed scheduling is to minimize hardware subject to the given time constraint by balancing the concurrency of operations, values to be stored, and data transfers. Whereas the resource constrained scheduling methods proceed from control step to control step selecting operations to be scheduled according to the priority function, force-directed scheduling successively selects among all operations and all control steps an operation and the control step it is to be assigned to according to a measure called *force*.

For the introduction of the force measure and the force-directed scheduling algorithm, again only the functional unit set is considered. Of course optimizing hardware in general requires consideration of storage and interconnect as well. Features like multi-cycling and chaining are presented in the original papers. Supporting overlapping functionality, however, is not discussed. In the following presentation this feature is not considered either. For convenience, it is also assumed that operation type sets are disjoint and all operations are single-cycle operations.

As an example the Differential Equation example is used again. For convenience, the same assumptions concerning the Differential Equation example as in the original papers are made: The global time constraint on the loop body is 4 cycles; the common subexpression $(u \times dx)$ is not eliminated; the add, subtract, and compare operations are to be executed on ALUs supporting $+$, $-$, and $<$ operations; and for the multiply operation a single-cycle multiplier is available.

The force measure relies on a similar statistical interpretation as delay reduction look-ahead does. Again ASAP and ALAP schedules are calculated first to derive the time frames ΔT for all operations (Equation 6.14) and the corresponding average number of operations covering a control step (Equation 6.17). In force-directed scheduling ASAP and ALAP schedules are derived with respect to the global time constraint τ_{exact} which may be larger than the critical path length S_{ASAP}. The resulting ASAP and ALAP schedules with respect to the assumptions made are shown in Figure 6.17.

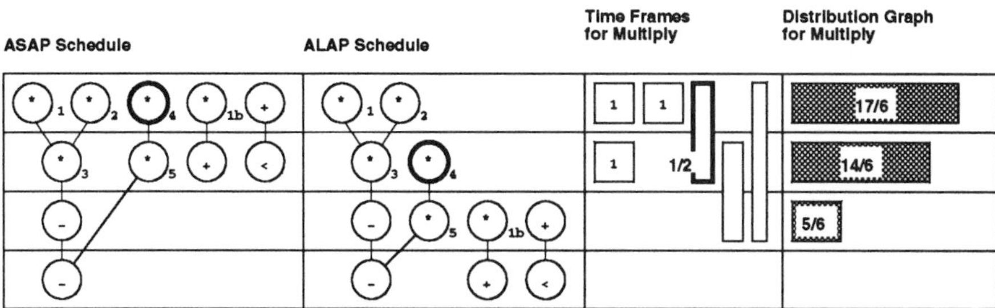

Figure 6.17: ASAP/ALAP schedules, time frames, and distribution graphs.

The collection of average values in the control steps is called *distribution graph*. Inspecting the distribution graph in Figure 6.17 shows that the average values are fairly unbalanced, indicating a large functional unit requirement in some control steps and a bad utilization of functional units in the other control steps. The basic idea is to balance the distribution graphs in order to minimize the hardware to be used.

What happens to the distribution graphs if the multiply operation marked with a 4 is tentatively assigned to the first control step? When, in other words, is the time frame of Multiply 4 reduced to a single control step? The resulting distribution graph—the left distribution graph in Figure 6.18—shows that this is a bad choice since the average values are even more unbalanced than in the original distribution graph. Tentatively assigning Multiply 4 to Control

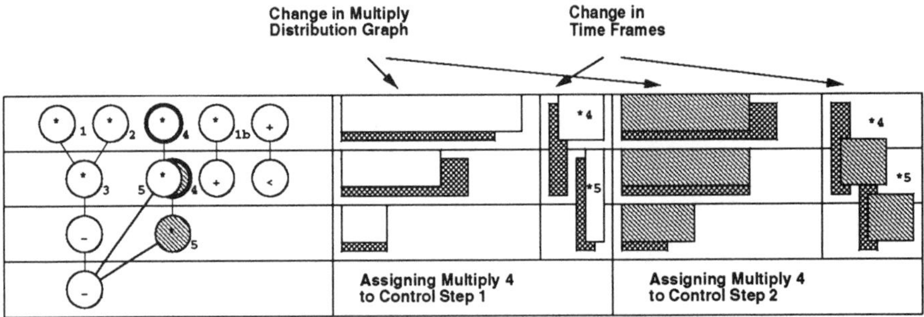

Figure 6.18: Changes in distribution graphs.

Step 2, in contrast, balances the distribution graph as shown in the right part of Figure 6.18.

Consequently, one has to measure the effect of an attempted assignment to the control steps within the time frame of the operation. Force is a rather intuitive measure for how desirable a certain operation-to-control step assignment is. A simple argument motivates the basic formula. Assume, the average number of operations matching with functional unit type k in control step s_0 $n_k(s_0)$ is large in comparison to the average distribution

$$\sum_{s=\sigma_{ASAP}(o_0)}^{\sigma_{ALAP}(o_0)} \frac{n_k(s)}{\Delta T(o_0)}$$

of operations in the control steps belonging to the time frame ΔT of operation o_0. Assigning o_0 to control step s_0 will unbalance the average distribution and the average number in control step s_0 even more; this is a unfavorable assignment. If the average number in s_0 is small in comparison to the average distribution in the time frame, assigning o_0 to s_0 is preferable since the distribution graph becomes more balanced. So, force simply is:

$$\mathcal{SF}_{(\sigma(o_0)=s_0)} = n_k(s_0) - \sum_{s=\sigma_{ASAL}(o_0)}^{\sigma_{ALAP}(o_0)} \frac{n_k(s)}{\Delta T(o_0)} \tag{6.21}$$

The smaller or more negative this expression becomes, the more preferable is an operation-to-control step assignment. The resulting forces for the multiply

Multiply Operation	Control Step	Associated Self-Force
4	1	0.25
4	2	-0.25
5	2	0.75
5	3	-0.75

Table 6.10: Self-forces for Multiply 4 and Multiply 5.

operations marked with 4 and 5 are listed in Table 6.10. Among the four forces calculated the smallest force results when assigning Multiply 4 to Control Step 2.

Of course, this assignment results in consequences that need to be considered. When assigning the Multiply 4 operation to Control Step 2, the succeeding multiply operation marked with a 5 is implicitly assigned to Control Step 3. Thus, fixing the time frame of an operation possibly affects the time frames of data dependent operations resulting in additional changes of the distribution graph. Since this effect has not been taken into account, the force contribution $\mathcal{SF}$ calculated so far is called *self-force*. The additional contribution from restricting the time frames of successor operations is called *successor-force*. Assigning Multiply 5 to Control Step 2, on the other hand, adds a contribution from the predecessor operation Multiply 4—called *predecessor-force*. In order to take these effects into account a force associated with an operation whose time frame is affected is introduced

$$\mathcal{AF}_{(\Delta T^{new}(o))} = \sum_{s=\sigma_{S_{new}}(o)}^{\sigma_{F_{new}}(o)} \frac{n_k(s)}{\Delta T^{new}(o)} - \sum_{s=\sigma_{S_{init}}(o)}^{\sigma_{F_{init}}(o)} \frac{n_k(s)}{\Delta T^{init}(o)}, \tag{6.22}$$

where $\sigma_{S_{init}}$ and $\sigma_{S_{new}}$ denote the ASAP control steps of the initial and the new time frame and $\sigma_{F_{init}}$ and $\sigma_{F_{new}}$ the ALAP control steps of the initial and the new time frame. The overall force becomes

$$\mathcal{F}_{(\sigma(o_0)=s_0)} = \mathcal{SF}_{(\sigma(o_0)=s_0)} + \sum_{o \in Succ(o_0)} \mathcal{AF}_{(\Delta T^{new}(o))} + \sum_{o \in Pred(o_0)} \mathcal{AF}_{(\Delta T^{new}(o))}. \tag{6.23}$$

In order to reduce the computational complexity, however, the final procedure first calculates and stores all the self-forces. Then the graph is traversed from bottom to top storing for each operation a force consisting of the sum of its self-force and the forces stored for its successors. Finally, the graph is traversed

Multiply Operation	Control Step	Associated Force
4	1	0.25
4	2	-1.00
5	2	1.00
5	3	-0.75

Table 6.11: Forces for Multiply 4 and Multiply 5.

from top to bottom and for each operation the final overall force is calculated as the sum of its force and the forces stored for its predecessors. The resulting forces for the Differential Equation example are listed in Table 6.11. One can observe that the assignment of Multiply 4 to Control Step 2 turns out to be even more preferable, since restricting the time frame of Multiply 5 to Control Step 3 improves the balance of the distribution graph. As an exercise the reader may apply the force-directed scheduling procedure as depicted in Procedure 6.1 to the Differential Equation example. The resulting schedule including the final distribution graphs that indicate a functional unit allocation of 2 parallel multipliers and 2 ALUs to satisfy the schedule is shown in Figure 6.19.

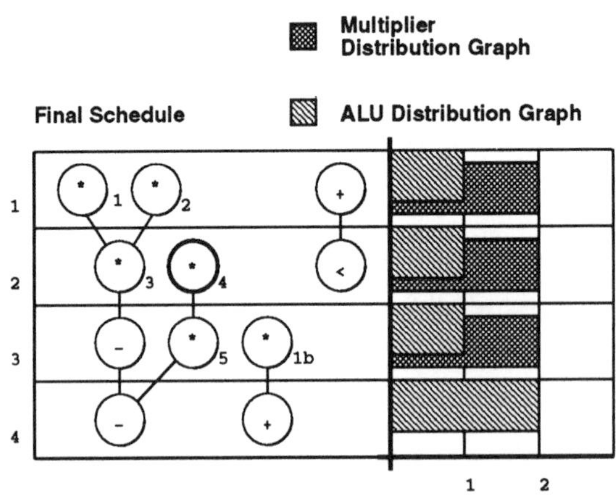

Figure 6.19: Result schedule and final distribution graphs.

Force-directed scheduling under a global time constraint τ_{exact}.
REPEAT
Evaluate or update actual ASAP and ALAP schedules and ΔTs
Create or update $n_k(s)$
For every unscheduled operation and for every feasible control step calculate $\mathcal{SF}_{(\sigma(o)=s)}$
Traverse the DFG from bottom to top for successor contributions
Traverse the DFG from top to bottom for predecessor contributions
Schedule operation with the lowest overall force $\mathcal{F}$
UNTIL *all operations scheduled.*

Procedure 6.1: Force-directed scheduling.

6.4.5 Scheduling and Control Flow

In the preceding sections only basic blocks have been considered in the presentation of scheduling algorithms. In general, one has to decide how to treat the specified control flow or, in other words, how to handle loops and conditional branches.

The classical high-level synthesis approach for handling nested loops is to schedule the innermost loop body and the corresponding test expression first, and then to treat the loop as a supernode associated with the schedule length S as operation delay (see, e.g., [PaKn89b, GuKR91]). In this case, however, a resulting schedule is optimized for only those program paths, in which every loop is executed at least once. Other program paths may result in suboptimal computation times. As an example consider a `WHILE` loop that is skipped since its test expression evaluates to false the first time. Operations succeeding the loop are delayed by S control steps although in practice they can start immediately.

Synthesis results are also affected by the general decision whether loops completing on different data dependencies are allowed to run in parallel. Different schemes of controlling can be suggested. Since the number of states in a single controller may explode, hierarchical control is often used consisting of a super-controller that starts the loop-controllers and resumes execution when it has received the ready signals of all loops running in parallel. Mapping an algorithmic description onto an architecture with hierarchical control, however, is usually considered system-level synthesis (cf. Chapter 7).

Conditional branches cause optional program paths in a way similar to loops. While the number of control steps may vary in different branches, it is fixed in each branch. There is no need to introduce hierarchical control. However, one may exploit resource sharing due to mutual exclusion of operations in different branches.

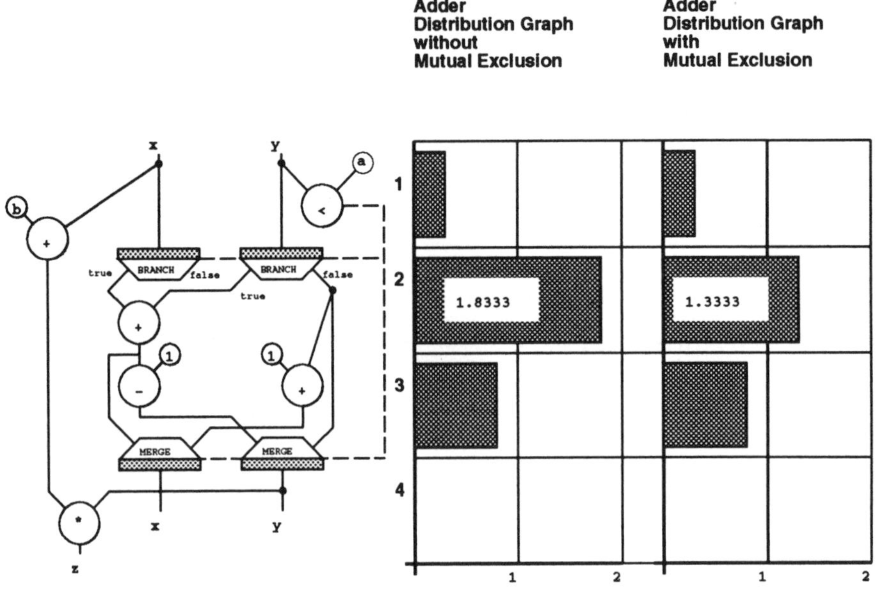

Figure 6.20: Distribution graphs with conditional branches.

The classical approach usually assumes that all branches have the length of the longest branch and that the control flow is not changed by implicit transformations of conditional branches into data flow using a select operation as proposed in [WaTh89]. As an example, the calculation of the distribution graphs covering control steps in conditional branches as used in [GuKR91] and [PaKn89b] is presented here. Assume a data flow graph as depicted in Figure 6.20.[14] The average number of adders in Control Step 2 is $1.8\overline{3}$. However, the adder usage is overestimated, since mutual exclusion has been neglected. To obtain a more precise distribution graph, in each control step only the

[14]For the derivation of the distribution graphs single-cycle operations and a schedule length $S = 4$ have been assumed.

maximum of all average values in mutual exclusive branches is added to the distribution graph as it is shown in the right part of Figure 6.20.

Experience has shown that modifications of the control flow of algorithmic descriptions by behavioral transformations are very important for high-throughput applications. Loop transformations such as loop unrolling, loop folding, loop merging, and conditional branch transformations such as the transformation to a select operation are examples. For a detailed discussion of behavioral transformations the reader is referred to Chapter 7. As an example for the integration of loop transformations with one of the core high-level synthesis tasks—namely scheduling—loop folding is presented in Section 6.4.6.3.

6.4.5.1 Path-Based Scheduling

Another approach to scheduling is path-based AFAP scheduling [CaBe90], [BeCP91]. In contrast to the classical approaches, a control flow graph in its strict sense is used as the underlying internal representation for scheduling. Using the data flow graph is suitable for synthesis from algorithms in which a suitable number of operations can be executed in parallel, as for instance in DSP applications. This assumption, however, is not necessarily applicable to control-dominated ASICs and especially not to processor-like applications where the algorithmic description mainly contains control constructs and only few arithmetic or logic operations that can be executed in parallel. Path-based scheduling considers all program paths and exploits optimization potential concerning state assignment especially in conditional branches. In contrast to the classical approaches in which only the length of the longest program path is minimized, AFAP scheduling minimizes the length of each program path. For this reason it has been called "as fast as possible" (AFAP) scheduling.

In order to demonstrate the effect of the AFAP scheduling approach, a simple conditional example is examined, whose control flow graph is shown in Figure 6.21. It is assumed that resource constraints and a cycle time constraint have to be obeyed. Resources include 1 adder, 1 subtractor both with 40 ns delay, 1 OR gate with 2 ns, 1 read and 1 read/write port both with 10 ns delay. Interconnect and register delay are estimated with 8 ns, and the cycle time is set to 60 ns. The example contains 3 paths (1, 2, 3, 4, 13, 14, 15, 16), (1, 2, 5, 6, 7, 8, 9, 12, 13, 14, 15, 16) and (1, 2, 5, 6, 7, 8, 10, 11, 12, 13, 14, 15, 16). A list schedule obtained on the corresponding data flow graph exploiting chaining and mutual exclusion is shown in Table 6.12. All paths in the list schedule require 7 cycles.

The basic idea of AFAP scheduling is to optimize the control step assignment for every path separately, and then to minimize the number of states needed for the complete program when the paths are combined. The method differs from the classical approaches since the control step assignment is ob-

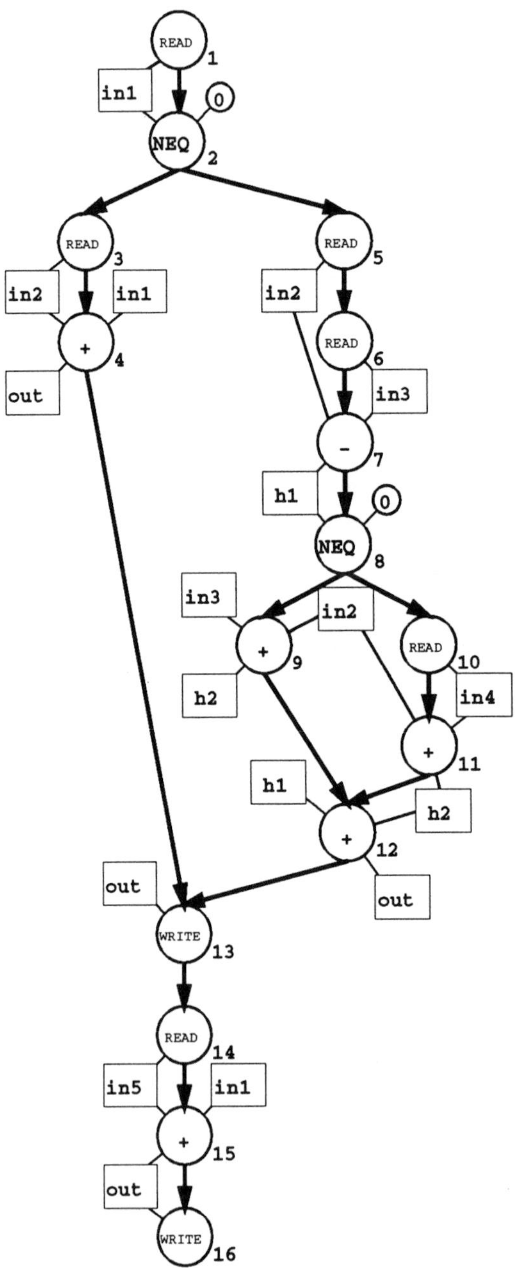

Figure 6.21: Control flow graph of the conditional branch example.

Control Step	Operations
1	1,2
2	3,4;5,6,7
3	8
4	9;10,11
5	12,13
6	14,15
7	16

Table 6.12: List schedule for the conditional branch example.

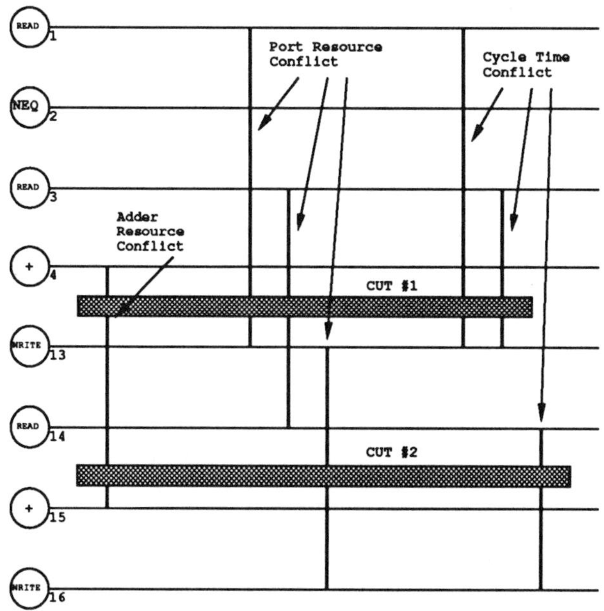

Figure 6.22: Constraint violation in the first path.

tained from cutting[15] a sequence of operations in the control flow graph if resource constraints or the cycle time constraint are violated. A drawback of this approach is that the result depends on the initial order of statements in the algorithmic description, which is not unique at all if many operations can

[15]Often, the term "state splitting" is used.

be parallelized. In fact, this is the reason why the method in its original form is restricted to processor-like or control-dominated applications.

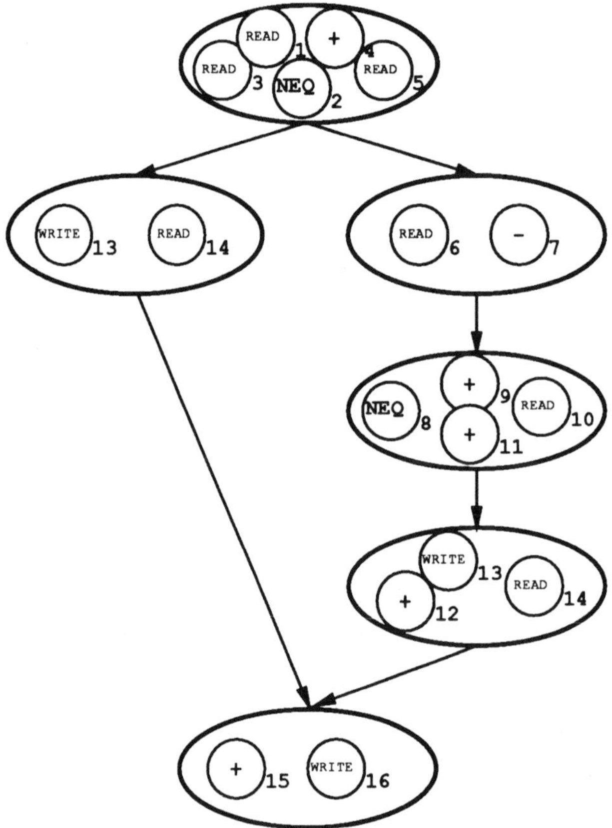

Figure 6.23: Final state assignment.

In order to understand the control step assignment in a single path the shortest path is examined. First, the operations belonging to one path are put in the specified sequence. Then, the operation sequence is scanned and a line is drawn for each pair of operations that would cause a constraint violation if the two operations were assigned to the same control step. The line is drawn so that it connects the two slots of the operation pair. The result of this construction is a diagram as it is shown in Figure 6.22. The problem is now to cut all lines in order to resolve all constraint violations. Since a cut corresponds to introducing a new control step by splitting an existing one, the minimum number of cuts in the diagram that cut all lines must be found. This problem is equivalent to finding the minimum clique cover of an undirected graph $G(V, E)$, where the constraint violations are the vertices $v \in V$ and where two vertices are connected by an undirected edge $e \in E$ if and only if the two constraint violations intersect. Fortunately, the resulting graph $G(V, E)$ is an interval graph, for which a minimum clique cover can be found in polynomial time. An introduction to the problem complexities of the minimum clique cover and the minimum graph coloring problems, and to the optimum algorithms to be used in case of interval graphs is given in Section 6.5.5.1. A minimum clique cover for the considered path results in 2 cuts (or 3 control steps respectively) which are shown as grey-shaded cross-bars in Figure 6.22. Repeating the procedure for the two other program paths yields 4 cuts or 5 control steps for both.

In the second phase the paths have to be combined. To minimize the number of states needed for the complete program, one tries to combine as many cuts as possible from different program paths. Again this is formulated into the problem of finding a minimum clique cover. In contrast to the first phase no polynomial-time algorithm exists for the second phase due to the resulting graph structure. In Figure 6.23 the resulting state assignment is shown which has been obtained from a near-minimum clique cover of the cuts in all paths. The original control structure as implied by the specified control flow has been relaxed. In contrast to the list schedule presented in Table 6.12 there are only two paths left, the number of control steps needed has been reduced, and the left path is now considerably shorter than the right path.

6.4.6 Pipelining

If throughput requirements are not a critical factor, the desired output can be computed by processing the $n\,th$ input sample when computation on input sample $n - 1$ has been completed. If, however, the required sampling rate at the input side or the production rate at the output side is very high, the computations on consecutive input samples need to be performed overlapping in time. This technique is called *pipelining*.

Typical applications that require heavily pipelined solutions are found in the video digital signal processing domain.

In this section the basic terminology and concepts used in pipeline scheduling are reviewed. The basic definitions used in pipelining theory are illustrated in Figure 6.24.[16]

Definition 6.4 (Computation time) *The computation time is the total number of time units needed to complete the computation on one input sample.*

[16]For more details refer to [Kogg81].

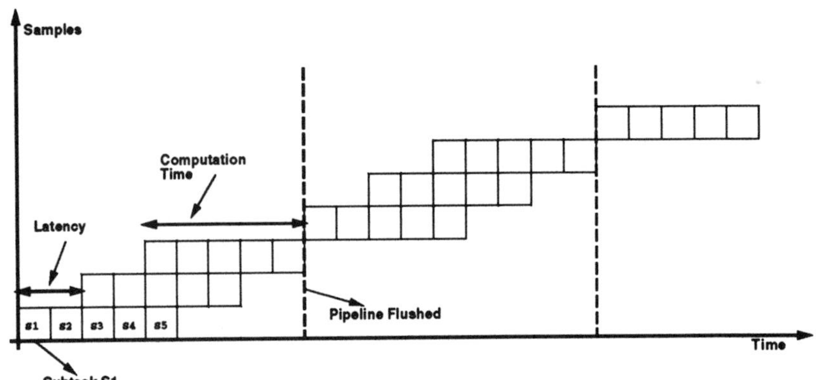

Figure 6.24: Pipelining.

The computation time C equals the schedule length S in conventional scheduling. In the following, the computation time is denoted by S. In Figure 6.24 a subtask sequence is shown that requires five time units to process one input sample; every second time unit a computation on a new input sample is initiated. The basic quantity for characterizing the performance of a pipeline is the *latency* L.

Definition 6.5 (Latency) *Latency is defined as the number of time units between two consecutive initiations, where initiation is the start of a computation on an input sample.*

The latency in Figure 6.24 is 2. Data dependencies may cause flushing and refilling of the pipeline. The average initiation rate $R_{init,N\to\infty}$ is therefore used as a more general measure for the pipeline performance:

$$R_{init,N\to\infty} = \frac{1}{L \times t_{stage} + r_{synchro} \times (C - L) \times t_{stage}} \tag{6.24}$$

where t_{stage} is the time one stage needs to complete its subtask and N is the number of computations to be performed or input samples to be processed. The resynchronization rate

$$r_{synchro} = \frac{N_{flush}}{N}$$

characterizes how often a pipeline needs to be flushed. N_{flush} is the number of input samples that cause flushing of the pipeline. In the example presented in Figure 6.24, $r_{synchro}$ equals $0.\overline{3}$.

The basic term for characterizing pipeline implementation is the *(pipe) stage*.

Definition 6.6 (Stage) *A (pipe) stage is a piece of hardware that is capable of executing certain subtasks of the computation.*

A pipeline implementation is not unique for a given pipeline scheme. For the scheme shown in Figure 6.24 one can define three stages, two of which are reused during the computation. Alternately, one can define five stages none of which is shared between the subtasks. The basic tool for dealing with pipeline implementation is the *reservation table* shown in Figure 6.25.

Definition 6.7 (Reservation table) *The reservation table is a two-dimensional representation of the data flow during one computation. One dimension corresponds to the stages, and the other dimension corresponds to time units.*

A place in the reservation table is marked if a stage of the corresponding row is used or active at the time unit of the corresponding column. To check whether a certain latency L or a latency pattern can be satisfied by a chosen implementation, one simply shifts the reservation table according to the latency pattern, until all shifts modulo S are made. The shifted reservation tables are then overlapped with the original reservation table. As soon as marked places in the overlapped reservation tables collide, the intended pipeline implementation cannot satisfy a latency pattern. In Figure 6.25 the concept of the reservation table is illustrated. The implementation satisfies a latency of 2 but does not satisfy a latency of 3. Latency patterns may be complex; the considerations in the following paragraphs, however, are restricted to a single, fixed latency.

6.4.6.1 Functional Pipelining

Overlapping several instances of the reservation table means overlapping computations on different input samples in time. This is the essence of pipelining. Drawing the analogy to high-level synthesis, a scheduled data flow graph corresponds to the reservation table, a time step of the schedule to a time unit of the reservation table, and the logical grouping of operations in one time step to a stage. In conventional pipelining, stages have physical equivalents, i.e., the stage hardware is either shared completely in different time units or not shared at all. Handling arithmetic operations that require large functional units, however, suggests sharing hardware in another sense than in conventional pipelining. Consider a multiplier and an adder both used in Stage 1. The potential for sharing differs for the multiplier and the adder with respect to other stages. The multiplier can be used in Stage 2 again, the adder in

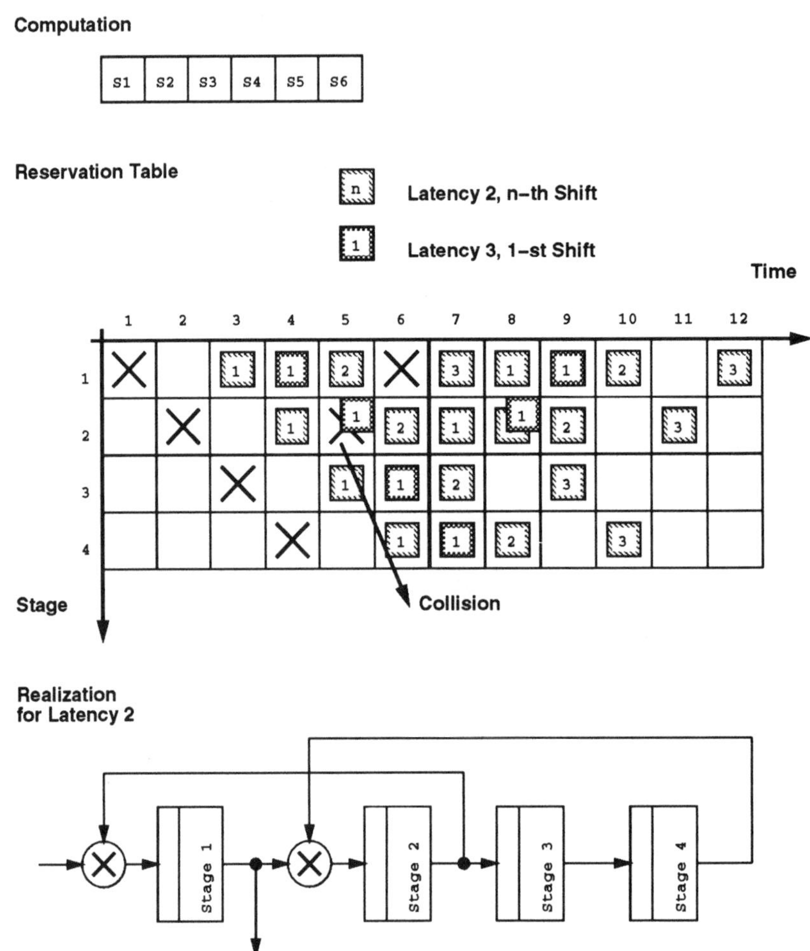

Figure 6.25: The use of the reservation table.

Stage 3. Thus there is no physical stage corresponding to the logical grouping of operations in a time step. This is referred to as *functional pipelining*.[17]

In order to illustrate functional pipelining, a simple example is introduced: Finite impulse response (FIR) filters are characterized by a sequence of response values $y(n)$ where the n *th* response value is a function of a finite

[17] This term is reserved for the case in which no data dependencies exist between computations on different input samples.

sequence of inputs $x_n, x_{n-1}, \ldots, x_{n-(N/2-1)}$ and no function of the response $y_{n-1}, y_{n-2}, \ldots, y_{n-k}$. A symmetric N-point FIR filter has the following filter equation:

$$y_n = \sum_{i=0,\ldots,N/2-1} (x_{n-i} + x_{n+1-(N-i)})h(i)$$

with the filter coefficients $h(i)$. In Figure 6.26 overlapped instances of the scheduled data flow graph of the 8-point FIR filter are shown assuming a latency of 2. One can see that the time steps have been partitioned into groups such that

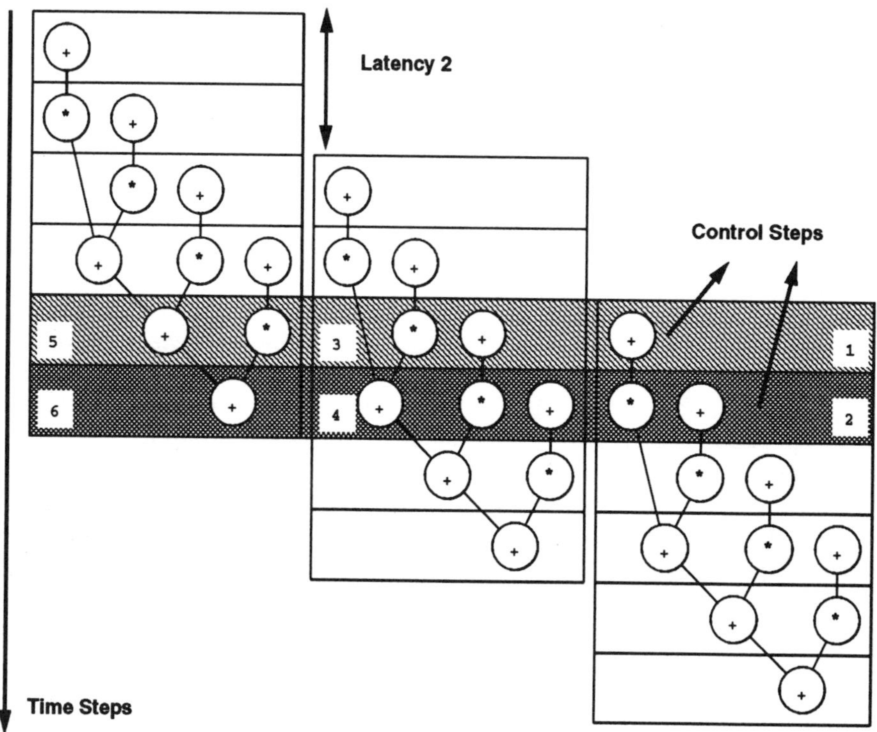

Figure 6.26: Overlapped instances of a scheduled data flow graph.

time steps belonging to different groups never overlap in time. Figure 6.26 illustrates also why time steps and control steps have to be distinguished in pipelining. A control step corresponds to a group of time steps that overlap in time. Again, operations belonging to different control steps may share functional units without conflict.

For a formalization of resource constrained scheduling with functional pipelining, the inequalities in the ILP formulation Equation 6.5 that represent the resource constraints have to be modified. Operations belonging to one control step, or in other words operations belonging to the time steps $s+n\times L$, for $n \geq 0$ are executed simultaneously and cannot share hardware. Thus, the inequality Relation 6.5 becomes:

$$\sum_{n=0}^{\lfloor \frac{S-s}{L} \rfloor} \sum_{o_i \in r_k} x_{i,s+n\times L} - |R_k| \leq 0; 1 \leq s \leq L,\ 1 \leq k \leq |K| \qquad (6.25)$$

As in conventional scheduling, the decision problem for pipeline scheduling is NP-complete [RaLi75]. In the following section the typical extensions of scheduling heuristics for pipeline scheduling are introduced.

6.4.6.2 The SEHWA Pipeline Scheduling

The basic foundations of pipelining as used in high-level synthesis are described in [PaPa88]. The resource constrained pipeline scheduling explained below is taken from the pipeline synthesis tool SEHWA introduced in [PaPa88] as well.

The resource constrained pipeline scheduling consists of a modified list scheduling. The main ingredients are the *urgency* priority function $\mathcal{U}$ and the *allocation table*. In SEHWA, forward urgency and backward urgency are distinguished. Forward urgency measures the longest path from an operation to the outputs of the data flow graph. Similarly, backward urgency of an operation measures the longest path from the inputs of the data flow graph terminating at the operation. In the case of single-cycle operations, forward urgency, as defined above, is identical to the labels used in Hu's algorithm (Algorithm 6.1).

The two-dimensional allocation table $\mathcal{AT}[k,c]$ is used to check for resource availability. The rows correspond to functional unit types and the columns correspond to control steps, i.e., groups of time steps feasible with respect to a given latency. The entries of the allocation table give the number of reserved functional units of type k in control step c.

A description of the basic pipeline list scheduling algorithm is given as pseudo-code in Algorithm 6.3. It is restricted to disjoint operation type sets and single-cycle operations. Urgency is used as a static priority function. For each time step s the list of operations is scanned from the most urgent to the least urgent operation. An operation is scheduled whenever all its predecessors have been scheduled and a functional unit of the proper type is available in control step $c = s \bmod L$.

Pipeline list scheduling with a static priority function, assuming disjoint operation type sets and single-cycle operations. Given are $|R_k|$ functional units of type k, $|K|$ functional unit types, a latency L, and an allocation table $\mathcal{AT}[k,c]$ where $1 \leq k \leq |K|$ and $1 \leq c \leq L$.

```
FOR all o ∈ V DO calculate 𝒰;
ℒ_sorted ← Sort V with decreasing 𝒰;
s ← 1;
c ← 1;
ℒ'_sorted ← NIL;
Σ ← ∅;
WHILE V ≠ Σ DO
BEGIN
      WHILE ℒ_sorted ≠ NIL DO
      BEGIN
            op ← HEAD(ℒ_sorted);
            ℒ_sorted ← TAIL(ℒ_sorted);
            IF Pred(op, DFG) ⊆ Σ ∧ 𝒜𝒯[k : op ∈ r_k, c] < |R_k| THEN
            BEGIN
               σ(op) ← s;
               Σ ← Σ ∪ {op};
               𝒜𝒯[k : op ∈ r_k, c] ← 𝒜𝒯[k : op ∈ r_k, c] + 1;
            END ELSE
               ℒ'_sorted ← APPEND(ℒ'_sorted, op);
      END;
      s ← s + 1;
      c ← s mod L;
      ℒ_sorted ← ℒ'_sorted;
END;
writeln('Schedule Length S:', s);
```

Algorithm 6.3: Pipeline list scheduling.

In conventional scheduling it is sufficient to provide at least one functional unit of each required functional unit type to ensure that a schedule exists. In case of pipeline scheduling, however, at least[18]

$$|R_k| = \left\lceil \frac{|\{o \in V : o \in r_k\}|}{L} \right\rceil$$

functional units of type k have to be provided to ensure that a schedule for the

[18] Restricting to single-cycle operations. For more details refer to [PaPa88].

given latency L exists.

Of course, each scheduling algorithm may be extended for pipelining. Instead of resource constraints one may specify both the schedule length S and the latency L. Force-directed scheduling can be used in that sense. In fact, specifying the schedule length is only useful if a high resynchronization rate (see Equation 6.24) is to be expected. The usual approaches to functional pipelining without resource constraints rely on the ASAP schedule length S_{ASAP}. Using force-directed scheduling, hardware utilization is balanced over groups of time steps instead of single time steps. The distribution graphs have to be modified accordingly. The adder distribution graphs for the conventional and for the pipeline scheduling for the 8-point FIR Filter are shown in Figure 6.27. Comparing schedule lengths and the functional units allocated for SEHWA and

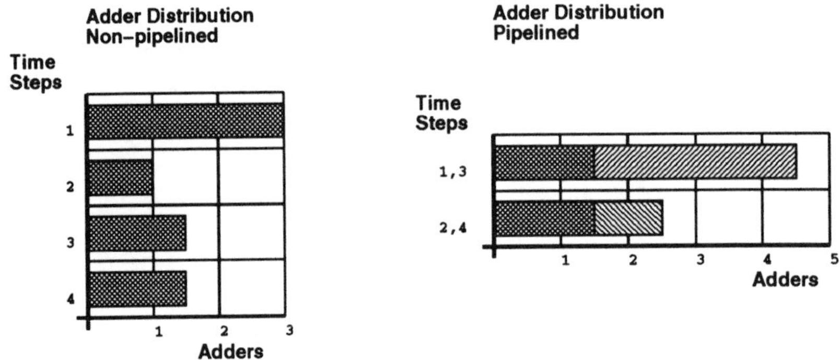

Figure 6.27: Modified distribution graphs for force-directed scheduling with pipelining.

for the force-directed scheduling (see Figure 6.28), shows that in the presence of resynchronization the force-directed result is preferable.

6.4.6.3 Loop Folding

Within functional pipelining as presented so far, there is no reason to have a lower bound on the latency L_{LB} except that resource constraints forbid small latencies. In general, of course, a lower bound on the latency exists if computations on different input samples are data dependent. A typical situation where data dependencies between different instances of a repeated computation may occur are loops. In this case, computation instances correspond to loop iterations. It is known from software compilers for vector processors that loops bear a large potential for improving throughput by reorganizing the loop structure.

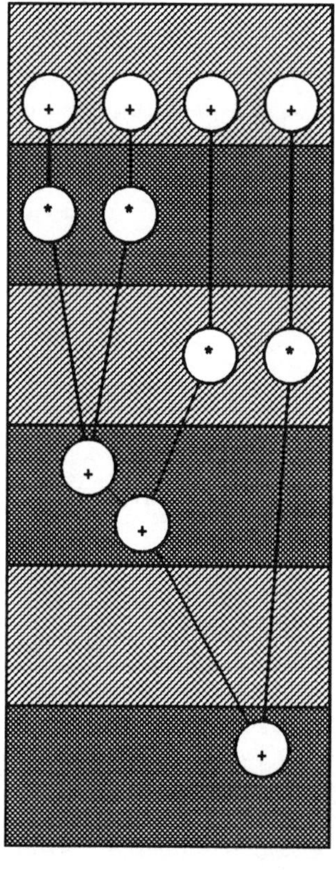

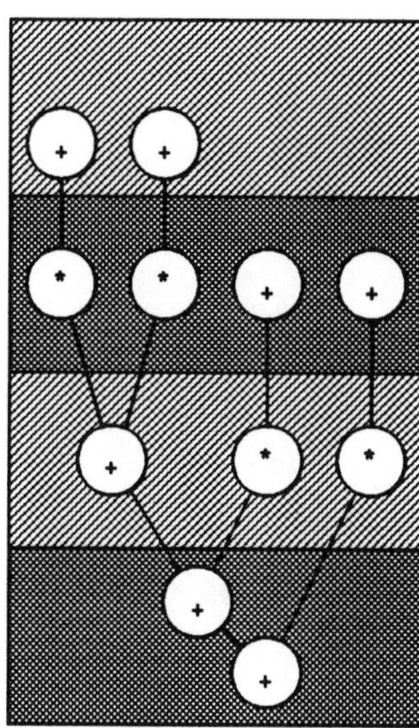

Figure 6.28: Forward urgency list schedule and force-directed schedule.

The same applies for a large class of DSP algorithms where nested loops with constant iteration count are typical.

In the following it is illustrated how reorganization of the loop or, in other words, loop folding can be used in high-level synthesis to improve throughput. First, loop folding is applied as a behavioral transformation and the effect on the synthesis result is derived from a subsequent resource constrained scheduling.

Then, an integrated approach in which loop folding is part of the scheduling task itself is explained.

The task of the example below is to compute the imaginary part of a 256-point Discrete Fourier Transform

$$\mathcal{I}(k) = \sum_{n=0}^{N-1} hw(n)\, in(n)\, sin(n\, k\, \frac{2\pi}{N})$$

where $hw(n)$ is the so-called *Hamming window*. As can be seen from the algorithmic description in Procedure 6.2, the values of the Hamming window and the sine values are read from external memory. The multiplication $n * k$ is performed by accumulating the k values.

```
PROCEDURE 256-point DFT
BEGIN
    FOR k ← 0 TO 63 DO BEGIN
        nk ← 0;
        sum ← 0;
        j ← 0;
        REPEAT
            sum ← sum+ read(in[j]) * (read(sin[nk]) * read(hw[j]));
            nk ← nk + k;
            j ← j + 1;
        UNTIL j = 255;
        write(sum, I(k));
    END;
END.
```

Procedure 6.2: 256-point DFT with original loop organization.

In Figure 6.29 the corresponding data flow graph is depicted. Applying ASAP scheduling under the assumption that the read and add operations are single-cycle operations and the multiply operation is a two-cycle operation, yields a critical path length of $S_{ASAP} = 6$. Analysis of the data dependencies between two loop iterations $n - 1$ and n shows that the longest feedback path between the new and old instances of values sum, nk, j is not longer than 1 time step. It is therefore possible to reorganize the loop so that the critical path is shortened. In Procedure 6.3 the 256-point Discrete Fourier Transform is shown with a folded loop. Two new loop variables *temp_1* and *temp_2* have been introduced in order to cut the critical path and the initialization and the termination code have been adjusted accordingly. Applying ASAP scheduling

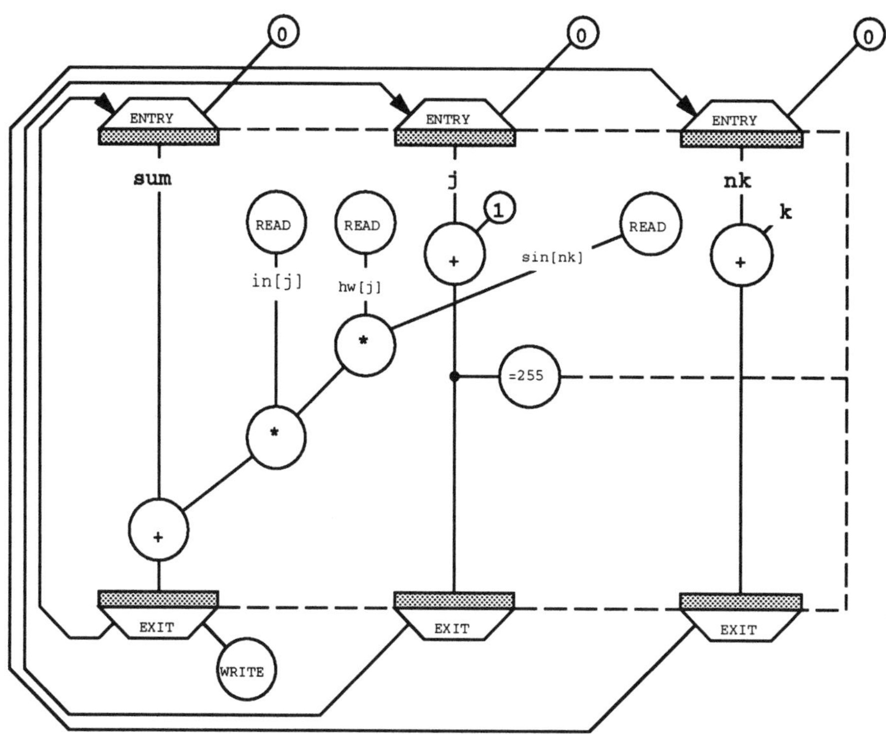

Figure 6.29: Data flow graph for original loop organization.

to the new data flow graph demonstrates that the critical path of the reorganized loop is shorter ($S_{ASAP} = 3$). In Figure 6.30 the new data flow graph is shown. Comparing the new and the old data flow graph shows that in the new loop organization the loop body of the old loop organization seems to be *folded*. Understanding the loop body of the old loop organization as one computation instance, computations overlap in time in the folded loop. This is equivalent to pipelining. The schedule length of the folded loop corresponds to the latency L. In the following, the latency L and schedule length of the loop S_{loop} are used as synonyms.

The effect on the synthesis result can now be studied by applying resource constrained scheduling to the loop body. For the original loop organization, two read/write ports, one adder and one parallel multiplier are sufficient to provide a schedule length of $S_{loop} = 6$. In the folded loop, the parallel multiplier has to be replaced by a pipelined multiplier in order to achieve the minimum schedule

```
PROCEDURE 256-point DFT
BEGIN
    FOR k ← 0 TO 63 DO BEGIN
        nk ← k;
        sum ← 0;
        j ← 1;
        temp_1 ← read(in[0]);
        temp_2 ← read(sin[0]) * read(hw[0]);
        REPEAT
            sum ← sum + temp_1 * temp_2;
            temp_1 ← read(in[j]);
            temp_2 ← read(sin[nk]) * read(hw[j]);
            nk ← nk + k;
            j ← j + 1;
        UNTIL j = 255;
        sum ← sum + temp_1 * temp_2;
        write(sum, I(k));
    END;
END.
```

Procedure 6.3: 256-point DFT with folded loop.

Loop	PipeMult	ParMult	Add	Port	S_{loop}	Execution Time
Original	-	1	1	2	6	98432
Folded	1	-	1	2	3	49408

Table 6.13: Throughput for original and for folded loop organization.

length of $S_{loop} = 3$. Comparing the total execution times[19] listed in Table 6.13 demonstrates that folding the loop has improved throughput drastically.

As described above, loop folding may be applied as a behavioral transformation [Girc87]. However, schedule length or latency is known only after operation-to-functional-unit-type assignment and ASAP scheduling; if resource constraints have to be obeyed it is even unknown until resource constrained scheduling is finished. Therefore, an integrated approach for loop folding is desirable.

In the example above, there are data dependencies between two successive

[19] For the outer loop it has been assumed that the initialization for the inner loop requires at least one clock cycle, and that the write operation is a single-cycle operation as well.

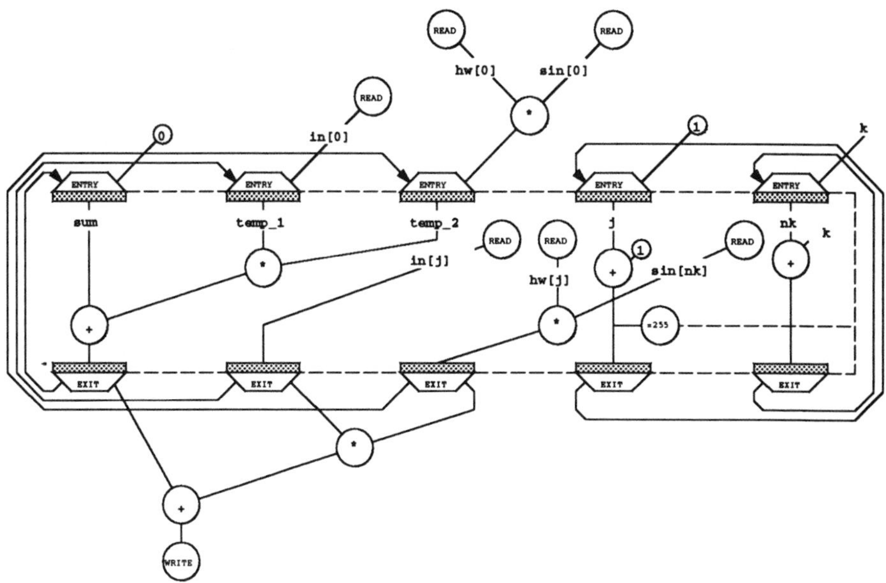

Figure 6.30: Data flow graph for folded loop organization.

loop iterations. In general, data dependencies between loop iterations $(n - \lambda)$ and n may exist that have to be accounted for during scheduling. Consequently, the precedence constraints in the ILP formulation have to be modified for data dependencies across loop iterations. In order to keep the following considerations consistent with the ILP formulation presented in Figure 6.11, only single-cycle operations are considered. Let o_j be an operation in loop iteration n that is data dependent on an operation o_i executed in iteration $(n - \lambda)$. λ is called the *looping degree.* The number of time steps Δt that have to pass between these two operations on the absolute time axis is:

$$\Delta t = t(o_j) - t(o_i) \geq 1. \tag{6.26}$$

Assuming that a schedule is known, the absolute times $t(o_i)$ and $t(o_j)$ can be expressed using the time steps of the schedule, and the latency L:

$$t(o_i) = \sigma(o_i)$$

$$t(o_j) = \sigma(o_j) + \lambda\, L$$

This allows substitution of the precedence constraints Relation 6.8 in Figure 6.11 with the precedence constraints Relation 6.27 which are obtained by reformulating Relation 6.26.

$$\sigma(o_i) - \sigma(o_j) \leq -1 + \lambda\, L \tag{6.27}$$

Scheduling with loop folding simply corresponds to modifying the looping degree λ. If o_j is moved from loop iteration n to loop iteration $(n + \phi_j)$, ϕ_j is called the *folding index* of o_j. For the description of scheduling with loop folding the folding index has to be introduced in the precedence constraints:

$$\sigma(o_i) - \sigma(o_j) \leq -1 + (\lambda + \phi_j - \phi_i)\, L.$$

The looping degree $(\lambda + \phi_j - \phi_i)$ has become variable. This shows the increase in algorithmic complexity of scheduling with loop folding. Due to the product $(\lambda + \phi_j - \phi_i)\, L$ of the effective looping degree and the resulting latency, a Non-Linear Programming (NLP) problem, known to be harder than ILP, is obtained.

Throughout this section it was assumed that it is sufficient to consider acyclic data flow graph. Feedback variables due to loops were cut for this purpose. A loop may, however, be specified so badly that the resulting scheduling length S_{ASAP} is (much) larger than the lower bound on latency L_{LB}. Since the lower bound on latency is determined by the cycles in the data flow graph, the assumption that the underlying data flow graph is acyclic is now given up. Given a latency L, the existence of a schedule for a cyclic data flow graph corresponds to the realizability criterion for retiming as introduced in Chapter 5. Forward edges of the data flow graph are weighted with the operation delay of the head operation $w(e = (o_i, o_j))$. Backward edges are weighted with the difference between the head operation delay and the product of the looping degree and the given latency. Assuming single-cycle operations only, yields:

$$w(e = (o_i, o_j)) = 1,\ \forall e \in E_{forward}$$

$$w(e = (o_i, o_j)) = 1 - \lambda\, L,\ \forall e \in E_{backward}$$

In analogy to realizability, a schedule for the given latency L exists if there are no cycles C in $DFG(V, E)$ whose edge weight sum exceeds zero:

$$\sum_{e \in C} w(e) \leq 0,\ \forall C \in DFG(V, E, w(e, L))$$

A lower bound for the latency L_{LB} can be found by iteratively decrementing the latency until the obtained graph $DFG(V, E, w(e, L))$ is no longer realizable.

Due to resource constraints, the bound L_{LB} may, however, not be reached. In other words, the scheduler has to avoid the violation of realizability. If the data flow graph of the loop is not to be changed, i.e., loop folding is not performed, the minimum latency L equals the ASAP schedule length S_{ASAP} that is obtained by ASAP scheduling ignoring the back edges. An iterative approach to loop folding can be sketched as follows [Goos89]. The idea is to increase iteratively the effective looping degrees $(\lambda + \phi_{o_j} - \phi_{o_i})$ such that L can be decreased and throughput can be increased. The approach basically consists of two phases. Since the edge weights depend on the result schedule or the latency, scheduling has to be solved iteratively.

To provide a starting point for the second phase, in the first phase, a schedule with an initial latency $L_{initial}$ is evaluated. This phase consists of two nested loops, the outer loop for the latency iteration and the inner loop for the scheduling iteration. A first estimate is taken from ASAP scheduling: $L^{(0)}_{initial} = S_{ASAP}$. Using this estimate the edge weights are calculated and the inner loop is performed in order to obtain a schedule. If the result schedule does not meet the current latency estimate ($S_{resource} > L^{(i)}_{initial}$) the latency is incremented ($L^{(i+1)}_{initial} = L^{(i)}_{initial} + 1$). The whole process is repeated until a solution with $S_{resource} = L_{initial}$ is found.

The purpose of the second phase is to successively change the loop organization in order to find a schedule with a near-minimum latency L_{final} that satisfies throughput constraints. Like the first phase, the second phase consists of two nested loops, the outer loop for the folding iteration and the inner loop for the schedule iteration. Starting with $L^{(0)}_{final} = L_{initial}$ the folding indices of operations that cover the last control step ($\sigma(o) = L_{initial}$) are increased by one, and scheduling is performed assuming that $L^{(i+1)}_{final} = L^{(i)}_{final} - 1$ can be satisfied. The whole process is repeated until in the next iteration L_{final} cannot be decreased even if more folding iterations are applied without decrementing the latency.

For the convergence of the iterative process it is essential that the list scheduling does not violate realizability. A priority function taking care of this condition is given by

$$\mathcal{P}(o_i) = \min_{C: o_i \in C} \frac{l_C^{remain}}{l_C^{required} + 1}, \tag{6.28}$$

where l_C^{remain} is the remaining length for the unscheduled part of the cycle, and $l_C^{required}$ is the length required for the unscheduled part of the cycle. The priority function becomes the smaller, the larger the unscheduled part of a cycle is and the more control steps have been consumed already by the scheduled part of the cycle. The priority function tries to avoid "overconstraining" cycles. The most constrained operation is the one with the smallest $\mathcal{P}$. The operation with the least $\mathcal{P}$ value has to be scheduled with highest priority.

6.4.7 Classification of Scheduling Approaches

Scheduling approaches are usually classified according to the algorithms and the specific cost functions used [McPC90].

Scheduling algorithms can be classified into transformational, iterative or constructive, and global algorithms. An example of *transformational scheduling* is found in the YORKTOWN SILICON COMPILER (YSC) [Bray88] in which, starting from an initial maximal parallel schedule, a final schedule is obtained by successively splitting states. Most of the scheduling algorithms are of *iterative* or *constructive* nature, usually proceeding from control step to control step. Examples are ASAP scheduling with conditional postponement, critical path scheduling as presented in the MAHA/ADAM [PaPM86], and the various list scheduling approaches. In contrast to the constructive scheduling approaches, in *global scheduling* approaches all control steps and all operations are considered simultaneously when operations are assigned to control steps. Examples of global scheduling algorithms are found in force-directed scheduling (see Section 6.4.4), the simulated annealing approach [DeNe89], the NISCHE neural net scheduling [HePo89], and in the various ILP scheduling approaches [HwHL90, HwHL91, HwHL91].

In addition to classifying scheduling methods according to algorithm and cost function, it is worthwhile to classify them according to the *scheduling scheme* [Stok91], i.e., according to the assumptions made concerning the data flow graph units handled by the scheduler.

Most of the algorithms presented in literature as well as the algorithms presented in the preceding paragraphs assume that basic blocks are given to the scheduler as a scheduling unit. These *basic block scheduling* schemes, however, are weak in optimizing at boundaries between basic blocks. They work well for examples such as the 5th Order Digital Wave Filter (see Section 6.4.3.4), or the loop of the Differential Equation example (see Section 6.3.1) in which many operations belong to a basic block and the critical path of the basic block is not too short compared to the length of the complete program paths. An exception to the basic block schemes is the *hierarchical scheduling* scheme of the HYPER system [RCHP91]. Like the basic block schedulers, HYPER starts with scheduling the innermost loops; in contrast to these, however, HYPER allows corrections of the scheduling when higher hierarchy levels are processed. As the counterpart to both, the *path-based scheduling* scheme presented in Section 6.4.5 has to be considered. This, in turn, is best suited for applications whose algorithmic description contain many control constructs and many different program paths.

6.5 Allocation and Assignment Tasks

As already stated, the three tasks scheduling, resource allocation, and resource assignment neither need to be performed in a certain sequence nor to be considered as independent tasks.

As an alternative to considering scheduling and functional unit allocation as the prior tasks, a *partitioning* approach consisting of bottom-up functional unit allocation and functional unit assignment before scheduling is introduced. This is followed by a systematic overview of the synthesis tasks that are left after scheduling. Following a short review of the various allocation and assignment methods used in high-level synthesis, algorithms for register allocation and register assignment are presented in detail as a representative for all allocation and assignment tasks. A sketch on bus allocation and bus assignment concludes this section.

6.5.1 Alternative Approach

Partitioning is an important task in overall synthesis strategies. The motivation for partitioning varies for the various synthesis strategies depending on the level of abstraction at which partitioning is to be performed. Partitioning at the system level can be used as a preparatory step for high-level synthesis (cf. Chapter 7).

In the following, partitioning as an approach to the high-level synthesis process itself is introduced [McFa86, SGLM90]. The motivation for an approach like this is to direct the synthesis process at an early stage, based on area and timing estimates obtained from a supplementary floorplanning step, in which topological information based on interconnections between clusters is exploited. Using topological information in this early stage corresponds to performing functional unit allocation and functional unit assignment as prior synthesis tasks. In addition, this approach provides a natural way for searching the design space, i.e., exploring the area/time trade-off.

6.5.1.1 Partitioning Approach

Partitioning approaches rely on assigning operations to functional unit clusters and on allocating these clusters according to a distance measure. A *functional unit cluster* is a bottom-up built multifunctional data path which either exists in the library or has to be built from several functional units available in the library. *Distance* reflects how preferable it is to have two operations executed on the same cluster.

The basic approach is to form a hierarchical cluster tree from the operations in the data flow graph, where the clustering process is driven by the distance measure. The operations form the leaf nodes of the tree. Internal nodes represent clustered operations. The cluster nodes are labeled with a height function that equals the distance measure. To select among different partitionings or clusterings the cluster tree is cut at a certain height. Clusters with a height smaller than the one chosen are either realized by a library component or by building a multi-functional data path, whereas clusters with a height larger than the one chosen are used to guide floorplanning, for example by organizing them into a slicing tree. The cluster tree for the Differential Equation example—the distance measure used is introduced below—and three cuts together with their idealized points in the design space are shown in Figure 6.31.[20] Of course, to derive the number of time steps needed, the subsequent scheduling step has to be invoked. The cluster tree is subsequently cut from the smallest to the largest height. Clusters below the cut are realized by library components capable of executing all operations in the cluster or with bottom-up built data paths. Accordingly, Cut #2 represents a functional unit set of 2 parallel multipliers, 1 adder, 1 subtractor and 1 ALU(+,−,<). The functional unit assignment is given by the grouping of operations in clusters.

Searching the design space can now easily be understood. By proceeding from cut to cut, one proceeds from fast and area-intensive designs to slower but area-efficient designs. Area and delays are estimated from a floorplan which is built by using the cluster tree above the cut as a topological tree. For floorplanning, the area and the shape of the clusters as well as their interconnections need to be known. Interconnections are implied by the obtained functional unit assignment. Area and shape are calculated according to the following model: Functional units from the library are stacked lengthwise (if more than one are needed). Local registers, and multiplexors, both estimated from the schedule, and two busses are added.

The distance measure consists of three contributions. The first contribution measures similarity of functionality. Two operations are similar in functionality, and therefore near to each other, if the relative cost gain $\Delta\mathcal{C}(o_1, o_2)$ from assigning the two operations to the same cluster is large in comparison to assigning them to two different clusters.

$$\Delta\mathcal{C}(o_1, o_2) = \frac{cost(o_1) + cost(o_2) - cost(o_1, o_2)}{cost(o_1, o_2)} \qquad (6.29)$$

In this formula $cost(o_1, o_2)$ represents a cost function in terms of area and delay of a cluster implementation that requires the minimum number of library

[20] The cluster tree has been evaluated under the following assumptions: The library contains an ALU(+, −) which is 20% larger and another ALU(+, −, <) which is 40% larger than an adder. User-defined parameters are $N = 1, M = 1$. Multiply operations are single-cycle operations.

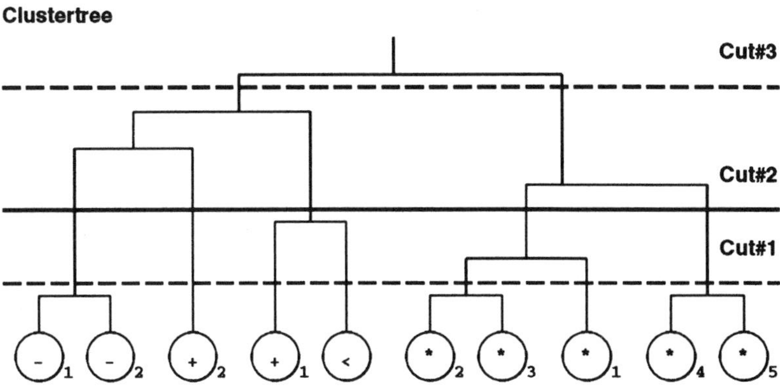

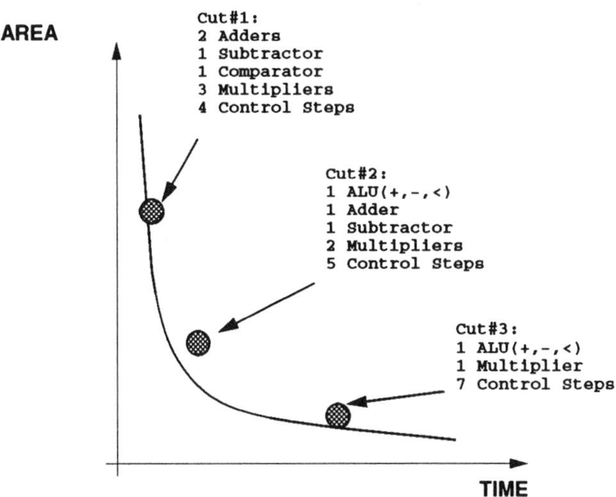

Figure 6.31: Cluster tree for Differential Equation example.

functional units. Mapping two operations to the same cluster is preferable if they cannot be executed in parallel. Therefore, a second contribution $\mathcal{P}$ is a simple expression to weight the influence of potential parallelism:

$$\mathcal{P}(o_1, o_2) = \begin{cases} 1 & \text{if } o_1 \text{ and } o_2 \text{ can be executed in parallel} \\ 0 & \text{else} \end{cases} \tag{6.30}$$

Finally, to control the amount of interconnections between the clusters, the influence of data flow connections is taken as a third contribution. The interconnect gain $\Delta\mathcal{I}$ is given by:

$$\Delta\mathcal{I}(o_1, o_2) = \frac{|\{(o_i, o_1), (o_1, o_i) \in E\} \cap \{(o_i, o_2), (o_2, o_i) \in E\}|}{\max(|\{(o_i, o_1), (o_1, o_i) \in E\}|, |\{(o_i, o_2), (o_2, o_i) \in E\}|)} \quad (6.31)$$

Using these contributions and the user-defined parameters N, M a distance measure $\mathcal{D}$ can be defined as follows [McFa86] :

$$\mathcal{D}(o_1, o_2; N, M) = \frac{(\mathcal{P}(o_1, o_2) + 1)^N}{\Delta\mathcal{C}(o_1, o_2) + M\ \Delta\mathcal{I}(o_1, o_2)} \quad (6.32)$$

The distance between two clusters is assumed to be the average distance between each pair of the operations belonging to the two clusters.

The distance measures used for potential parallelism and for interconnect can be refined. In the approach presented in [SGLM90] especially restrictions imposed on the subsequent scheduling are taken into account more precisely than by the simple expression given above.

6.5.2 Completing the Data Part

For a systematic introduction of the tasks that are left after scheduling, the typical synthesis approaches mentioned in the last sections are recalled:

- Functional units are allocated by specifying resource constraints for the subsequent resource constrained scheduling as presented in Section 6.4.3.
- A time constraint is specified and time constrained scheduling follows as presented in Section 6.4.4.
- The partitioning approach presented in Section 6.5.1.1 determines allocation and assignment of functional units first. Then scheduling is invoked.

At this stage of synthesizing the register-transfer level structure, a lower bound on the number of resources that have to be provided to satisfy the schedule has been obtained. For those types of resources that have not been allocated before scheduling, the minimum number or a number close to the lower bound is derived when completing the data part. From that point of view the following subtasks are left to complete the data part:

- Allocation of functional units if functional units have not been allocated before scheduling. Allocation of functional units remains after time constrained scheduling (cf. Section 6.4.4).

- Assignment of operations to functional unit instances if functional unit assignment has not been solved before scheduling as in the partitioning approach presented in Section 6.5.1.1.
- Allocation of storage if storage elements have not been allocated before scheduling.
- Assignment of values that have to be stored to storage elements.
- Allocation of busses, if busses are required and if they have not been allocated in advance.
- Assignment of data to be transferred to busses if busses are used.

Approaches differ not only significantly as to when and how scheduling is performed. The sequence and the combination of the subtasks needed for completing the data part also vary due to different data part architectures and different optimization goals.

6.5.3 Approaches to Allocation and Assignment

Allocation and assignment approaches are usually classified according to the algorithms or heuristics used. These include rule-based schemes, greedy or iterative methods, branch and bound, linear programming, and graph-theoretical formulations [Stok91]. Rule-based schemes are typically used to perform resource allocation, especially type selection and resource assignment, prior to scheduling. An example is the CATHEDRAL II/2ND system [Goos88] in which rules are used to decide about expanding multiplication or division into adder shifter sequences, to allocate loop counters for iterations, or to select addressing schemes for background memory. Greedy or iterative approaches, branch and bound, and linear programming approaches typically proceed from control step to control step, solving all allocation and assignment tasks for the current control step before the next control step is processed. Graph-theoretical formulations, on the other hand, are usually used to perform the three allocation and assignment tasks separately and consecutively, but globally with respect to the schedule.[21]

An example of a greedy approach is found in the MABAL tool of the ADAM system [KuPa90a]. MABAL treats operation-to-functional unit and register assignment equivalently and considers the trade-off between resource and interconnect cost for each assignment decision. In addition, two forms of interconnect realization, multiplexor library cells and synthetic multiplexors consisting of tristate drivers and decode logic are considered as trade-offs.

[21] Graph-theoretical formulations therefore often have been called *global*.

Branch and bound search is used for interconnect synthesis in the S(P)LICER system [PaGa87]. The approach relies on a four-level interconnection model: connections between registers and functional unit input busses constitute the first level, connections between the functional unit input busses and the functional unit inputs themselves form the second level, connections between functional unit outputs and corresponding output busses are the third, and connections between output busses and registers,[22] finally, belong to the fourth level. Interconnect and operation-to-functional unit assignment are performed by processing consecutively the four interconnection levels and giving priority to assignments that use existing busses or bus connections.

Linear programming minimizing an objective function that represents interconnect cost is used in the MIMOLA system [BaMa90] to solve operation-to-functional unit and register assignment for all operations and values that belong to the current control step in list scheduling.

Most of the allocation and assignment approaches, however, rely on graph-theoretical formulations, especially clique partitioning and graph coloring to perform allocation and assignment globally with respect to the schedule. The price that has to be paid with these approaches is that functional unit allocation and assignment, register allocation and assignment, and bus allocation and assignment have to be performed separately although they are strongly interrelated and mutually dependent. A typical approach to overcome this problem is to incorporate means for interconnect optimization in register allocation and assignment and functional unit allocation and assignment. Examples for graph-theoretical formulations are found in the FACET system [TsSi86] (see also Section 6.5.4.2), in the HAL system [PaKn89a], in the CADDY program [KrRo90], in the LYRA/ARYL tools [HCLH90], and in the ASYL system [MiSa91b]. Figure 6.32 shows the different approaches to completing the data part for these systems. Within the FACET system, Tseng's clique partitioning heuristic is used to perform register allocation and assignment prior to functional unit allocation and assignment—clustering operations on ALUs in this case—and finally bus allocation and assignment. A similar technique enhancing clique partitioning with interconnect optimization is used in the HAL system and the complementary formulation as graph coloring problem (see also [Pfah87]) is used in the CADDY tool. In addition, the systems mentioned so far exploit commutative operations: *Operand alignment* swaps input pairs to lower the multiplexor cost. The LYRA/ARYL tools allow for separation of the allocation from the assignment task (see Subsection 6.5.4.4) and also allow exploration of the design space by interchanging allocation and assignment of functional units and registers. Finally, the ASYL system starts from a user-specifiable

[22] The busses are unidirectional. The interconnection scheme corresponds to the multiplexed architectural style (see Section 6.1.2).

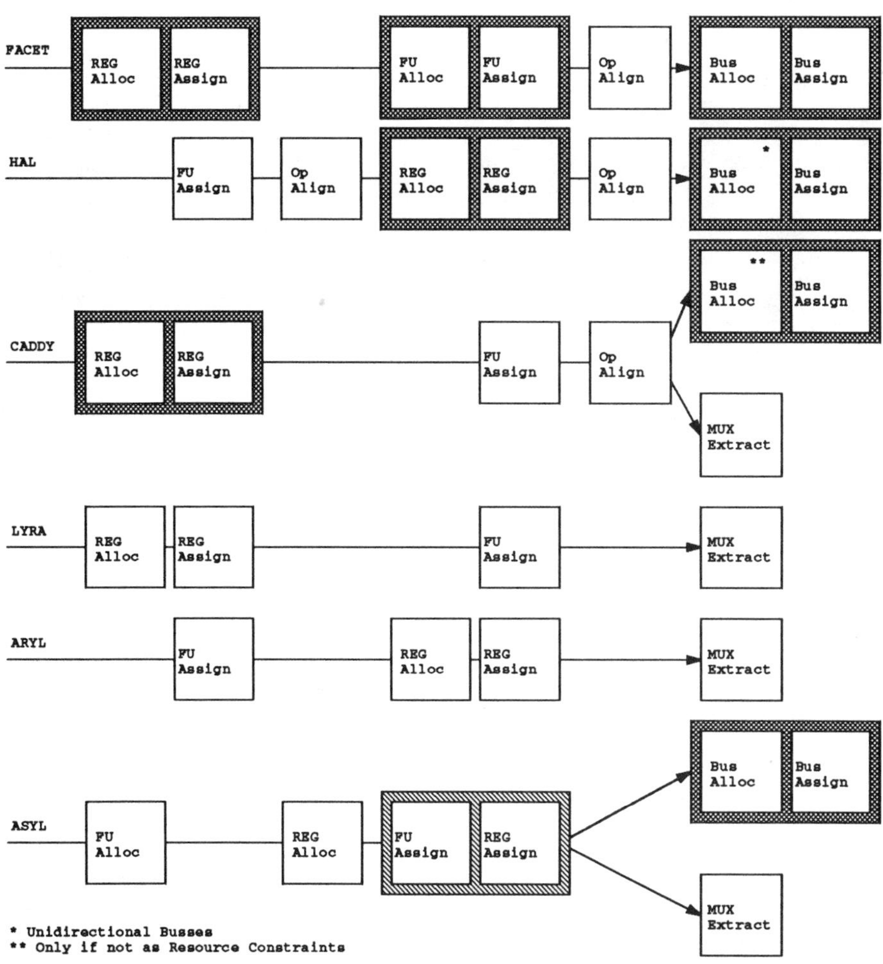

Figure 6.32: Various approaches to data part completion.

allocation of functional units and registers and performs functional unit and register assignment concurrently.

The graph-theoretical techniques are usually implemented in a generic form to be used for all allocation and assignment tasks. The following sections address register allocation and assignment as a representative for the data part allocation and assignment tasks.

6.5.4 Register Allocation and Assignment

6.5.4.1 Lifetime

The starting point for register allocation and register assignment is a scheduled data flow graph. For the moment it is assumed again that the scheduled data flow graph $DFG(V, E)$ represents a basic block. The reader may recall Figure 6.8 in which the schedule of the Differential Equation example obtained from mobility list scheduling is shown. Assuming a multiplexed data part architecture, a value needs to be stored if the corresponding data flow edge(s) cross a control step boundary.[23] A value created by operation o corresponds to the set of edges to the successor operations of o: $val = \{o\} \times Succ(o, DFG)$.[24] A control step σ is associated with the time interval $[T_\sigma, T_{\sigma+1})$ where T corresponds to a control step boundary. The *lifetime* i_{val} or life interval of a value is defined as

$$i_{val} = [\,T_{\sigma(o)} + \Delta(o), \max_{o' \in Succ(o,DFG)} (T_\sigma(o') + \Delta_{use}(o'))\,),$$

where $\Delta_{use}(o')$ equals the operation delay $\Delta(o')$ if operation o' is assigned to a combinational functional unit. If operation o' is assigned to a pipelined functional unit, $\Delta_{use}(o')$ equals the delay of the first stage of the pipelined functional unit. If i_{val} contains a control step boundary T, the value *val* needs to be stored. Values having disjoint lifetimes may share a register. The minimum number of registers needed is fixed by the schedule. In Figure 6.33, the number of life intervals crossing or starting at the $4th$ control step boundary is 6, hence at least 6 registers are needed.

Register allocation means determination of the number of registers to be provided. However, to find the *minimum* number required is not always solvable in polynomial time. The conditions under which the determination of the minimum number is polynomially solvable are explained in Section 6.5.5.1. Moreover, allocation of the minimum number of registers is not necessarily the ultimate goal, since allocating more than the minimum number may be preferable if interconnect area is taken into account.

Register assignment means to assign values to register instances. Since interconnect area is strongly dependent on the grouping of values in registers, it is recommended to control register assignment by the effects on interconnect area.

In the following, various algorithms and heuristics for combined[25] and for

---

[23] For the bidirectional bus architecture slightly different assumptions are made.

[24] For simplicity it is assumed that operations only have one output.

[25] This is the reason why often "allocation" is used as synonym for "allocation *and* assignment."

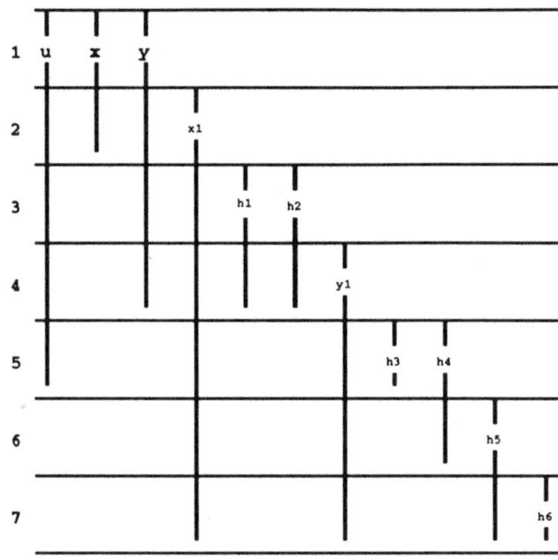

Figure 6.33: Lifetime table for Differential Equation example.

separate register allocation and register assignment, and aspects of interconnect optimization are reviewed.

6.5.4.2 Left-Edge Algorithm

One way to look at register allocation and assignment is to draw an analogy to the track assignment problem in channel routing with the goal of minimizing the number of tracks needed. The wires (values) that do not intersect in space (lifetime) may share a track (register). Accordingly, the *left-edge algorithm* [HaSt71] was adapted to register allocation and assignment [KuPa87]. The birth time of a value, $birth(val) = T_{\sigma(o)} + \Delta(o)$, corresponds to the left and the death time, $death(val) = \max_{o' \in Succ(o,DFG)} (T_\sigma(o') + \Delta_{use}(o'))$,[26] to the right edge of a wire.

Algorithm 6.4 works as follows: First, the values V are sorted in increasing order of their birth times. The first value in the list is assigned to the first register. The list is scanned for the next value v_{next} whose birth time is larger than or equal to the death time of the value v_{cur} assigned before. v_{next} is also assigned to the first register, and the list is scanned for the next value until no more values are found to share the first register. A new register containing the (new) first element of the list is created and again the list is scanned for further values that can share the current register. The whole process is repeated until the list is empty or, in other words, all values are assigned to a register.

[26]Note that the death time is *not* contained in the life interval.

Given a set V of values v with the life interval i_v, a birth time $birth(v)$ and a death time $death(v)$, where $I = \{i_v : v \in V\}$ is the set of intersecting intervals on a line, *assign v to registers $\mathcal{R}$ such that the minimum number of registers is allocated.*

```
BEGIN
    I_sorted ← Sort V with increasing birth(v);
    number ← 0;
    WHILE I_sorted ≠ NIL DO
    BEGIN
        number ← number + 1;
        v_cur ← HEAD(I_sorted);
        I_sorted ← TAIL(I_sorted);
        R[number] ← {v_cur};
        I'_sorted ← NIL;
        WHILE I_sorted ≠ NIL DO
        BEGIN
            v_next ← HEAD(I_sorted);
            I_sorted ← TAIL(I_sorted);
            IF birth(v_next) ≥ death(v_cur) THEN
            BEGIN
                R[number] ← R[number] ∪ {v_next};
                v_cur ← v_next;
            END ELSE I'_sorted ← APPEND(I'_sorted, v_next);
        END;
        I_sorted ← I'_sorted;
    END;
    writeln('Registers allocated:', number);
END.
```

Algorithm 6.4: Left-Edge algorithm for register allocation and assignment.

The algorithm has polynomial time complexity and guarantees to allocate the minimum number of registers. Unfortunately, there are two disadvantages to the left-edge algorithm.

First, not all lifetime tables may be interpreted as intersecting intervals on a *line* which is the necessary condition for using the left-edge algorithm without heuristic extensions. The reader may recall Figure 6.33. Value u is created at the end of Control Step 7 for the next loop iteration and values $x1$ and $y1$ have to be transferred to values x and y for the next loop iteration. If the transfers of value $x1$ to x and value $y1$ to y would take place in different control steps, or in other control steps than Control Step 7, the lifetimes had to be interpreted as intersecting intervals on a *circle* in order to represent all conflicts. The presence of loops causes such lifetime tables. Also, the existence of conditional branches prohibits the interpretation of intersecting intervals on a line, since values occurring in mutual exclusive branches may share a register although they seem to overlap in lifetime.

Second, the assignment produced is neither unique nor necessarily optimal. There may be an assignment that yields fewer multiplexors, or fewer multiplexor inputs in the final design. To find an optimal assignment in terms of interconnect area is still a topic for research.

6.5.4.3 Tseng's Heuristic

In this and the next section clique partitioning for register allocation and assignment is introduced.

From the value lifetimes a *compatibility graph* $G = (V, E)$ is built, in which V is the set of values, and where two vertices v, w are connected by an edge $e = (v, w) \in E$ if and only if the two values are compatible in lifetime, that is their lifetimes do not intersect ($i_v \cap i_w = []$). Register allocation and assignment is equivalent to finding the minimum clique partitioning of the compatibility graph. A clique corresponds to a register. The compatibility graph for the Differential Equation example (see Figure 6.33) is shown in Figure 6.34.

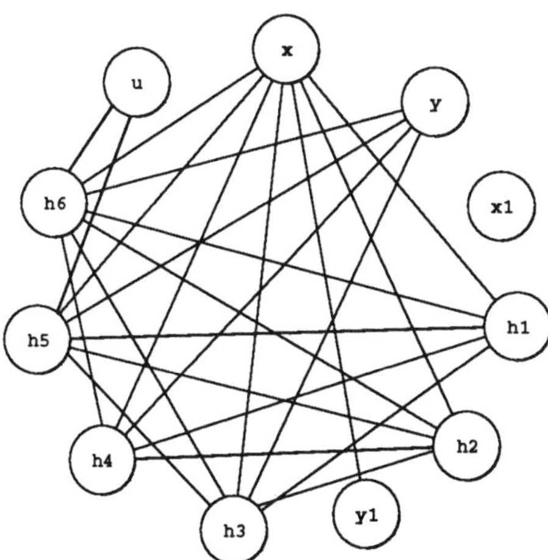

Figure 6.34: Compatibility graph for register allocation and assignment of the Differential Equation example.

Allocation and assignment of functional units as well as allocation and assignment of busses can be solved the same way. Operations are compatible if they are executed in disjoint control steps; data transfers are compatible if they do not intersect.

Definition 6.8 (Clique, clique cover) *Let $G = (V, E)$ be an undirected graph with a set V of vertices and a set E of edges.*

A clique is a set of vertices that form a complete subgraph of G, i.e., a set where every pair of distinct vertices is adjacent. The clique is maximal if it is not properly contained in another clique of G. The clique is maximum if there is no clique of larger cardinality.

A clique cover of G is a set of cliques A_i for $i = 1, 2, \ldots c$ that covers the entire graph: $V = A_1 \cup A_2 \cup \ldots A_c$. A clique partitioning of G is obtained if vertices are deleted from A_i such that $A_i \cap A_j = \emptyset, \forall i \neq j$. A clique cover is called minimum if c is the smallest possible.

Minimum clique partitioning has been proven to be NP-complete [GaJo79] for general graph structures. Therefore, Tseng et al. used a polynomial time heuristic that constructs a near-minimum clique partitioning based on exploiting the graph structure [TsSi86]. Although it does not produce optimum results for special graph structures for which optimum polynomial time algorithms exist, Tseng's heuristic has become a classical approach.

Algorithm 6.5 works as follows: For all edges $e = (v, w)$ the number of common neighbors ν

$$\nu(e) = |\{v' \in V : (v', v) \in E \wedge (v', w) \in E\}|$$

and the number of edges η to be deleted if the vertices v and w are merged into vertex v

$$\eta(e) = |\{(v, v') \in E : (w, v') \notin E\} \cup \{(w, v') \in E\}|$$

are calculated. Merging vertices v and w into vertex v corresponds to assigning values v and w to register $\#v$. The list $\mathcal{LR}_v$ stores the values assigned to register $\#v$. Common neighbors ν and edges η to be deleted constitute the graph structure information that is exploited to select among competing merges. The vertices connected by the chosen edge e_{best} are merged into vertex v and the edges that are no longer valid are deleted. For affected edges, ν and η are updated. Then the edges incident to vertex v are traversed. Accordingly, the value belonging to e_{best} is again assigned to register $\#v$. This is repeated until there are no edges left that are incident to v. The edges of the whole graph are inspected again, and values are assigned to the next register. The whole

Tseng's basic clique partitioning algorithm for register allocation and assignment. Given is a compatibility graph $G = (V, E)$ of arbitrary structure where $E = \{(v, w) : v, w \in V \wedge i_v \cap i_w = [\,]\}$.
Remark 1: $E_{sub}(v)$ is the edge set of $G_{sub}(V_{sub}, E_{sub})$ with $v \in V_{sub}$ where G_{sub} is a disconnected subgraph of G.
Remark 2: In [TsSi86] η is only calculated if more than one edge with the same ν occurs.

```
BEGIN
      FOR all v ∈ V DO 𝓛𝓡_v ← v;
      FOR all e ∈ E DO calculate ν and η;
      REPEAT
          e_best ← e = (v, w) : η(e) = min_{E_maxν}(η)
                  where E_maxν = {e' ∈ E : ν(e') = max_E(ν)};
          𝓛𝓡_v ← APPEND(𝓛𝓡_v, w);
          E ← E \ {(v, v') ∈ E : (w, v') ∉ E} \{(w, v') ∈ E};
          V ← V \ {w};
          FOR all e ∈ E_sub(v) DO  calculate ν and η;
          E_v ← {(v, v') ∈ E};
          WHILE  E_v ≠ ∅ DO
          BEGIN
                  e_best ← e = (v, w) : η(e) = min_{E_maxν}(η)
                          where E_maxν = {e' ∈ E_v : ν(e') = max_{E_v}(ν)};
                  𝓛𝓡_v ← APPEND(𝓛𝓡_v, w);
                  E ← E \ {(v, v') ∈ E : (w, v') ∉ E} \{(w, v') ∈ E};
                  V ← V \ {w};
                  FOR all e ∈ E_sub(v) DO calculate ν and η;
                  E_v ← {(v, v') ∈ E};
          END;
      UNTIL  E = ∅;
      writeln('Registers allocated:', |V|);
      FOR all v ∈ V DO
      BEGIN
          write('Values assigned to Register #v:');
          REPEAT
                  write(HEAD(𝓛𝓡_v));
                  𝓛𝓡_v ← TAIL(𝓛𝓡_v);
          UNTIL 𝓛𝓡_v = NIL;
          writeln;
      END;
END.
```

Algorithm 6.5: Tseng's heuristic for clique partitioning.

process is repeated until all edges have been removed from the compatibility graph.

When assigning compatible elements to resource instances, whether values to registers, operations to functional units, or data transfers to busses, there are always competing assignments that differ in terms of interconnect area. In the next section it is shown how best use can be made of these additional degrees of freedom.

6.5.4.4 Weight-Directed Clique Partitioning

To illustrate how optimization for interconnect area can be added to clique-partitioning, the multiplexed data part architecture is chosen. It is assumed that functional unit assignment has already been performed. The area due to interconnect hardware usually is abstracted to the number of multiplexor inputs. A refined measure takes the number of point-to-point connections as a measure for the area contribution of wiring into account.

The typical situations that occur when compatible values, or intermediate registers, are merged into one register and the corresponding effects on the interconnect area are shown in Figure 6.35. Weights $\omega(v_1, v_2)$ are assigned to these situations, expressing how preferable a merge of a compatible pair of values or intermediate registers is.

$\omega = 4$: Merging compatible values/intermediate registers having both a source functional unit and a destination functional unit in common reduces the number of point-to-point connections by two and the number of multiplexor inputs at least by one.

$\omega = 3$: Merging compatible values/intermediate registers having a source functional unit but no destination functional unit in common reduces the number of point-to-point connections by one.

$\omega = 2$: Merging compatible values/intermediate registers having only a destination functional unit in common reduces the number of point-to-point connections by one but adds a multiplexor input in front of the register.

$\omega = 1$: Merging compatible values/intermediate registers where a source functional unit of one value is the destination functional unit of the other, clusters the register to the corresponding functional unit but adds a multiplexor input in front of the register.

$\omega = 0$: The worst case occurs when the two compatible values have no functional unit in common. This results in adding a multiplexor input in front of the register without any savings in wiring.

Register allocation and assignment after functional unit assignment. Given is a compatibility graph $G = (V, E)$ *of arbitrary structure in which* $E = \{(v, w) : v, w \in V \wedge i_v \cap i_w = [\,]\}$.

BEGIN
    FOR *all* $v \in V$ **DO**
    BEGIN
        Calculate $\mathcal{S}_v$ *and* $\mathcal{D}_v$;
        $\mathcal{LR}_v \leftarrow v$;
    END
    FOR *all* $e \in E$ **DO** *calculate* $\omega(e)$;
    REPEAT
        $\tau \leftarrow \max_{e \in E}(\omega(e))$;
        $E_{red} \leftarrow \{e \in E : \omega(e) \geq \tau\}$;
        FOR *all* $e \in E_{red}$ **DO** *calculate* ν *and* η;
        REPEAT
            $e_{best} \leftarrow e = (v, w) : \eta(e) = \min_{E_{max_\nu}}(\eta)$
                where $E_{max_\nu} = \{e' \in E_{red} : \nu(e') = \max_{E_{red}}(\nu)\}$;
            $\mathcal{LR}_v \leftarrow \text{APPEND}(\mathcal{LR}_v, w)$;
            $E \leftarrow E \setminus \{(v, v') \in E : (w, v') \notin E\} \setminus \{(w, v') \in E\}$;
            $V \leftarrow V \setminus \{w\}$;
            Update $\mathcal{S}_v$ *and* $\mathcal{D}_v$;
            FOR *all* $e \in \{(v, v') \in E\}$ **DO** *calculate* $\omega(e)$;
            $E_{red} \leftarrow \{e \in E : \omega(e) \geq \tau\}$;
            FOR *all* $e \in E_{red}$ **DO** *calculate* ν *and* η;
            $E_v \leftarrow \{(v, v') \in E_{red}\}$;
            WHILE $E_v \neq \emptyset$ **DO**
            BEGIN
                $e_{best} \leftarrow e = (v, w) : \eta(e) = \min_{E_{max_\nu}}(\eta)$
                    where $E_{max_\nu} = \{e' \in E_v : \nu(e') = \max_{E_v}(\nu)\}$;
                $\mathcal{LR}_v \leftarrow \text{APPEND}(\mathcal{LR}_v, w)$;
                $E \leftarrow E \setminus \{(v, v') \in E : (w, v') \notin E\} \setminus \{(w, v') \in E\}$;
                $V \leftarrow V \setminus \{w\}$;
                Update $\mathcal{S}_v$ *and* $\mathcal{D}_v$;
                FOR *all* $e \in \{(v, v') \in E\}$ **DO** *calculate* $\omega(e)$;
                $E_{red} \leftarrow \{e \in E : \omega(e) \geq \tau\}$;
                FOR *all* $e \in E_{red}$ **DO** *calculate* ν *and* η;
                $E_v \leftarrow \{(v, v') \in E_{red}\}$;
            END;
        UNTIL $E_{red} = \emptyset$;
    UNTIL $E = \emptyset$;
END.

Algorithm 6.6: Weight-directed clique partitioning.

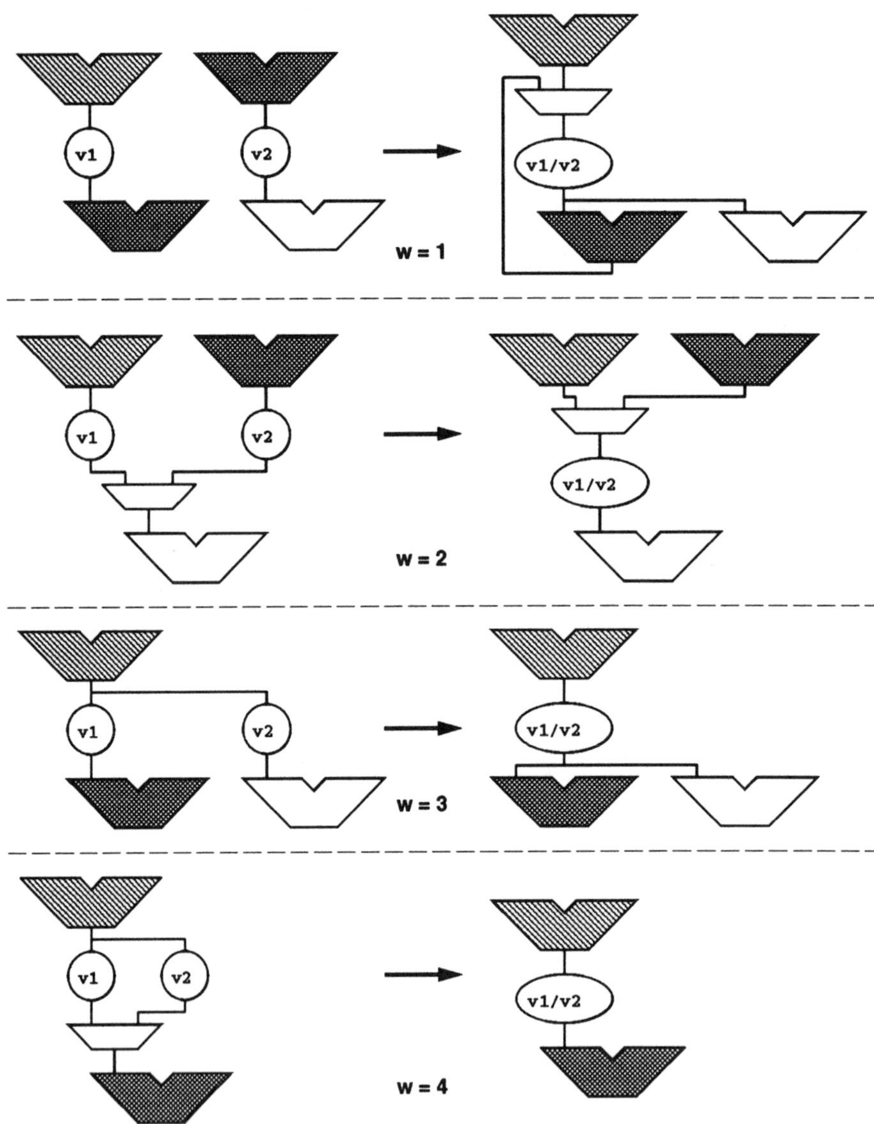

Figure 6.35: Four different situations after functional unit assignment.

Using structural weights as introduced above, clique-partitioning heuristics can now be modified in order to give priority to merges that are profitable in terms of interconnect area (see, e.g., [Paul88]).

As an example, a modification of Tseng's heuristic for *weight-directed clique-partitioning* is introduced. The weight-directed clique-partitioning algorithm (Algorithm 6.6) relies on two compatibility graphs, the original compatibility graph G and a dynamically built, reduced compatibility graph G_{red} where E_{red} contains only edges exceeding the weight threshold τ. First, all edges that have the highest weight are collected in E_{red}. Then, the graph-based heuristic for finding the best edge for merging is applied to G_{red}. The source functional unit and destination functional unit sets $\mathcal{S}_v$ and $\mathcal{D}_v$ of the new vertex v, the weights ω of edges containing v, and G are updated. Edges that have obtained the highest weight are included in E_{red} and the corresponding number of common neighbors ν and the number of edges η to be deleted are calculated. Again, the vertices adjacent to the clique v currently built are traversed and successively inspected for merging until no more edges connected to this clique can be found. This is repeated for the next clique until all edges with highest weight have been processed. The process is restarted with the next smaller maximum weight. This is repeated until no more edges remain in E.

6.5.5 Separate Scheme for Allocation and Assignment

Much effort has been spent in high-level synthesis to improve graph-theoretical techniques for allocation and assignment tasks. This section addresses two aspects of this research [SpTh90, RoKr91, HCLH90, Stok91, MiSa91b].

- The class of problems for which minimum allocation can be done in polynomial time is extended. This refers to graph coloring of *triangulated* instead of *interval* graphs.
- The allocation and the assignment tasks are solved separately. This is advantageous if trade-offs between register cell area and area due to interconnect hardware are to be made. The corresponding variants can be searched by starting with the allocation of the minimum number of registers followed by register assignment, and successively incrementing the number of registers allocated.

In order to use *graph coloring*—the complementary formulation of clique-partitioning—the *conflict graph* is constructed by deriving an undirected graph $G = (V, E)$ where V is the set of values, and where two vertices are connected by an edge $e \in E$, if and only if the two values intersect in lifetime. Register allocation and assignment is equivalent to finding the minimum vertex coloring of the conflict graph. A color corresponds to a register.

Definition 6.9 (Stable set, proper c-coloring, minimum vertex coloring, chromatic number) *Let $G = (V, E)$ be an undirected graph. A* <u>*stable set*</u>

is a subset of vertices X that contains no vertices that are adjacent to each other. (Remark: If all vertices within a stable set are painted with one color no two vertices adjacent in G receive the same color.) A proper c-coloring is a partitioning of G such that $V = X_1 + X_2 + ... + X_c$, $X_i \cap X_j = [\,], i \neq j$ *where each* X_i *is a stable set. A proper c-coloring with smallest possible c is called minimum vertex coloring and its cardinalty c is called chromatic number* $\chi(G)$ *of the graph G.*

Finding the minimum vertex coloring of a general graph has been proven to be NP-complete [Golu80]. For specific conflict graph structures, however, vertex coloring can be performed in polynomial time. For this reason, it is worthwhile to examine the conflict graphs that occur in high-level synthesis.

6.5.5.1 Conflict Graphs in High-Level Synthesis

The conflict graph resulting from the value lifetime table of a scheduled basic block is an *interval graph.*

Definition 6.10 (Interval graph, interval representation) *An undirected graph* $G = (V, E)$ *is called an interval graph if its vertices can be put into one-to-one correspondence with a set of intervals I of a linearly ordered set such that two vertices are connected by an edge* $\in E$ *iff*[27] *their corresponding intervals* $i \in I$ *intersect. I is called the interval representation of G.*

Several problems which are NP-complete for general graphs are solvable in polynomial or even linear time on the number of vertices and edges in an interval graph [Golu80]. For example, the left-edge algorithm (Algorithm 6.4) allocates the minimum number of registers in polynomial time.

Feedback variables in loops have birth times that are later than their death times. If such variables but no conditional branches are present, one obtains a superset of the interval graphs, the *circular arc graphs*, as shown in Figure 6.36. An intersection graph that results from a family of arcs placed around a circle is called a circular arc graph. Finding the minimum coloring of general circular arc graphs is NP-hard [SpTh90, Golu80]. Specialized heuristics for circular arc graphs have been used [HaEl89]. Most of the approaches in high-level synthesis are, however, restricted to interval graphs. For repetitive schedules, interval graphs are achieved by splitting all loop variables into two values, resulting in two intervals. These values are marked as *loop-dependent.* If they are assigned to different registers, a register-to-register transfer must be established to ensure proper function. A method for eliminating some of the register-to-register transfers is presented in [Stok91].

[27] The form "iff" is used as an abbreviation for "if and only if."

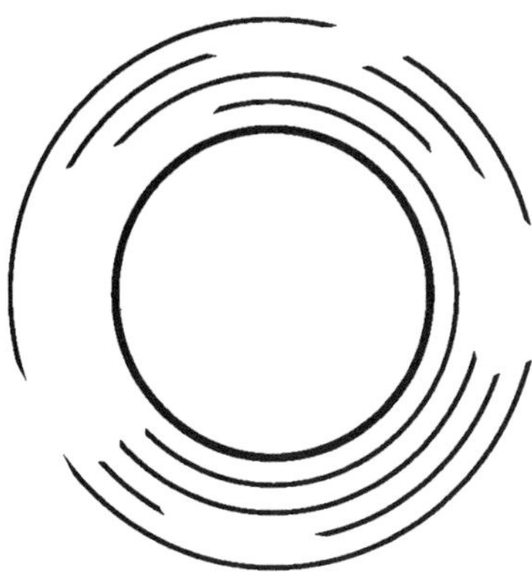

Figure 6.36: Interval representation of a circular arc graph.

There is an important class of algorithmic descriptions containing conditional branches that result, after some modifications, in *triangulated* conflict graphs. The class of triangulated graphs is a proper superset of the interval graphs.

Definition 6.11 (Triangulated/chordal graph) *Let $G = (V, E)$ be an undirected graph. A sequence of vertices $v_1, v_2, v_3, ..., v_l, v_1$ where $v_i \neq v_j$, $\forall i \neq j$ is called a cycle of length l if (v_{i-1}, v_i) for $i = 1, 2, ..., l \in E$ and also $(v_l, v_1) \in E$. It is called chordless cycle if $(v_i, v_j) \notin E$ for i and j differing by more than 1 mod l.*

A graph G is called triangulated or chordal if it does not contain any chordless cycle with length > 3.

Theorem 6.1 (Walter, Gavril, and Buneman [Golu80]) *The following two statements are equivalent:*

(i) $G = (V, E)$ is the intersection graph of a family of subtrees $\{T_v\}_{v \in V}$ of a tree T.

(ii) $G = (V, E)$ is triangulated.

Triangulated graphs can be recognized and colored in polynomial time [Golu80]. Many lifetime tables arising from algorithmic descriptions with conditional branches require the notion of intersecting subtrees of a tree in order to represent all conflicts. This is illustrated in Figure 6.37 depicting the lifetime table[28] for the small conditional branch example listed below.

```
IF (x < c) THEN
BEGIN
    b11 := x + 1;
    b12 := x + b11;
    bout := b11 * b12;
END ELSE
BEGIN
    b21 := x - 1;
    b22 := b21 + x;
    bout := b21 * b22;
END;
```

In order to capture all conflicts, value x has to be represented as a tree. The conflict graph which is also shown in Figure 6.37 is a triangulated graph. Using an interval graph leads to the allocation of 4 registers, because the maximum number of values whose lifetimes cut the same control step boundary is 4. However, two registers are sufficient since the values in different conditional branches do not actually overlap.

Conditional branches may still cause conflict graphs that are not triangulated. This situation occurs if a value is created in one of the mutually exclusive branches and crosses the conditional branch MERGE vertices. Again, these values have to be cut into two values exactly at the conditional branch MERGE vertices in order to obtain a triangulated graph.

6.5.5.2 Two-Step Approach

The weight-directed clique-partitioning heuristic, the left-edge algorithm, and the basic clique-partitioning approach derive the number of registers to be allocated simultaneously with the grouping of values in registers. Besides the fact that weight-directed clique-partitioning works for arbitrary compatibility graphs, the advantage of this approach is that it keeps a global view to all values during register allocation and assignment. The disadvantage, however, is that the number of registers that are finally allocated cannot be controlled directly. Control over the number of registers allocated is needed to influence the trade-off between register cell area and interconnect area. It is therefore attractive to treat register allocation and register assignment, including interconnect optimization, separately in a *two-step approach*.

[28] Assuming a schedule with two-cycle multiplications.

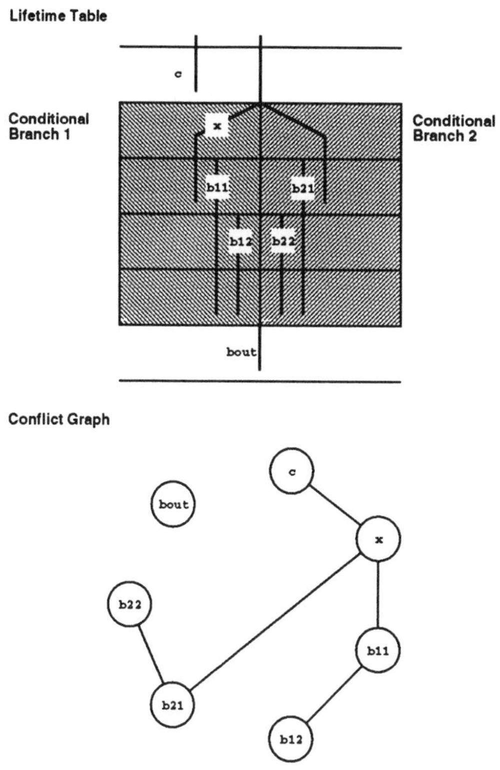

Figure 6.37: Lifetime table and triangulated conflict graph.

Given a scheduled data flow graph, the minimum number of registers needed may be calculated first in order to get a lower bound on the number of registers to be allocated. The minimum number of registers needed corresponds to the number of vertices in the maximum clique or the chromatic number of the conflict graph. Assuming interval or triangulated graphs, this can be computed in polynomial time. In case of interval graphs, the number of vertices in the maximum clique corresponds to the maximum number of values cutting a control step boundary. The set of values cutting the same control step boundary is called *cut set*. The reader may recall the considerations for the Differential Equation example in Section 6.5.4.1. Since the transfers of values $x1$ to x and $y1$ to y are performed in Control Step 7, the resulting conflict graph is an interval graph. Accordingly, the minimum number of registers needed can be derived from the maximum cut set. Moreover, in interval graphs, any maximal clique corresponds to some cut set(s). All maximal cliques can be found by inspecting all cut sets and filtering the non-maximal cliques out.

The idea of the two-step approach is to allow for interconnect optimization as described for the weight-directed clique-partitioning by exploiting the fact that the values belonging to a maximal clique can be colored simultaneously or, in other words, can be assigned to register instances simultaneously. A bipartite graph $G(A, R, E)$ is constructed where one vertex set represents the allocated registers R and the other one represents the values belonging to a maximal clique A. Directed edges connect a value $v \in A$ with a register $r \in R$ if value v can be assigned to register r without conflict. A conflict occurs if the lifetimes of values that have been assigned to register r earlier intersect with those of v. Again, weights for the edges may be introduced in order to represent how profitable an assignment is. Finding an assignment for all values in A to the registers R that yields the least overall cost is equivalent to *bipartite weighted matching*. An algorithm for bipartite weighted matching can be found in [Tarj83].

Since all maximal cliques have to be colored in order to obtain the complete register assignment, it is crucial that the maximal cliques are ordered in such a way that a value occurring in two or more maximal cliques cannot obtain different colors. Interval graphs as characterized by Theorem 6.2 allow for an ordering of maximal cliques such that maximal cliques containing the same vertex occur consecutively [Golu80].

Theorem 6.2 (Gilmore and Hoffman) *Let G be an undirected graph. The following two statements are equivalent:*

(i) G is an interval graph.

(ii) The maximal cliques of G can be linearly ordered such that for every vertex x of G the maximal cliques containing x occur consecutively.

It is obvious that for the cut sets such an ordering is given by the sequence of the control steps. A two-step approach as presented in [HCLH90] can be sketched as follows:

- Register allocation: First the minimum number of registers needed is calculated by searching for the maximum cut set.

- Register assignment: Then the maximal cliques are scanned according to the sequence of control steps. All values of cut sets that correspond to maximal cliques are assigned at once to the allocated registers using structural weights for interconnect optimization and solving bipartite weighted matching as described above.

Searching for trade-offs between register cell area and interconnect area can now easily be understood. Instead of using the minimum number of registers needed, several different register allocations followed by register assignment are tried, and the solution resulting in the smallest overall area is chosen.

For interval graphs, it is sufficient to use the sequence of the control steps in order to have a proper order for coloring the maximal cliques. In case of triangulated graphs the order of maximal cliques for coloring has to be determined differently. This is obtained by constructing a perfect vertex elimination scheme by lexicographic breadth-first search [SpTh90].

Being triangulated is a hereditary property defined through *simplicial* vertices. A vertex v of $G = (V, E)$ is simplicial if its *adjacency set* $Adj(v)$ induces a clique in G. The adjacency set is the set of all vertices that are connected to v. Each triangulated graph possesses at least two simplicial vertices. One can, therefore, check for triangulated graph structure by finding a simplicial vertex, eliminating it and the incident edges from G, finding the next vertex, again eliminating it and the incident edges, until either no vertices remain in G (the graph is triangulated), or no more simplicial vertex can be found (the graph is not triangulated). An ordering of the vertices, such that each vertex is a simplicial vertex of the remaining graph, is called *perfect vertex elimination scheme*.

Vertex elimination schemes can be constructed in different ways. In the following, a vertex elimination scheme constructed backwards by *lexicographic breadth-first search* is used. If the graph is not triangulated the vertex elimination scheme constructed according to Algorithm 6.7 is not perfect. A vertex

Given an undirected graph $G(V, E)$, construct a vertex elimination scheme σ by lexicographic breadth-first search. (Leuker, Rose, and Tarjan [Golu80])

```
BEGIN
    FOR all v ∈ V DO ℒ(v) ← 0;
    Σ ← 0;
    FOR n ← |V| DOWNTO 1 DO
    BEGIN
        v ← v : ℒ(v) = max_{w∉Σ}(ℒ(w));
        σ(n) ← v;
        Σ ← Σ ∪ {v};
        FOR all w ∉ Σ ∧ w ∈ Adj(v) DO ℒ(w) ← ℒ(w) + n;
    END;
END.
```

Algorithm 6.7: Vertex elimination scheme using lexicographic BFS.

elimination scheme has to be tested for whether it is a perfect scheme or not; an algorithm for testing a vertex elimination scheme can be found in [Golu80].[29] A modification of the testing algorithm together with a perfect vertex elimination scheme as constructed in Algorithm 6.7 can be used to produce all maximal cliques and calculate the number of vertices in the maximum clique χ. The resulting polynomial-time algorithm (Algorithm 6.8) is due to [Golu80].

Some of the cliques found are not maximal and must be filtered out, since only the maximal cliques are needed. In order to filter out the non-maximal cliques, the number $\mathcal{S}$ is introduced which counts the vertices to be checked for being part of a clique. The cliques that are not maximal miss vertices that have

Given a perfect vertex elimination scheme σ, list the maximal cliques and find the chromatic number χ.

```
BEGIN
    χ ← 1;
    FOR all v ∈ V DO S(v) ← 0;
    FOR n ← 1 TO |V| DO
    BEGIN
        v ← σ(n);
        X ← {x ∈ Adj(v) : σ⁻¹(v) < σ⁻¹(x)};
        IF Adj(v) = ∅ THEN writeln({v});
        IF X ≠ ∅ THEN
        BEGIN
            u ← σ(min{σ⁻¹(x) : x ∈ X});
            S(u) ← max(S(u), |X| − 1);
            IF S(v) < |X| THEN
            BEGIN
                writeln({v} ∪ X);
                χ ← max(χ, 1 + |X|);
            END;
        END;
    END;
    writeln('The chromatic number is', χ);
END.
```

Algorithm 6.8: Maximal cliques and chromatic number χ.

been eliminated earlier according to σ. The cardinality stored in $\mathcal{S}$ is therefore equal to or larger than that of the current clique, and the clique is not to be listed.

[29] See in [Golu80], Figure 4.6, Page 90.

A perfect elimination scheme constructed with breadth-first search provides a sequence for a coloring which is optimum with respect to the number of colors, but greedy with respect to other optimization aspects, e.g., interconnect area. More than a greedy approach to minimum coloring can, however, be achieved if one orders the maximal cliques *in reverse* with respect to the order produced by Algorithm 6.8. Again, all vertices of a maximal clique can be colored simultaneously using structural weights and solving bipartite weighted matching. Hence, a minimum coloring *including* interconnect optimization is achieved.

6.5.6 Synthesis Results

In order to illustrate the capabilities of the algorithms and heuristics introduced in the preceding sections, the 5th Order Wave Digital Filter is used again. As a starting point, the cluster approach from [SGLM90] and mobility scheduling are applied.

The data flow graph contains seven delay elements which correspond to seven feedback values. If the feedback values are not cut, the lifetime table results in a circular arc conflict graph.

In order to apply the left-edge algorithm (Algorithm 6.4) or the two-step approach (see Section 6.5.4.4), the feedback values have to be cut at the delay elements into two loop-dependent values. This results in an increase from 35 to 42 values to be handed over to register allocation and assignment. After register assignment for those loop-dependent value pairs that have been assigned different registers, register-to-register transfers are established in the last cycle.

No modifications of the value lifetime table are required when Tseng's clique partitioning (Algorithm 6.5) or the weight-directed clique-partitioning (Algorithm 6.6) are applied. Both heuristics can deal with general graph structures.

The minimum number of registers achievable for this design is 12. Whereas the left-edge algorithm randomly clusters the values into registers—which is reflected in the number of multiplexor inputs, or the number of two-to-one multiplexors respectively—the two-step approach (see row 4 in Table 6.14) results in much fewer multiplexor inputs and multiplexors. This saves not only cell area but also wiring area since the number of point-to-point connections is reduced as well.

Rows 2 and 3 of Table 6.14 illustrate that with the weight-directed clique partitioning the minimum number of registers cannot be reached. In order to reduce interconnect cost, the price of one extra register has to be paid. Without interconnect optimization 13 registers and with interconnect optimization 14 registers are allocated. Although the minimum number of allocated registers obtained in rows 1 and 4 is 12, 13 registers, obtained with Tseng's heuristic, may be the minimum achievable number when the feedback values are *not* cut.

Algorithm	Number of Registers	Number of Mux Inputs	Number of 2:1 Multiplexors
Left-Edge	12	50	35
Tseng	13	51	38
Weight-Directed	14	28	22
Two-Step	12	28	20

Table 6.14: Register allocation and assignment results.

6.5.7 Bus Allocation and Assignment

In Section 6.3.1 the multiplexed data part architecture was used to study the synthesis of the Differential Equation example. In order to illustrate bus allocation and assignment a bidirectional bus architecture is used now.

It is assumed that

- functional unit allocation
- resource constrained scheduling
- functional unit assignment
- register allocation
- register assignment

have already been performed. The functional unit set contains one two-stage pipelined multiplier and one ALU. Functional unit assignment is already solved by functional unit allocation for this case. The difference between the functional units used in the multiplexed data part architecture (see Figure 6.10) and the functional units used now is that besides the internal storage of the pipelined multiplier, a register is associated with each functional unit output. In Figure 6.38 the scheduled data flow graph and the storage provided are shown. Output registers, internal registers of functional units, and additional registers for intermediate storage are indicated as small squares.

In order to complete the design, busses have to be allocated and data transfers have to be assigned to them. According to the assumptions made, values may be transferred from an output register of a functional unit or from an additional register; values are transferred to the input of a functional unit or to an additional register. Data transfers that take place in the same control step conflict. The data transfers conflicting in Control Step 2 are highlighted in Figure 6.38. The maximum number of conflicting data transfers is 4, accordingly at least 4 busses have to be allocated. Again, structural information similar

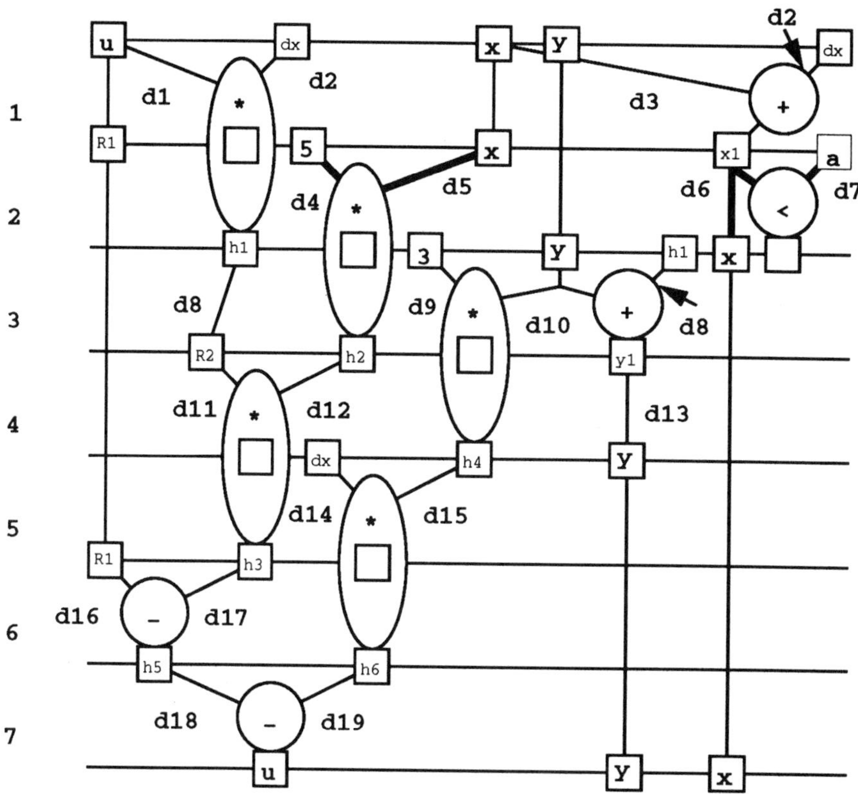

Figure 6.38: Data transfers for bus allocation and assignment.

to the one used for register assignment is exploited when bus assignment is performed. The resulting assignment of data transfers to busses is listed in Table 6.15.[30] The corresponding netlist of the data part is shown in Figure 6.39.

[30]The following assumptions have been made: The weight for assigning a data transfer to a bus corresponds to the number of sources and destinations the data transfer and the bus have in common. Data transfers were assigned according to the sequence of control steps.

Bus	$B1$	$B2$	$B3$	$B4$
Data Transfer	d1	d2	d3	d6
	d4	d7	d5	d13
	d9	d8	d10	d18
	d11	d12	d16	
	d14	d15		
		d17		
		d19		

Table 6.15: Bus assignment for bidirectional bus architecture.

6.6 Outlook

The focus of this chapter is on methods and approaches for solving scheduling, resource allocation, and resource assignment. But, there are many topics left to be mentioned in order to present a complete overview on high-level synthesis.

An important issue deserving more detailed consideration is *domain-specific* synthesis strategies. Many approaches have been published that are tailored to specific application domains and that differ significantly in the emphasis of synthesis tasks from the classical approach. Examples are the HIS system [BeCa89, CaBe90, BeCP91, Camp91b] targeting processor applications; the control-dominated applications path of the CALLAS system [März89, Duzy89, KoGD90, März90] and the HERCULES system [DeKu88, KuDe90]; the PHIDEO system [LMWV91, VdWe91] tailored to video applications; and the HYPER system [PoRa89, RCHP91] targeting DSP applications.

There are several tasks which have not been dealt with in detail in this chapter which need more consideration.

First, when large amounts of data need to be stored *memory management* becomes an essential synthesis task. Memory management deals with the allocation of memories, the assignment of data to memories, and the generation of address computation units.

In order to provide application specific units, a second essential synthesis task called *high-level data path mapping* has been identified. High-level data path mapping consists of partitioning the data part into application specific units [NGCD91] and their definition.

Another important issue is that of encoding problems which occur at several stages in high-level synthesis. In addition to state encoding performed in the subsequent FSM synthesis, encoding tasks and related optimization potential exist at the algorithmic level for data type encoding and—at the interface

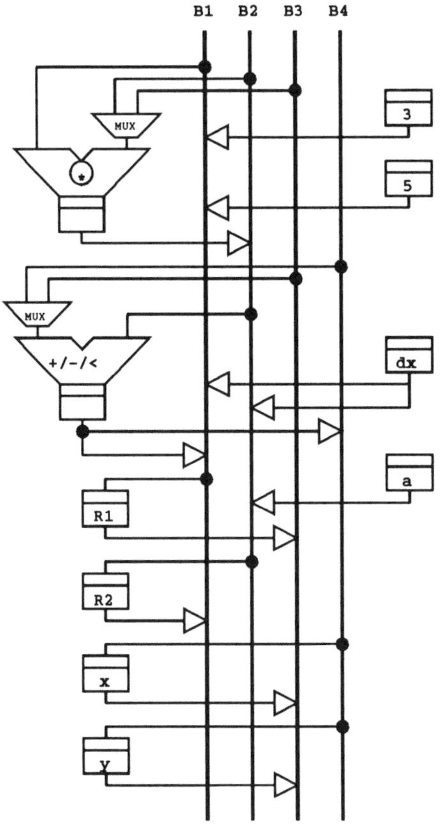

Figure 6.39: Bidirectional bus architecture of the Differential Equation example.

between data part and controller—for the encoding of control signals.

Assumptions made and methods used for deriving the internal behavioral representation from an algorithmic description, especially the role of data flow analysis in synthesis [RoKr91], have been omitted. The corresponding tasks are closely related to the semantics and features of the algorithmic description and to the specific form of the internal behavioral representation used.

High-level synthesis overlaps to some extent with other synthesis steps, namely register-transfer level synthesis at the abstraction level below, system level synthesis at the level of abstraction above. Typical register-transfer level synthesis tasks such as RT-level retiming as presented in Chapter 5, multiplexor optimization [WeBP91], and interconnect synthesis including mapping to library cells [KuPa90a, LyEG90] are needed to complete the creation of an appropriate register-transfer level structure. System-level synthesis tasks (see Chapter 7) such as partitioning [LaTh89], transformations in order to create processes or pipe stages [WaTh89], and retiming [HaEl89] all performed at the algorithmic level or above, may have a stronger impact on the synthesis results than the use of refined algorithms for scheduling, resource allocation, and resource assignment.

Considerable effort has been spent to tighten the link between the classical high-level synthesis tasks and physical design in order to meet hardware constraints. The integrated scheduling and floorplanning approach presented in [WePa91] is one example.

6.7 Problems for the Reader

1. Explain why list scheduling with mobility priority (see Section 6.4.3.3) prefers the two critical path add operations in the 5th Order Digital Wave Filter example (see Section 6.4.3.5) for the assignment to Control Step 12.

2. Derive the following alternate formula for the self-force measure used in force-directed scheduling (see Section 6.4.4):

$$\mathcal{SF}_{(\sigma(o_0)=s_0)} = \sum_{s=\sigma_{ASAP}(o_0)}^{\sigma_{ALAP}(o_0)} \mathcal{SF}_{(\sigma(o_0)=s_0)}(s)$$

where

$$\mathcal{SF}_{(\sigma(o_0)=s_0)}(s) = \sum_{s=\sigma_{ASAP}(o_0)}^{\sigma_{ALAP}(o_0)} n_k(s)(p_{o_0}^{new}(s) - p_{o_0}^{old}(s))$$

and

$$p_{o_0}^{new}(s) = \delta_{s,s_0}$$

with

$$\delta_{s,s_0} = \begin{cases} 1 & \text{if } s = s_0 \\ 0 & \text{else} \end{cases}$$

3. Show that the assignment of subtasks to stages given in Table 6.16 allows pipeline schemes with latency 2, 3 and 4. Use the reservation table as introduced in Section 6.4.6 for this purpose.

Subtask	$S1$	$S2$	$S3$	$S4$	$S5$	$S6$
Stage	$St1$	$St2$	$St3$	$St4$	$St5$	$St1$

Table 6.16: Subtask-to-stage assignment.

4. Extend pipeline list scheduling given in Algorithm 6.3 for separate chaining and multi-cycling as introduced in Section 6.4.3.3. Do not consider structural pipelining.

5. Apply the left-edge algorithm 6.4 presented in Section 6.5.4.2 to the lifetime table presented in Section 6.5.4.1. Note that the values are already sorted according to increasing birth times.

6. Recall from Section 6.5.5.1, why the conflict graph shown in Figure 6.40 is not triangulated.

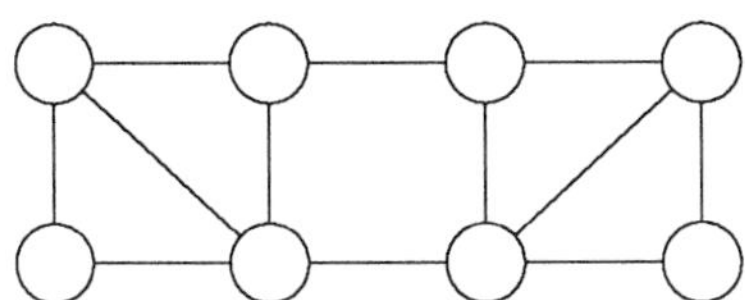

Figure 6.40: Conflict graph.

Chapter 7

System-Level Synthesis

Wolfgang Glunz, Andreas Pyttel, Gerd Venzl

7.1 Introduction

The objective of system-level synthesis is to synthesize complex real world systems from a system-level specification. Starting from a specification which is *abstract* and *minimal*, i.e., containing only the necessary requirements and constraints, the result is an implementation in terms of existing components or specifications of subsystems, which in turn can be synthesized. The goal is an implementation which is quickly available, cost-effective both in terms of design and manufacturing costs, and fully supports the required functionality and performance. Of course, fully automated synthesis at the system level is not the goal, at least with the present understanding of synthesis and design automation technology. Given the fact that systems are continuously becoming more complex, it may, actually, never be possible.

In this chapter, selected problems and solutions to system-level synthesis are presented. The selection is highly subjective, because research in system-level specification, synthesis, and optimization is still in its embryonic phase and the picture is still changing. This introductory section presents a general view on system-level synthesis. Some commonly used notions like "system," "system level," and "system-level synthesis" are explained.

7.1.1 What is a System?

Above, the term "system" has been used in two different contexts, i.e., complex real world systems and system-level specification. These different contexts suggest two different connotations of the term "system." An object which is termed a *system* is commonly understood to be large and complex with respect to some measure of size, degree of internal couplings, or difficulty of design. The first connotation, therefore, is complexity.

On the other hand, the term "system level" suggests more than just size or complexity. Millions of gates on a chip or hundreds of standard parts assembled on various printed circuit boards, obviously, do not represent the system-level view. This view, at least, implies *structure*, i.e., physical partitions or hierarchical composition. Moreover, a meaningful structure comes along with an abstract concept of behavior. Consider the example of a microprocessor. Its behavior is described through instructions which manipulate data. The second connotation, therefore, is abstraction.

The reader might suggest that complexity and abstraction are not completely orthogonal dimensions. Complexity has always driven engineers and scientists to consider abstraction as a means of understanding. In turn, abstraction has helped to manage complexity and to solve the practical problems of building complex systems. Current work on system-level synthesis reflects the two sides of the term "system." Some of the work mainly deals with complexity. Partitioning tries to impose structure on behavioral descriptions (Section 7.2). Behavioral transformations modify an existing behavioral description (Section 7.3). Other work is concerned with abstraction and abstract system-level descriptions in the behavioral domain (Section 7.4).

Before the term "system-level synthesis" is explained, it is instructive to highlight some properties of real world systems. Primarily, a system is a question of perception. An engineer, who is involved in the design of a factory automation system, views workstations as components of a system. On the other hand, a designer of a workstation views the workstation as a system and microprocessors as components. From an engineering point of view, systems are the problems of today, the objects currently under planning, construction, marketing, or maintenance. Obviously, the perception of systems has a historical dimension. The systems of today may be well-understood simple components of tomorrow.

Systems share a few common features: To a large extent they are separated from the rest of the world, meaning the interactions are rare and well-defined (*semi-isolation*). The term "embedded system" seems to contradict this. But, even those systems can only interact through well-defined interfaces. Systems solve sets of often complicated but, usually, related problems posed by a specific problem domain (*domain specificity*). Usually systems are composed of subsystems each solving a particular subproblem. For a given non-trivial system a hierarchy of subsystems exists (*hierarchical composition*). These consist of systems themselves, or are seen as simple components, depending on the point of view of the engineer.

Non-trivial systems share the property that they are not only large in a certain sense (chips, printed circuit boards, cabinets, etc.), but heterogeneous, also. They comprise heterogeneous hardware (digital and analog electronics, mechanical elements such as sensors and actuators), and various levels of firmware and software (***heterogeneity***). Obviously, systems are not built from parts at random. Instead, a system exhibits a coherent plan which describes its concepts on a functional and structural basis (*coherence*). In addition, this plan, often called the system's architecture, usually determines the gross features of its implementation. The computer architecture which defines the hardware-software interface of a computer is a well-known example [Goor89].

Summarizing, complete and automatic synthesis of those systems which are the industrial challenges of today is not feasible. System-level synthesis tackles certain subproblems in the specification, design, and implementation of system architectures.

7.1.2 System-Level Synthesis

From its basic meaning synthesis, as opposed to analysis, is the process of combining objects to form new objects which are more complex or more powerful in a qualitative sense. This means that the new objects can achieve tasks which its constituent parts are not capable of. A microcomputer, for instance, can repeatedly execute a stored program which a central processor unit (CPU) or a microprocessor without memory cannot perform.

Through the activities and results of research on the synthesis of electronic systems, synthesis has been annotated with other meanings and expectations. Synthesis is expected to proceed automatically or, at least, in an interactive manner with significant computer assistance. Synthesis is driven by a specification of the functionality (behavior) of the object to be synthesized. The idea of driving synthesis from a specification of the behavior which is qualitatively different from the behavior of the components is tied to the concept of levels of abstraction. In this perspective, synthesis is the implementation of an object on a given level of abstraction through objects on the levels of abstraction below. Consider the transformation of the instruction set of a computer (the computer architecture) into algorithms which interpret those in a behavioral hardware description language. This is an example of system-level synthesis: The starting point is an abstract data type "computer" with certain values (the I/O data) and operations (instructions) on these values.

However, specifications of systems are often structural. The components can be described at the system level as abstract processors implemented at the algorithmic level. From a practical point of view, there is good reason to consider the corresponding global transformations and optimizations at the algorithmic level as system-level activities. Most current high-level synthesis systems use as their target architecture single processors consisting of a data part controlled by a finite state machine controller. Hence, system-level synthesis identifies those activities which deal with two or more concurrent processes. System-level synthesis starts from concurrent algorithms and generates communicating processors, whereas high-level synthesis implements these processors. Communicating processors require new methods for specification and open up additional degrees of freedom for implementations which in turn can successfully be used for optimizations.

After this introduction a selection of approaches and results are presented which have influenced the current view of system-level synthesis. In Section 7.2 partitioning of behavioral descriptions at the algorithmic level is considered. The importance of partitioning for system-level design is emphasized. Two general methods are described: clustering and iterative improvement. In Section 7.3 interactive transformations of algorithmic-level behavioral descriptions are explained. These transformations can be used to create different behavioral descriptions for a given design and, thereby, to allow the designer to explore the design space for his or her problem. These two sections do not, in the strict sense, deal with the system level as it has been defined in Chapter 1. Instead, they address transformations and optimizations at the algorithmic level. Section 7.4, finally, addresses the system level more directly and refers to corresponding methods of specification and implementation. The topics are concurrency, abstract data types (ADT), and communication.

7.2 System-Level Partitioning

Partitioning is the decomposition of a system into a set of subsystems. It is a necessary step in order to handle complexity and to manage the design process (divide and conquer). In the early stages of design, partitioning is closely related to the development of a system concept and to the definition of an architecture which supports the requirements and is realizable, subject to given constraints.

Efficient partitions are essential for the quality of designs at any stage in the design process and at any level of abstraction. Creating good partitions, however, is a challenging task. Usually, partitioning must be performed when the system exists only "on paper," and reliable data are not available to assess the consequences of certain decisions. Trying out many different solutions and implementing them at a level of abstraction where reliable physical data can be obtained, in general, is not feasible. Design times, in this case, explode beyond acceptable limits.

In current design practice, partitioning at the system level is done manually. Often, designers consider system-level partitioning a simple task. They tend to do it "on the back of an envelope" within a few hours or days by utilizing proven architectures or modifications thereof. This is, at least, true for standard non-explorative designs that constitute a major part of industrial design work. As a matter of fact, no system-level partitioning tool is available which can compete with the global view and the body of knowledge of an experienced designer.

Explorative designs require innovative architectures or carefully optimized modifications of known architectures in order to meet requirements. In this case, various candidate architectures have to be created manually utilizing experience and creativity of designers. High-level modeling can be employed in order to estimate performance figures via simulation or analysis. A substantial body of knowledge on this topic has been accumulated in the area of multiprocessor analysis and performance estimation [ABCo86]. For an application to general hardware design see [RoBT85].

High-level synthesis has opened up new perspectives on partitioning. Designs which can be generated automatically using high-level and subsequent low-level synthesis tools can serve as *prototypes*. They allow a detailed evaluation of design alternatives with respect to, e.g., area and timing. Using this method, a heuristic or even exhaustive search through the design space may still not be possible. However, the knowledge gained from synthesis research can be expected to provide estimation algorithms which run much faster than synthesis and produce sufficiently exact predictions. This can help to solve one of the most urgent problems of system-level design, namely the reliable evaluation of manually created partitions. An example of this type of estimation tools is contained in BUD (Bottom-Up Designer) [McFa83, McKo90] which is presented in Section 6.5.1. The approach in [KuPa90b] has a similar objective.

Another impact of high-level synthesis on the partitioning problem arises from the understanding of behavioral modeling. This opens up a way not only to evaluate but also to generate behavioral partitions for special purposes, e.g., multi-chip partitions. A recent approach to this problem is presented in Section 7.2.2. In general, however, a long time will pass before automatic partitioning and design space exploration become feasible. The method of choice might be the computer-assisted transformation of behavioral descriptions. Thereby, different implementations of a design can be explored, since the outcome of synthesis depends on the form of the input description. Behavioral transformation is addressed in the following Section 7.3.

Partitioning serves as a means to manage complexity and it imposes structure on specifications and implementations. Therefore, it can be used to ease and guide the process of synthesis itself. This aspect of partitioning is covered in Section 7.2.1 in which various ways of clustering operators and functions in behavioral specifications for high-level synthesis are presented. A similar approach applied to logic-level synthesis can be found in [CaBr87]. An application of behavioral partitioning as a preprocessing step to the knowledge-based selection of customized multiprocessor architectures is proposed in [Temm89].

7.2.1 Multi-Stage Clustering and Architectural Partitioning

This section is based on the work on architectural partitioning which has been implemented in the architectural partitioner APARTY [LaTh89]. The objective of APARTY is to provide information about physical aspects of a design (area and timing) by analyzing the behavioral specification for high-level synthesis. This can be done prior to high-level or register-transfer level synthesis, in order to guide the synthesis process. In this respect, the approach is similar to the Bottom-Up Designer (BUD) [McKo90] (see also Section 6.5.1).

As opposed to BUD, however, the emphasis of APARTY is more on the high-level aspects of partitioning. BUD performs clustering of data flow operators considering functional proximity, interconnect proximity, and low-level parallelism [McKo90]. APARTY refines these criteria and adds clustering of procedures (blocks of data flow operators) to exploit high-level parallelism and sharing of procedures. Another difference is that for particular clustering criteria (schedule clustering and operator clustering, cf. Sections 7.2.1.4 and 7.2.1.5) preliminary results from high-level synthesis (scheduling) are used to guide partitioning. Great care is taken to keep the approach flexible and to support a variety of different architectural design styles. Therefore, more clustering criteria than in BUD have been proposed. The approach is based on multi-stage clustering which, in turn, is a refinement of hierarchical clustering. In order to keep the presentation as self-contained as possible the following subsection provides a brief introduction into clustering.

7.2.1.1 Multi-Stage Clustering

The starting point for *hierarchical clustering* [John67] is a set of objects with a certain measure of proximity assigned to each pair. Objects (and existing clusters) are successively merged into new clusters such that those with largest proximity are treated first. If more than one pair with maximum proximity exists, a random choice is made. The algorithm terminates when a single cluster has been created. In this way a hierarchical cluster tree is generated. The basic objects are the leaves of the tree. Each node of the tree represents a merge operation with a proximity value assigned to it. Since clustering is a sequence of binary operations, the tree is binary. However, linked nodes that represent identical proximities can be combined to generate non-binary trees. The nodes of the cluster tree can be ordered according to their proximity values.

In this case, a definite sequence of partitions can be derived progressing from coarse ones (low proximity) to fine ones (large proximity).

The algorithm requires proximities between clusters. For each merge operation, the proximities between the new cluster and all other existing clusters have to be calculated. In order to keep the complexity of the algorithm low, approximations are usually employed for the computation of proximities. For each new cluster either the maximum, minimum, or an average value is chosen from its member objects. This procedure has the property that proximities can never increase. In general, it can not be guaranteed that an exact evaluation of a given proximity function defined for arbitrary clusters has this property. On the other hand, the approximations introduce errors which accumulate when clusters grow. A solution is provided by multi-stage clustering, for which the proximities are recalculated for the existing clusters when the clustering criterion is changed.

Multi-stage clustering is an extension of the hierarchical clustering algorithm. Two or more clustering processes are applied in series, each of them emphasizing a different aspect of the design, and using a different clustering criterion. Each clustering process generates a complete cluster tree. Intermediate cluster trees are cut at a certain value of proximity and the resulting clusters form the objects for the subsequent clustering stage.

There are various degrees of freedom which tend to make multi-stage clustering an art, rather than a science, namely

- the choice of the clustering criterion for each stage, including procedures to compute proximities between clusters,
- the sequence in which these clustering stages are applied,
- the choice of proximity thresholds for cutting the intermediate cluster trees.

These degrees of freedom add to the flexibility of the clustering method. An interactive partitioning tool gives the designer control of the various decisions. It appears that multi-stage clustering simulates the approach of a human designer better than a single clustering stage which combines different aspects into a single criterion. To be really useful, however, the tool must aid the designer in the evaluation of the different partitions which can be created.

There are a few other advantages of multi-stage clustering over conventional single-stage hierarchical clustering.

- Multi-stage clustering provides a general framework for clustering and allows for decoupling of the various criteria. This makes the approach easier to implement, extend, and modify.

- Multi-stage clustering provides a method to compute and, whenever a new stage is started, to recalculate proximities of clusters. The error induced by the approximation of proximities which is used in the course of a single clustering stage can be limited.

- Multi-stage clustering helps in certain pathological situations when, according to a single criterion, many objects have identical proximities. This situation produces very few large clusters. A low-level clustering stage which is based on another criterion can break this inconvenient symmetry and can produce better partitions.

In the following subsections the various clustering criteria and their combination in the APARTY tool are presented.

7.2.1.2 Control Clustering

The goal of this clustering criterion is to keep together those data flow operators which are executed in series without intervening fork/join and conditional branching constructs. Partitioning across long threads of control implies transfer of control between partitions and creates further complicated control structures. In addition, it can decrease performance, due to communication and synchronization overheads. The resulting clusters can be viewed as the basic "instructions" of the design which are selected by controlling the branches in the data flow. This clustering stage can only be applied as a first stage in the clustering, since its criterion is defined for directed acyclic data flow graphs only. In general, preceding clustering stages produce acyclic graphs.

For two data flow operators a and b the proximity function $cprox(a,b)$ used in control clustering is defined as:

$cprox(a,b) = 1$	if a path from a to b, or vice versa, with no branching operator in between exists in the graph, i.e., the operators are in the same sequential branch of the graph,
$cprox(a,b) = 0$	if no path from a to b, or vice versa, exists in the graph, i.e., the operators are in different branches of the graph,
$cprox(a,b) = p$	if a path from a to b, or vice versa, with one or more branching operators in between exists in the graph.

The value p $(0 < p < 1)$ is a measure of the probability that both a and b will be activated in a single execution of the procedure represented by the graph. Probabilities can be derived from static analysis, e.g., an n-fold conditional branch yields a probability $1/n$ for each individual path, or through dynamic analysis by means of traces of the procedure.

As stated earlier, this clustering stage can be applied as a first stage. It can be useful for creating instruction sets from processor or controller specifications.

However, the resulting clusters can be very large, in terms of area requirements, if many complex operators like multiplication and division are assigned to one cluster. Therefore, for applications in digital signal processing with a large degree of parallelism at the level of (data)-operations, control clustering may not be useful.

7.2.1.3 Data Clustering

In data clustering the amount of data to be transferred between operators or clusters is considered. The goal is to reduce interconnect cost in terms of layout area. In addition, reduced data dependencies have the potential to increase the overall performance of a design. Propagation delays can be kept small for those data which are not transferred between partitions. Furthermore, the potential for parallelism at the level of the partitions is increased. However, data clustering does not consider these performance issues explicitly. It can happen by chance that those data which are critical for the overall performance of a design are transferred between partitions. Similarly, in order to detect parallelism, global properties of the behavior have to be considered, e.g., when inter-block data are produced and consumed in the blocks.

The proximity function $dprox(a, b)$ for data clustering is defined by the number of common data entering or leaving two operators or clusters a and b divided by the total number of carriers connected to a or b.

$$dprox(a,b) = \frac{CommonData(a,b)}{TotalData(a) + TotalData(b)}$$

In $CommonData(a, b)$ the common data are counted separately for clusters a and b in order to ensure that $dprox(a, a) = 1$.

7.2.1.4 Schedule Clustering

Operators that can be executed in parallel must not be in the same cluster if high performance through low-level parallelism is a goal. In order to measure the potential parallelism, APARTY uses a preliminary as-soon-as-possible (ASAP) schedule. For two operators a and b incompatibility $inc(a, b)$ is defined such that "$inc(a, b) = 1$" if a and b are scheduled into the same control step *and* have the same type, and 0 otherwise. The additive inverse "$1 - inc(a, b)$" is used as a measure of proximity of operators.

For clusters of operators $a = (a_1, \ldots, a_n)$ and $b = (b_1, \ldots, b_m)$ incompatibilities are generated as follows. Incompatibilities of operators a_i and b_j with clusters b and a, respectively, are defined as:

$$inc(a_i, b) \quad = \quad max(inc(a_i, b_1), inc(a_i, b_2), ..., inc(a_i, b_m))$$

$$inc(a, b_j) = max(inc(a_1, b_j), inc(a_2, b_j), ..., inc(a_n, b_j))$$

The maximum can be interpreted as equivalent to the logical *OR* operation. An operation is incompatible with a cluster, if and only if it is incompatible with at least one operation in the cluster. For two clusters a weighted sum of the operator-to-cluster incompatibilities is chosen. The weights are the relative costs of the operations a_i and b_j in terms of the total cost of the two clusters.

$$inc(a, b) = \frac{\sum_i inc(a_i, b) \times cost(a_i) + \sum_j inc(a, b_j) \times cost(b_j)}{cost(a) + cost(b)}$$

A properly normalized expression ($inc(a, a) = 1$) is obtained if the costs of the clusters are computed by adding the costs of their elements. The proximity of clusters for schedule clustering is, finally, chosen in [LaTh89] to be

$$sprox(a, b) = dprox(a, b) \times (1 - inc(a, b))$$

In [LaTh89] operator cost is measured in terms of physical size. Therefore, the schedule proximity *sprox* represents a subtle combination of low-level parallelism (*inc*), interconnect (*dprox*), and size (*cost*). The primary concern in schedule clustering is performance, i.e., schedule length. Pushing incompatible operators into the same partition, without increasing the schedule length, requires additional hardware. The corresponding cost is kept small through the introduction of the weights *cost*, reflecting operator area. On the other hand, a small hardware overhead is not a sufficient driving force for clustering. Therefore, the pure schedule proximity of clusters ($1 - inc(a, b)$) is weighted by the corresponding data proximity considering interconnect. To summarize, schedule clustering addresses schedule length as a constraint, area overhead as a penalty, and interconnect as the driving force.

7.2.1.5 Operator Clustering

This clustering step is related to the one from the previous section. The goal is to reduce area by clustering groups of operators which can share hardware without destroying the possibility of operator-level parallelism. Again, preliminary schedule information is used to estimate the potential low-level parallelism of operators. However, instead of incompatibility, compatibility of operators is considered. It is defined such that "$comp(a_i, b_j) = 1$" if a_i and b_j have the same type, but are *not* scheduled into the same control step, and 0 otherwise. The reader can verify that $comp(a_i, b_j)$ is not equal to "$1 - inc(a_i, b_j)$" since both functions vanish for operators of different types. Compatibility of operators with groups of operators (clusters) is calculated as

$$\begin{aligned} comp(a_i, b) &= max(comp(a_i, b_1), comp(a_i, b_2), \ldots, comp(a_i, b_m)) \\ comp(a, b_j) &= max(comp(a_1, b_j), comp(a_2, b_j), \ldots, comp(a_n, b_j)), \end{aligned}$$

and the operator proximity $oprox(a, b)$ is defined as

$$oprox(a, b) = \frac{\sum_i comp(a_i, b) \times cost(a_i) + \sum_j comp(a, b_j) \times cost(b_j)}{cost(a) + cost(b)}.$$

Operators that find an operator of the same type without schedule conflict in the other cluster contribute to the operator proximity. In contrary to the schedule proximity, no overall weighting factor is needed since operator proximity is the direct clustering goal.

7.2.1.6 Inter-Procedural Clustering

In the previous sections, it was neglected completely that the specification of a behavior can be provided in a modular fashion, i.e., by using an assembly of interrelated procedures. From a software engineering point of view, good reasons exist for such a programming style. On the other hand, it is well known that the outcome of synthesis depends on the way specifications are written. Inter-procedural clustering reduces the dependency on the structure of a specification. Since details depend on the way procedures are treated in specific hardware description languages used for synthesis, the following explanations are very general. They are based on [Thom90] in which ISPS is used as the hardware description language and the *Value Trace* (VT) is used as the corresponding internal representation.

Previous clustering stages may have generated a set of clusters completely contained within blocks of the original modular specification. Let capital letters denote blocks and lowercase letters denote clusters, i.e., a and b are clusters contained in blocks A and B, respectively. It is assumed that a detailed analysis of the data and control flow graphs of the system cannot be performed. The information available is the inter-block call graph or, more precisely, the calls from clusters to blocks, and the composition of blocks in terms of clusters. In analogy to the control and data clustering stages presented above in inter-procedural clustering both control and data aspects are considered.

Control-oriented procedural clustering minimizes the transfer of control between clusters contained in different blocks. The probability that an activation of cluster b in block B is due to a call of B from a can be approximated by

$$P(b|a) = \frac{NumberOfCalls(B, a)}{TotalNumberOfCalls(B) \times TotalNumberOfClusters(B)}$$

The function $NumberOfCalls(B, a)$ denotes the total number of calls of the block B from a. The meaning of the other functions is evident. Since the proximity between clusters has to be symmetric in its arguments, the following proximity for control-oriented procedural clustering can be chosen

$$cpprox(a, b) = max(P(a|b), P(b|a))$$

This criterion tends to cluster procedures with those clusters, from which they are called exclusively. Those procedures that are called from many different clusters stay separate.

Data-oriented procedural clustering evaluates the proximity of clusters based on their calls to common procedures. The clusters can belong to the same block, or to different blocks. A call of a procedure does not only involve transfer of control, but also transfer of data. If two clusters frequently call the same procedure, it is likely that they can share communication channels and control mechanisms. The corresponding proximity function can be

$$dpprox(a, b) = \frac{CommonCalls(a, b)}{Calls(a) + Calls(b)}$$

where $CommonCalls(a, b)$ is the numbers of calls (from a and b) to those procedures that are called by both a and b. $Calls(a)$ and $Calls(b)$ are the total number of any procedure calls in a and b, respectively. If interconnect in terms of area is the major concern, bit-widths of data transfers have to be considered too.

7.2.1.7 Architectural Partitioning

Multi-stage partitioning as provided by the program APARTY has been applied to a variety of designs [LaTh89, Thom90]. The sequence of stages which has turned out to produce reasonable results is

- *control clustering* which is the first step, if applied at all,
- *data clustering* which has been applied in all examples reported and appears to be central to the method of APARTY,
- *inter-procedural clustering* which can be applied in the case of modular specifications, the partitioning of the IBM/370 architecture being the only example reported [Thom90],
- *operator clustering* which is frequently applied as the final clustering step in order to clean up the design and to minimize hardware.

As was pointed out earlier, multi-stage clustering offers many degrees of freedom. They can be used by a designer to experiment with the partitioning tool, to generate various partitions, and to choose the best one for the implementation of his or her system. Evaluating the different partitions is a problem since running a complete synthesis down to the logic level for each case is too time consuming. Fast measurement and evaluation routines, like those supplied by BUD [McKo90], combined with a *design assistant* to suggest suitable modifications of the partitioning scheme provide an ideal solution. This remains an open problem for future research in system-level partitioning.

7.2.2 Constrained Partitioning of Behavioral Models

In the previous sections, clustering is used to produce a complete cluster tree for systems specified in the behavioral domain. A cut through the cluster tree defines a certain partitioning of the system. The primary goal of clustering is not to create real physical partitions but to guide synthesis routines. Constraints on properties of individual clusters and of the total system after partitioning are not considered explicitly. They are taken into account only indirectly by means of the clustering criterion (proximity matrix). The result of the clustering process has to be checked against the constraints afterwards. If the constraints are not met, a different cut of the cluster tree has to be chosen, or clustering has to be iterated with modified proximities or different clustering stages. Thus, an outer loop has to be set up around the clustering algorithm. There is no way to guarantee convergence in this approach.

The approach presented in this section considers constraints explicitly within the inner loop of the partitioning algorithm [GuMi90]. Behavior is described using the *sequencing abstraction model* which captures the operations to be performed and their data and control dependencies. These dependencies are either explicit in the hardware description, or they result from data flow restrictions or hardware resource sharing. This implies that not only basic optimizing transformations of the specification but also data flow analysis and decisions on hardware sharing have to be performed in advance. The model is represented by the sequencing graph *SIF* (*Sequencing Intermediate Format*), an acyclic directed graph which is used in the OLYMPUS synthesis system [DKMT90]. Edges additional to those representing the sequencing information are used to represent hardware sharing between operators. The hardware model is a hypergraph with

- vertices: operations and blocks of combinational logic,
- edges: ordered vertex pairs (dependencies),

- hyperedges: subsets of vertices (operators which share common hardware).

The following properties are assigned to the vertices and edges in order to measure cost:

- vertex area cost: area equivalent for combinational logic blocks or functional units which implement operations.
- vertex delay cost: propagation delay of functional units in terms of the constant cycle time,
- edge interconnect cost: number of bits for interconnect including transfer of data and control.

In addition, the following global costs of the hypergraph and the individual partitions are defined:

- hypergraph area cost A: sum of the area costs of vertices counting one vertex of each hyperedge,
- hypergraph pin-out cost C: weighted sum of the edges which enter or leave the graph (weights are the edge interconnect costs),
- hypergraph delay cost T: longest weighted directed path through the graph (weights are the delay costs of vertices passed) plus a constant synchronization delay for each edge cut by partition boundaries.

As was pointed out in [GuMi90] the hypergraph can be hierarchical such that complex vertices represent sub-hypergraphs. The constraints on the partitioning problem are the following:

- The area cost A and pin-out cost C for each partition are smaller or equal to A_{max} and C_{max}, respectively.
- The total delay cost T of the hypergraph is smaller or equal to T_{max}.
- Hardware resources shared between operators must remain in the same partition, i.e., hyperedges must not be cut.

The constants A_{max}, C_{max}, and T_{max} are user provided and reflect the properties of the implementation medium (area and pin count) as well as the timing goals.

As mentioned earlier, basic decisions on parallel/serial trade-offs have to be made to define the sequencing graph and resource sharing. Synchronous inter-block communication is assumed with a cost of one clock cycle for synchronization. Care must be taken, however, because the synchronization delays cause a shift in the schedule which in turn can influence the amount of possible hardware sharing. At the beginning of partitioning, complex vertices are not considered as composed objects. If no acceptable solution can be found this way, complex vertices can be considered for partitioning. Depending on the properties of the underlying hypergraph, the complex vertex can be either expanded or duplicated.

Two related versions of the partitioning problem can be formulated.

1. Hypergraph partitioning: Partition the hypergraph H into a minimal number n of hypergraphs H_i $(i = 1, 2, \ldots, n)$ such that for each block $A_i \leq A_{max}$ and $C_i \leq C_{max}$, and for the total hypergraph $T \leq T_{max}$ holds.

2. Hypergraph bipartitioning: Partition the hypergraph H into two hypergraphs H_1 and H_2 such that for each block $A_i \leq A_{max}$ and $C_i \leq C_{max}$ $(i = 1, 2)$, and for the total hypergraph $T \leq T_{max}$ holds.

The two problems obviously are interrelated, because a solution of the second problem is a solution of the first one, provided that the trivial partition (no partition at all) is no solution of Problem 1. The solution of the bipartitioning problem is easier than the solution of the general partitioning problem; efficient heuristics exist for the former. Therefore, the general problem is reduced to a series of modified partitioning problems. If no solution of problem 2 exists a bipartition is generated which relaxes the constraints for one block, say $i = 2$, and maximizes the area utilization of Block 1. This can be induced by a cost function

$$f(A_1, C_1) = a \times (C_1 - C_{max}) + b \times (T - T_{max}) + c \times (A_{max} - A_1)$$

which is to be minimized. The positive constants a, b, and c are used to tune this process. The minimization of the cost function f tends to drive A_1 close to the limit A_{max}, whereas C_1 and T are kept small. For Block 2, a solution to Problem 2 can then be tried. If, again, this is not possible the same procedure is iterated.

The classical heuristic for the solution of the partitioning Problem 2 is that due to Kernighan and Lin [KeLi70]. It starts with an initial partitioning which is improved iteratively by swapping vertices between partitions. Updating of the area and pin-out costs of the blocks can be achieved incrementally.

An incremental update is not possible for the delay cost. This poses a problem, because the invocation of a longest path algorithm for each move is too costly for large graphs. Without going into details, various heuristics can be applied to circumvent this problem. One approach is to skip delay computation in the inner loop of the algorithm and to select moves based on the area and pin-out costs only. Delays are updated in the outer loop of the algorithm discarding the moves that do not lead to an acceptable delay. Another solution computes an approximation of the delay in the inner loop. With this approximation it is assumed that the critical path does not change and that it represents a lower bound to the delay. Again, delay cost is computed exactly only in the outer loop of the algorithm.

Constrained partitioning is geared towards multi-chip partitioning. If it turns out that a given behavioral description with timing constraints cannot be realized on one chip, partitioning can be used to overcome the area limitations while still meeting the timing constraints. A special application is hardware prototyping, i.e., implementation of the same behavior on different target architectures (ASIC, FPGA, PLD). For this application more information is needed about the behavior compared to the clustering approaches, i.e., scheduling and hardware sharing have to be solved first. The advantage is that if a partition is found, it meets the specified timing.

7.3 Behavioral Transformations

The hardware model generally used in algorithmic-level synthesis is the synchronous digital machine, which consists of a data path and a controller. The data path consists of functional units such as ALUs, adders, registers, etc., while the controller is realized as a FSM or as a microprogrammed controller.

Most hardware description languages (e.g., VHDL) support structuring elements such as procedures, functions, loops and conditional control structures, e.g., *if* and *case*. Procedures, functions, loops, as well as nested *if*- and *case*-blocks mainly serve to achieve better readability of the programs. Hierarchical control structures are implemented in hardware using multiplexors to select the desired branch of the data path (*if*, *case*) and to select the states of the FSM or microprograms that control the data path via registers and multiplexors (loops). Procedures and functions require substates of the FSM or submicroprograms, respectively.

In contrast to the compilation task for general programming languages several ways to map a behavioral description to a target architecture are available. A large number of hardware structures may fit the desired functionality. This situation is referred to as *design space*.

The designer needs, therefore, to be supported in exploring different possibilities to find the most appropriate architecture. The generation of these possibilities is supported by transformations.

Transformations are divided into two groups. The first group comprises those similar to transformations used in optimizing compilers. These are referred to as *optimizing transformations.* They reduce the control graph and simplify the controller, what often leads to higher performance. Furthermore they can structure the code in a way to allow for more parallel computations than with the original implementation.

Another group of transformations has a more extensive effect on the synthesized structure. Transformations that generate processes running in parallel or pipelined, as well as transformations that partition a single process into more processes belong to this group; they are referred to as *behavioral transformations.*

7.3.1 Optimizing Transformations

Some important optimizing transformations are presented in the following subsections. These transformations are described in the System Architects Workbench (SAW) [Thom90, Thom88, WaTh87], the HERCULES system [DeKu88] and the Flamel hardware compiler [Tric87].

7.3.1.1 Procedure In-line Expansion

In behavioral hardware description languages procedures or functions are used as in conventional programming languages to achieve a better structuring of the program code.

One way to implement a procedure in hardware is to use a hardware block that performs the desired function. This block can consist of functional elements, registers and multiplexors. The inputs and outputs of this block correspond to the formal parameters of a procedure in programming languages. The connection of the actual values of a certain call to the inputs and outputs of the procedure block is performed by multiplexors. The hardware within the block is controlled by substates of the FSM or by a submicroprogram (Figure 7.1).

Another way to implement the procedure call is to expand the code of the procedure body. In this case, the hardware of the described block is implemented for each call individually. The multiplexors at the inputs and outputs as well as the substates for the FSM are not required. This reduces the control flow graph and those parts of the program that can be parallelized are easier to recognize.

Furthermore, the number of control steps can be decreased, too, which may result in a better schedule. This effect is more significant for short procedures.

On the other hand, procedure expansion often increases the amount of hardware. Therefore, the expansion should be kept optional and the expansion of selected calls of a certain procedure should be possible.

7.3.1.2 Loop Unrolling

The loop construct in hardware description languages allows for the repeated use of certain hardware functions that have to be realized only once. The loop is controlled by the FSM in accordance with the loop test condition. Additional hardware such as registers and multiplexors is needed to process intermediate values.

Loops with fixed bounds can be fully or partially unrolled. A fully unrolled loop decreases the nesting level of the control graph. The controller is simplified as well, because a loop test is no longer necessary.

The amount of hardware increases, but, in most cases, the performance improves too. There is also a potential for parallelizing the unrolled loop iterations. Loop unrolling is often done before performing transformations that create a pipeline.

However, if operators are shared, loop unrolling is not advantageous.

7.3.1.3 SELECT Transformations

The conditional execution of certain code can be achieved by an *if* or *case* construct. The hardware realizing an *if* or *case* construct uses multiplexors to choose the desired path. The transformations mentioned are described in [Thom90, Thom88, WaTh87] and are referred to as SELECT transformations.

Combining IF- and CASE-blocks. Nested blocks can be combined into one block by replacing two sequential condition checks by one. This reduces the nesting level. The select condition, however, becomes more complex. Instead of nested if blocks, a case with several branches is generated.

A lower nesting level facilitates the operations described in the following sections, because fewer multiplexors are needed. The Boolean select conditions which may become more complex can be minimized more easily. *If*-statements without *else*-branches can be combined especially well.

Motion of operators into or out of SELECT blocks. Operators that exist in all branches can be removed from the SELECT-Block. Likewise, operators from outside the SELECT block can be transferred into all branches of the block. Motion into the SELECT block is often used together with flattening of IF- and CASE-blocks to prepare for pipeline transformations.

7.3.2 Behavioral Transformations for System-Level Synthesis

7.3.2.1 Processes and Communication

A special feature of system-level synthesis is the synthesis of multiple processes that communicate with each other.

Processes are well known in the field of computer science. A process refers to the execution of a number of sequential statements. The parallel execution of two or more processes is called *concurrent execution*. The data transfer between the processes is referred to as *interprocess communication*. *Synchronization* induces that interprocess communication has to occur at a mutually agreed upon time.

Many mechanisms exist for interprocess communication and synchronization of processes. In some of them, communication is performed by shared data, which implies that only one process may access these data at any point of time. Semaphores, conditional critical regions, monitors and path expressions belong to this group.

Other mechanisms are based on *message passing*. Here, processes send messages to other processes and, thus, perform synchronization and data exchange. Message passing may use *direct naming*, i.e., the name of the process to receive data is specified. Another possibility is to communicate via mailboxes.

7.3.2.2 Implementing a Process Description in Hardware

Processes and their communication can be synthesized in different ways. First, a distinction must be made between synchronous and asynchronous communication. Second, the method to perform synchronization must be chosen. Synchronization can be achieved by control step scheduling or by the use of additional hardware or microcode. In the example from the SAW-system, two alternatives are illustrated.

Synchronization by control step scheduling. Data is transferred to a queue, register, or wire using a SEND operation, and execution follows the instruction after the SEND. The RECEIVE operation reads data from the queue, register, or wire, and execution continues with the next instruction after the RECEIVE.

The control step scheduler is responsible for the SEND and RECEIVE working in lockstep, i.e., their respective control steps must be executed at the same time. This, however, is not always possible. If the operations processing the data to be sent must be scheduled earlier, or if the operations computing the received data must be scheduled later, a register will be necessary to hold the data being transferred.

The scheduling of operations in different processes must not only consider the control steps of the single processes (local scheduling), but also the relationship between control steps of all the processes (global scheduling). To achieve lockstep execution the use of no-operations (NOPs) can be necessary.

It is not always possible to achieve synchronization of processes by control step scheduling. If a process contains *if-* or *case*-branches, that contain different numbers of operations, it is necessary to adapt the length of the shorter branches to the longest one. This is not very favorable.

If, however, every process has only one control path, or all paths have the same length, synchronization through control step scheduling can save additional hardware.

Hardware supported synchronization. As in synchronization by control step scheduling, the data transport is performed by a SEND and RECEIVE via a queue, register, or wire. If data are not yet available, the RECEIVE must block. This has to be insured by the implementation of the SEND and RECEIVE operation in hardware or microcode.

There are low-level techniques to limit access to a section of code to only one process at a time. High-level mechanisms can be implemented in microcode based on these low-level primitives.

The synchronization can also be achieved by a process, whose controller controls the controllers of the other processes. This is advantageous if the processes are synthesized separately.

7.3.2.3 Process Creation from Procedures

As mentioned in Section 7.3.1, a procedure can be implemented in hardware using a hardware block that is controlled by substates of the controller. In Figure 7.1 it is shown how the data transfer between the caller and the procedure is solved for two procedure calls (call A, call B).

The *process creation transformation* creates a process from a specified procedure. It changes the hierarchical structure of a procedure call into the concurrent structure of communicating processes, which contain their own controllers.

To achieve this, the data transfer performed by the procedure call is replaced by interprocess communication. Equally, the data transfer of the return-statement, is changed. Furthermore, the newly created process has to be restarted so that it will run continuously. Then the new process executes concurrently with the other processes.

In the hardware implementation, concurrent processes have their own data path and controller (Figure 7.2). The communication is implemented using the mechanisms described in Subsection 7.3.2.2.

Procedure creation can be used to split a design into two or more smaller data paths and controllers.

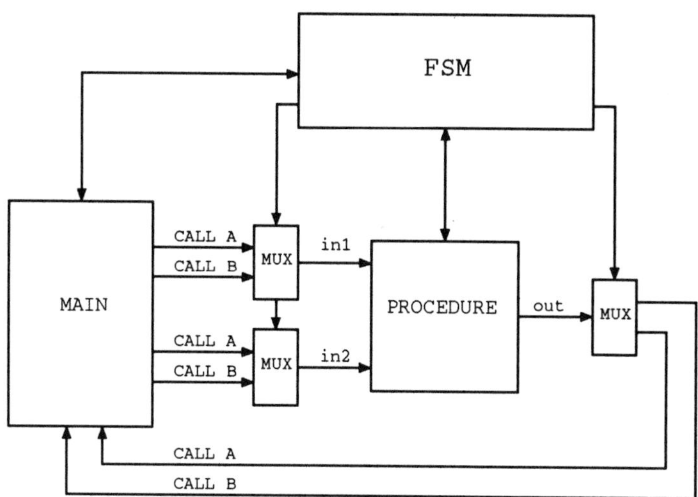

Figure 7.1: Hardware implementation of a procedure call.

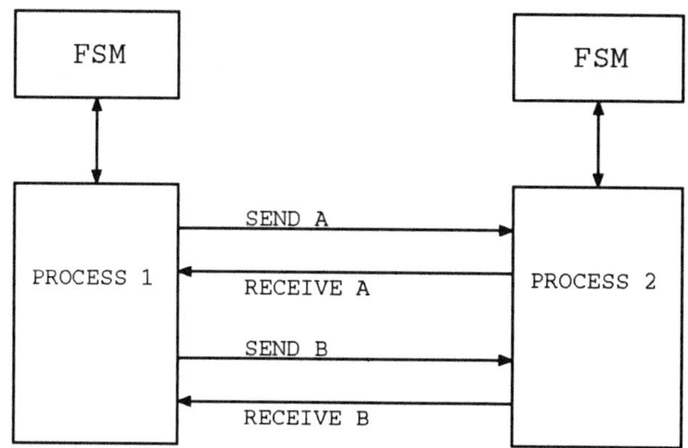

Figure 7.2: Transformed procedure call.

Potential for more concurrency. After the hierarchical control has been transformed into concurrent processes, operations in the former caller that are not dependent on the data received from the newly created process can be scheduled before the receive operation. Thus, a more parallel implementation is admired (cf. code B in Figure 7.3).

Less hardware per process. If a process is too large to be realized on a single chip, it is necessary to partition the process. Although additional hard-

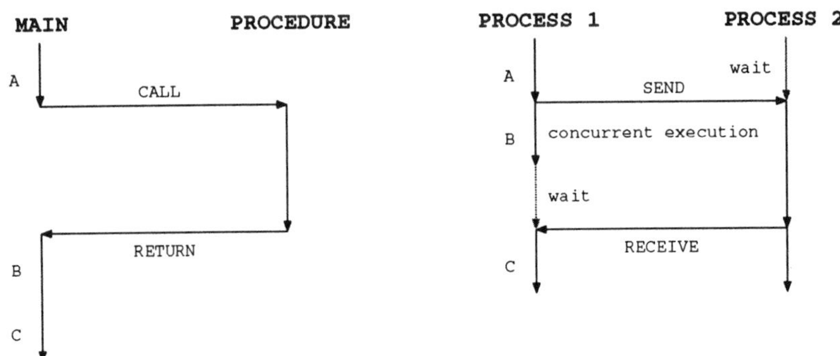

Figure 7.3: Possibility for more concurrency.

ware will often be necessary for the communication between the partitions, the size of a single process will often be smaller than the original one. Furthermore, a gain of speed may arise due to the smaller processes. Additionally, the controllers of the single processes decrease in complexity, when FSM-states or microcode can be reduced.

Process allocation. The process creating transformation described before creates one process from a procedure, and replaces each call by a SEND / RECEIVE mechanism.

Alternatively, a process can be created for each call which requires more hardware, but avoids delays caused by waiting for a process to become available.

A third possibility is to create several processes to be shared by the calling processes; this transformation is similar to the scheduling and allocating task in structural (RT-level) synthesis, but differs in that the processes created are more complex than the typical functional units.

7.3.3 Pipelining and Functional Pipelining

Pipelining and functional pipelining for high-level synthesis are described in Chapter 6. In this chapter, transformations that generate a pipeline or a functional pipeline from concurrent processes are presented.

There are several ways to design a pipeline. One possibility is the functional decomposition into stages (such as found in most microprocessors: reading operand, reading address, etc.). Another method divides a function, e.g., multiplier, signalprocessor, into more balanced stages.

7.3.3.1 Pipestage Creation

Pipestage creation is a transformation which is performed after process generation. It transforms concurrent processes into pipestages. To achieve this, the interprocess communication has to be changed such that every process sends data only to the subsequent process in the pipeline. Only the last stage is allowed to send data back to the first process.

It has to be made sure that the processes work in *lockstep*, i.e., the data transfers occur at the same time. This is realized using methods described in Section 7.3.2.2.

For the SAW-system a way to generate a pipeline with functionally decomposed stages is shown using an instruction set processor as an example. Consider the behavioral description containing the tasks IFETCH (instruction fetch), DECODE, OPFETCH (operator fetch) and EXECUTE, which are transformed into pipestages.

The initial description may be highly modular using procedures for every task and three nested SELECT levels for the instruction group decoding, the instruction opcode decoding and the address mode decoding. This serves to improve readability, but does not constitute a good basis for direct transformation into a pipeline.

The first step is to improve the control step schedule by expanding all procedures inline as well as by combining the nested SELECTs for the instruction group and the instruction opcode into one SELECT. The address mode decoding retains a second SELECT level.

The next step is to define the control (SELECT) structure for all stages in the pipeline. In the example, DECODE, OPFETCH, and EXECUTE depend on the instruction (forcing an instruction SELECT). The stages DECODE and OPFETCH depend on the instruction and the address mode (forcing an instruction SELECT and an address mode SELECT). IFETCH does its work unconditionally and needs no SELECT. The operators from the optimized description are then distributed to the appropriate stages.

Subsequently, these stages are transformed into processes and further transformed into pipestages (Figure 7.4).

An alternative is the creation of a two stage pipeline in which the task IFETCH is executed by the first stage and the tasks DECODE, OPFETCH, and EXECUTE are executed by the second stage.

7.3.3.2 Functional Pipelining

In functional pipelining the pipestages do not have physical equivalents as in conventional pipelining (cf. Chapter 6).

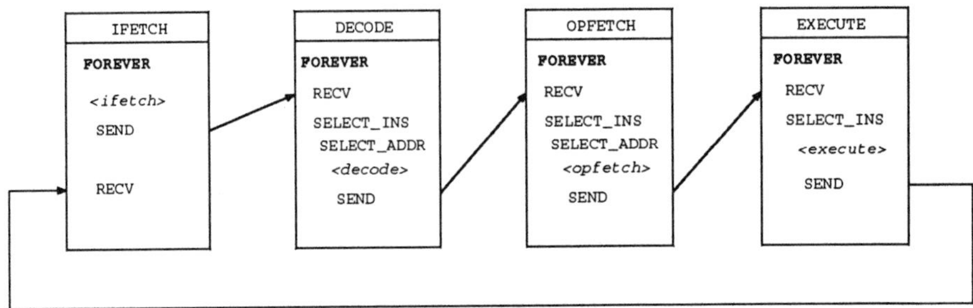

Figure 7.4: Pipeline.

The method to achieve functional pipelining from an algorithmic description is called *loop winding* [Girc87]. In loop winding the data flow of a single process (Figure 7.5) is partitioned into functional stages (Figure 7.6).

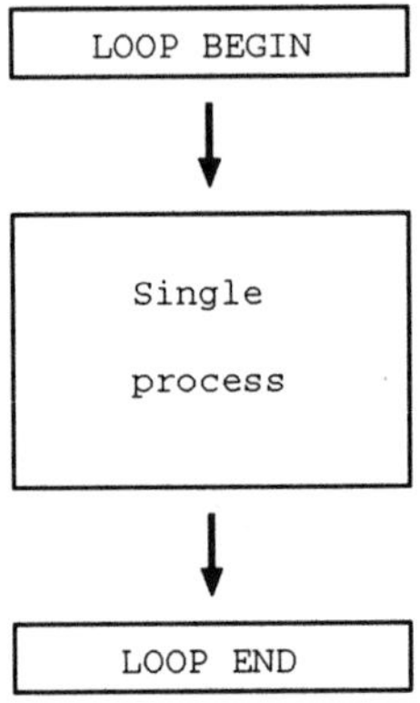

Figure 7.5: Single process before loop winding.

Because loop winding does not require a specific hardware structure, the synthesis of the generated process descriptions has some degrees of freedom, e.g., the processes can be synthesized separately, fulfilling only the data transfer controlled by the loop. The synchronization of the data transfers between the stages can be controlled by a special controller process. Loop winding is performed in three steps in a preprocessing phase.

In the first step, the data flow representation has to be prepared for functional pipelining. This consists of unrolling nested loops and determining the critical path length. Nested loops must be unrolled because they can interfere with the execution of the parallel subiterations that represent the functional stages.

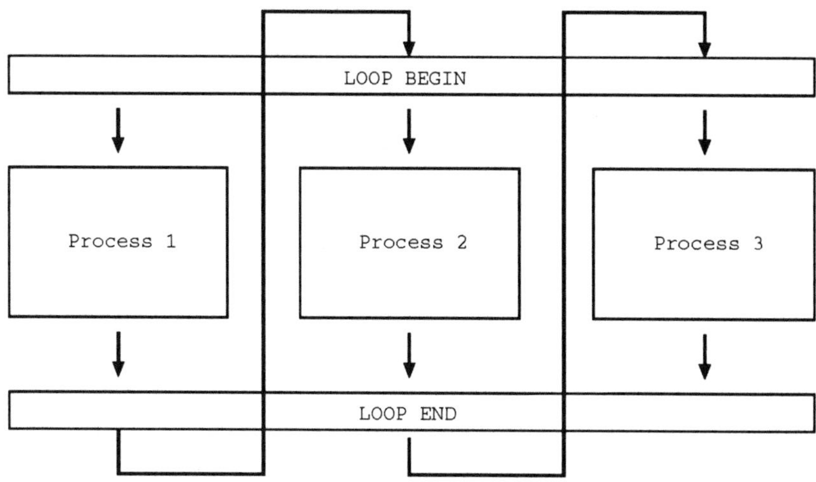

Figure 7.6: Functional pipeline after loop winding.

In the second phase, the data flow of the critical path is partitioned into an appropriate number of subiterations, which form the functional stages. The critical path of the data flow first has to be partitioned into an appropriate number of paths with equal length. Operations that are not on the critical path are then distributed to stages according to their weight. After that, an operation load balancing is performed for the stages.

The final phase of loop winding forms the wound loop. The data flow edges are changed so that data that have to be sent to the following partition can only be transferred via the loop end of the sender and the loop begin of the receiver (Figure 7.6). This insures proper pipeline maintenance (filling and emptying) for the data flow.

7.4 Synthesizing from System-Level Descriptions

As explained earlier, the goal of system-level synthesis is to support synthesis from a level of abstraction that is beyond that of high-level synthesis. In the following, some description techniques are discussed that raise the level of abstraction compared to the starting point of high-level synthesis. In addition, some global ideas are presented on how these techniques can be supported by synthesis tools. Generally, some of these features can simply be supported by translation to the algorithmic level, others need specific extensions of high-level synthesis tools.

First, two different approaches to the description of control and concurrency at a higher level of abstraction are considered. These approaches are:

- Use of fork/join constructs
- Use of extended finite state machines

Next, object oriented techniques are examined. Their advantages for hardware design are shown, and the possibilities for synthesis support are explained. At the end of this section, several aspects of communication are presented.

7.4.1 Fork/Join

The scope of most currently known high-level synthesis systems is the synthesis from descriptions written as sequential algorithms. This means that every system to be synthesized has to be decomposed into concurrently active and purely sequentially operating processes *prior* to synthesis. For system-level descriptions, however, it is necessary to provide language features that support the description of concurrency better than through the use of structural decomposition. Languages that support arbitrary mixtures of concurrent and sequential control flow include, e.g., OCCAM [Newp86], CAP/DSDL [Ramm82], and SCHOLAR [BeAl88]. CAP/DSDL (cf. Figure 7.8), for example, provides the language construct "`seqbegin S1; ...; Sn end;`" which encloses statements that have to be executed strictly sequentially. The statements included in the "`conbegin S1; ...; Sn end;`" construct, on the other hand, can be executed concurrently, without any implicit synchronization. The "`conbegin ... end`" block is finished, when all statements enclosed are finished with their execution.

These two language constructs can be nested arbitrarily and hence allow for an abstract description of the system behavior. This kind of description does not imply any specific implementation, in contrast to a system description that has already been decomposed into purely sequential processes.

As an example for the application of these constructs, a 3x3 array of communicating transputer nodes (Figure 7.7) is chosen [Newp86].

Each node in this array periodically performs the following actions: Two data values are received, from the north and the west channel. These two actions can be performed concurrently if the two values received from the north and west port are stored into different local variables (x and y). The product of the two values is added to a sum variable and the values x and y are sent, via the south and east channel, to the neighbors of the node. These three actions can also be performed concurrently. The 3x3 array thus performs a matrix multiplication on two 3x3 matrices, one being fed into the top channels while

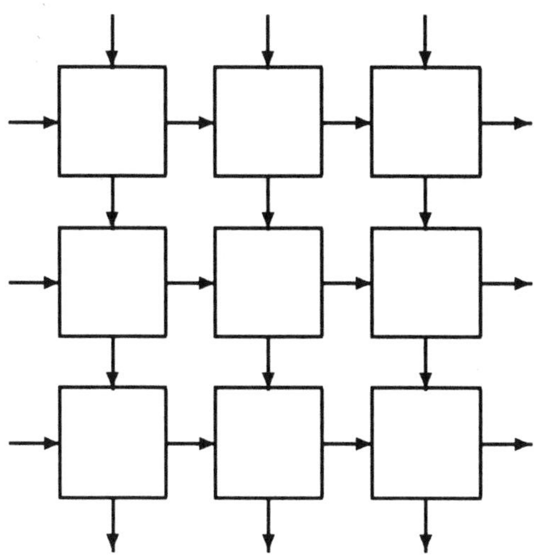

Figure 7.7: Network of transputers

the other enters the channels on the left hand side. In DSDL the behavior of each node can be described easily using the **seqbegin** and **conbegin** language constructs, as shown in Figure 7.8. For the communication two procedures **send** and **receive** are defined.

To express the same behavior in languages that do not provide fork/join or similar features, e.g., VHDL, a decomposition into multiple concurrently active and communicating, strictly sequential processes is necessary. In VHDL, for example, the behavior has to be written as shown in Figure 7.9. Each process in this description is strictly sequential, and thus can be synthesized by high-level synthesis tools. However, this results in a rather inefficient implementation, containing five FSMs and five data paths (one for each process), and therefore sharing of operators among the data paths of the individual processes is not possible.

For the direct implementation of fork/join constructs two similar approaches have been published. In [BKKR86] the translation of the DSDL constructs **seqbegin** and **conbegin** into a Petri net and techniques for the implementation of the resulting net are presented. In [BeAl88] a similar control architecture is presented, but without mapping the description onto a Petri net first. The controller is implemented based on a synchronous token passing mechanism (*sequencer controller*), which corresponds to a one-hot state encoding. The fork is implemented by distributing the control token (the hot value) to subsequencers.

```
type channel = record
                 (* definition omitted *)
                 end;
procedure send(in datain: integer; out dataout : channel);
seqbegin
(* body omitted *)
end;
procedure receive(out dataout: integer; in datain : channel);
(* receive dataout from channel datain *)
seqbegin
(* body omitted *)
end;
procedure tnode (in  n:channel; in  w:channel;
                 out e:channel; out s:channel);
VAR sum,x,y : integer;
seqbegin
  sum := 0;
  while "1" do
  seqbegin
    conbegin
      receive(x,n);
      receive(y,w);
    end;
    conbegin
      sum := sum + (x*y);
      send(x,s);
      send(y,e);
    end;
  end; (* while *)
end; (* procedure tnode *)
```

Figure 7.8: DSDL description of a node.

The join is implemented by waiting until the control token has reached the end of each subsequencer. The basic principle of the sequencer controller is shown in Figure 7.10. This approach is similar to the control strategy used in the HERCULES [DeKu88, KuDe89] synthesis system. In HERCULES, controllers are built from so-called *control elements* (see also Section 5.3) that support the control of the concurrent execution of independent operations, which need a data dependent number of cycles for each operation. An interesting side effect of this control architecture is that it directly supports pipelining. Pipelining can be achieved by putting multiple control tokens into the sequencer. However, care has to be taken that two succeeding tokens can never collapse, for example, if the control flow contains loops or branches.

With this control architecture the example given above can be synthesized in a single step, hence providing more potential for optimizations, e.g., the sharing of operators during scheduling and allocation. A problem of this architecture can be the larger number of state registers, but this is partially compensated by the trivial next state function. Further optimizations may aim for the reduction of the number of registers, e.g., by realizing sequences (and subsequences)

```
package types is
  type channel is record
                  -- omitted here
                   end record;
end types;
use work.types.all;
-- description of one node
entity tnode is
 port( north,west,east,south : inout channel);
end tnode;
architecture behavior of tnode is
  signal x,y : integer;
  -- signals for synchronization
  signal activate, r1ready, r2ready, both_ready : boolean;
  -- resolution functions omitted !!
  procedure receive (signal dataout: inout integer;
                     signal datain : inout channel) is
  begin
  -- body omitted
  end;
  procedure send    (signal datain : in integer;
                     signal dataout: inout channel) is
  begin
  -- body omitted
  end;
  begin -- architecture
  west_receive : process (activate)
  begin
    -- receive data from west
    receive(x,west);
    r1ready <= true;
  end process;
  north_receive : process (activate)
  begin
    -- receive data from north
    receive(y,north);
    r2ready <= true;
  end process;
  both_ready <= r1ready  and r2ready;
  main_control : process
  variable sum : integer := 0;
  begin
    loop
      activate <= not activate; -- to trigger the receivers
      r1ready <= false;
      r2ready <= false;
      wait until both_ready;
      :
      :
      sum := sum + (x*y);
      -- same for sending
    end loop;
  end process;
end behavior;
```

Figure 7.9: VHDL description of one node of the transputer network.

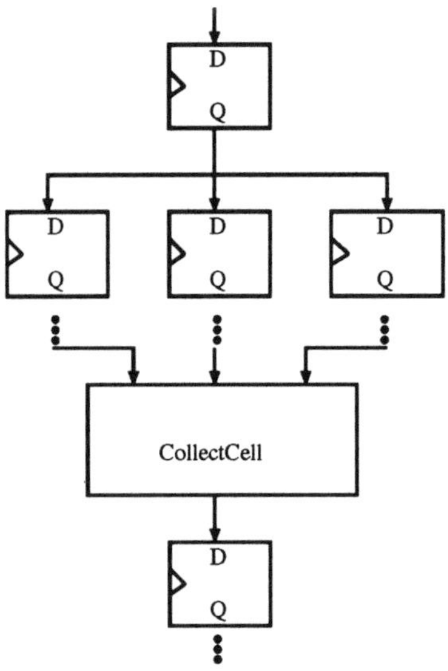

Figure 7.10: Principle of fork/join with sequencer controller.

without fork/join as conventional FSMs with optimized state encoding, or by merging several sequences into one FSM [Ramm89].

7.4.2 Extended Finite State Machines

Conventional finite state machines are widely used in hardware design and are typically supported by register-transfer level synthesis. However, the notion of states can be found also at the system level, e.g., a CPU having the (system-level) states FETCH, DECODE and EXECUTE. Conventional finite state machines do not support hierarchy, abstraction, and concurrency. Therefore, they are not suitable for system-level specifications since the number of states to be described gets too large. The decomposition into communicating finite state machines does not help much, since this often implies a specific implementation of the system. Several approaches [Hare87, VaNG91, STAT90] are known to extend the basic idea of finite state machines to include features that support the system level. The fundamental idea how the usability of finite state machines can be increased at the system level is presented in [Hare87] and named *statecharts*. Statecharts are a visual formalism for the specification of the behavior of reactive systems. They extend the classical FMSs by providing means to express concurrency, state-history, and hierarchy. Hierarchy in this context refers to the principle that states can be refined into substates.[1] Two types of decomposition are possible:

AND-clustering: A state S is refined into *orthogonal* substates O_i. To be in state S, a system must be in all states O_i at the same time. This also expresses concurrency.

OR-clustering: A state S is refined into *exclusive* substates E_i. To be in state S signifies that the system is in exactly one of the states E_i.

Transitions can be performed between states at any level of hierarchy. To leave a high-level state means to leave all corresponding substates the system is in. The basic states that are entered after entering a high-level state are defined by *default arrows*. A simple statechart is shown in Figure 7.11. The

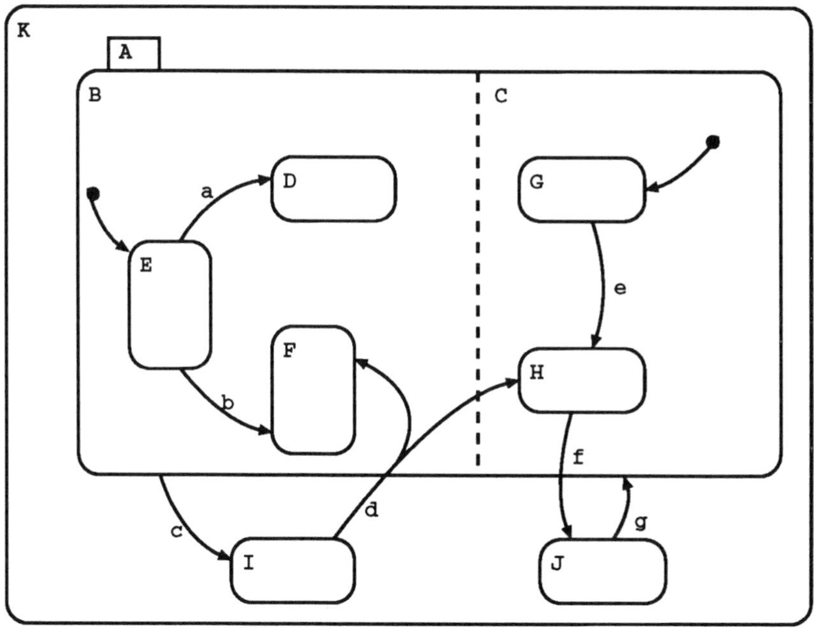

Figure 7.11: Simple statechart.

system in this example contains three states at the highest level of hierarchy (A,I,J). The state A is refined into two orthogonal states, B and C. If A is entered, i.e., if the condition g is true in state J, the default arrows indicate

[1] In the sequel of this section the terms "high-level state" and "low-level state" denote states in the state hierarchy.

that the substates E of B and G of C are entered. If condition c is true in any of the substates of A, a transition into state I is performed. The transition from state I, when condition d holds, is an example for a transition between different levels of hierarchy.

The synthesis from statechart descriptions was presented in [DrHa89] and [Hard91]. The basic idea is to synthesize a FSM for each level of hierarchy in the statechart. Each FSM controls the transitions at the corresponding level of hierarchy. The result of this synthesis strategy is a tree of interacting FSMs. The advantages and disadvantages of synchronous and asynchronous interaction between the FSMs are outlined in [DrHa89]. This proposed architecture is only one extreme of possible implementations. Another extreme is to generate a single FSM from statecharts, which results in a large number of states. This leaves a lot of room for optimizations and trade-offs, similar to the ones known from the work on FSM decomposition [DeNe88].

Related approaches. A combination between Harel's statecharts and VHDL is proposed under the name SpecCharts in [VaNG91, NaVG91]. The translation to VHDL for simulation purposes is explained in detail in [NaVG91]. Some global ideas about synthesis from SpecCharts are mentioned in [VaNG91]. In [JePC91] an extension to VHDL called MetaVHDL (MV) is described to support high-level modeling of controllers. The extension includes special support for hierarchical composition of FSMs, exception handling, and global transition between FSMs. MV can be translated to standard VHDL for simulation purposes or to SIF [JePa91] for synthesis.

7.4.3 Object Oriented Techniques

Object oriented design techniques are becoming more and more popular in software engineering. The main reasons are:

- Good support for software reuse
- High level of abstraction
- Good readability
- Easy maintenance.

The main ideas behind object orientation are:

- Data encapsulation. This refers to the principle that data structures are defined in combination with the operations manipulating the data structure. These operations are called *methods* or *member functions.* Access to private data in the data structure is only allowed using the defined methods, whereas access to public data can be gained directly. The definition of the data structure in combination with the methods is called a *class*. *Objects* are instantiations of classes.

- Objects can send messages to other objects in order to activate certain operations in the receiving object.
- Classes with similar properties can be grouped into a new class.
- A class derived from other classes inherits data and methods from those classes.

These ideas are also interesting for hardware design, specifically for the design at the system level. For hardware designers, the paradigm of object orientation is even more natural than for software engineers since hardware consists of objects (chip, modules, cells, ...) that send messages to each other. A processor, for instance, sends the message **write_data** to the memory. Data encapsulation again is very natural from a hardware designer's point of view. The internal data, e.g., of a processor, can only be accessed using defined functions (methods). Object oriented languages, therefore, support the description at the system level by identifying the global components (objects) of the system (global system structure) without specifying the physical connections between the components in detail. At this level of abstraction, only *use*-relations, e.g., usage of the RAM by the processor, are of interest. No interest is given to how the processor is linked physically to the RAM, and which protocol has to be used for communication. This is sometimes referred to as *partial structure*.

Maintenance is supported since it is possible to exchange an object of one class by an object of an other class, as long as both classes provide the same methods, i.e., if they are derived from the same base class. For example, a serial access RAM can be replaced by a parallel access RAM, since both provide the function **read** and **write**. This can be done without changing the description of the objects that use this RAM object. With hardware description languages that do not provide object oriented features this is very difficult. In this case it is necessary to change the protocol in all processes from which the RAM is accessed.

Several approaches have been published ([MuRa90, Ramm82, Oczk90], [GlUm90]) that incorporate object oriented ideas into hardware description languages, or that propose to use object oriented programming languages for hardware description [Wolf89, PaWl87, Take81]. Objects in these approaches correspond to system components with an abstract view to their functionality. Examples are a processor with its instruction set or a stack with its operations *push* and *pop*.

The synthesis from descriptions at this level of abstraction is considered in [MuRa90] and [VaNG91]. In the following, some basic ideas are presented on how synthesis from object oriented hardware descriptions can be performed, and which kind of optimizations are possible during the implementation phase.

For illustration of the ideas a language is used, which is based on VHDL enhanced with elements (classes) from C++ [Stro86].

In Figure 7.12 the class definition for a RAM is shown. This is an abstract

```
class ram is
private
  storage : array[1 to 1024] of byte;
public
  procedure read (address : integer; data : byte) is
  begin
    data := storage[address];
  end;
  procedure write(address : integer; data : byte) is
  begin
    storage[address] := data;
  end;
end class;
```

Figure 7.12: Definition of class RAM.

view of a RAM. No information about ports is contained in the class definition. This means that it is still open how the RAM is to be implemented, only the functionality of the RAM is defined. This type of class is referred to as *general* class. The class definition can be used to declare individual RAM objects, or it can be used as a base class for other, more specific, RAM class definitions.

In the approaches known to include object oriented features into HDLs, classes are used to describe components with an abstract (functional) view. This means that information about the ports of the components has to be included in the class definition. Additionally, the protocols (sequences of stimuli at certain ports) have to be defined that must be used by a process using this class in order to perform the different methods of the class. These protocols can either be specified within the class, if for example they are a property of the component to be modeled (cf. Figure 7.13), or they are specified independently, as it is done, e.g., in [MuRa90]. These types of classes are referred to as *component* classes in the following. A special RAM class `serial_ram` can be derived from the general RAM class. This is shown in Figure 7.13. The class `serial_ram` inherits all variables and methods from its base class `ram`. Since it is intended to describe a real component, the port information is added and the protocols for each method are defined.

These classes are used in a simple processor in Figure 7.14. For instantiation of objects a new clause `object` is used.

```
class serial_ram : ram is
port
  clock: in bit;
  RW: in bit;
  d: inout bit;
  a: in integer;
protocol read (address : integer; data : byte) is
  begin
    RW := '1';
    for i in 1 to 8 loop
      data[i] := d;    -- read data from d port
      cycle(clock);    -- wait one clock cycle
    end loop;
  end;
protocol write(address : integer; data : byte) is
  begin
    RW := '0';
    for i in 1 to 8 loop
      d := data[i];    -- write data to d port
      cycle(clock);    -- wait one clock cycle
    end loop;
  end;
end class;
```

Figure 7.13: Class definition of a serial access RAM.

```
-- to be synthesized
entity processor_board is
 port(.....);
end processor_board
architecture synthesis of processor_board is
  object myram : ram;
begin
 processor : process
  variable word : byte;
  variable op1,op2 : word;
  variable result : word;
  variable PC : integer;
 begin
  myram.read(PC,word);
  decode(word,instruction,op1,op2,address);
  case instruction
   when add =>  result := op1 + op2; PC := PC + 1;
     :
  end case;
  myram.write(address,result);
 end process;
end synthesis;
```

Figure 7.14: Using the RAM class.

For the translation of such a description to the level of abstraction acceptable for high-level synthesis tools several possibilities exist. First, the translation of the objects that are declared in the description is considered. One possible translation that is similar to what a C++ preprocessor performs on

C++ source code is the following:

1. Inheritance is expanded.
2. The class definition is translated to a record definition containing all class variables (public, private, all inherited).
3. The member functions are translated to ordinary functions with an additional argument of the above-mentioned record type.
4. All references to class variables are translated to references to the new additional argument.
5. Object instantiations are translated to data declarations.
6. All references to objects are translated accordingly.

As an example, in Figure 7.15 the resulting code is shown for the **ram**.

```
type ram_record is record
  storage : array[1 to 1024] of byte;
end record;

procedure ram_read (ram_data : ram_record; address : integer;
                    data : byte) is
  begin
    data := ram_data.storage[address];
  end;

procedure ram_write(ram_data : ram_record; address : integer;
                    data : byte) is
  begin
    ram_data.storage[address] := data;
  end;
```

Figure 7.15: Result of translation to algorithmic level.

Considering the example of Figure 7.14 this results in:

signal myram : ram_record; instead of **object myram : ram;**
and
ram_read(myram,PC,word); instead of **myram.read(PC,word);**

If the synthesis system can deal with procedure calls, e.g., by performing inline code expansion (cf. also Section 7.3), and if it can handle global signals, such a description can be handed over to a high-level synthesis tool. This corresponds to distributing the functionality of the class among the using processes.

Another possible solution is to translate the general class to a component class by adding port and protocol information to the class. This can be done by the user if he/she has already existing components in mind or generated

automatically, e.g., by defining all formal arguments of the member functions as ports together with a port specifying the desired instruction. The protocol part of the class again can be translated into normal procedures. The member variables of the class are *not* translated into signals, in this case; instead, each object declaration is converted to a component instantiation plus the declaration of signals connected to the ports of the component. All accesses to an object are translated to the activation of the corresponding protocol for communication with the real component. This results in
`ram_read_protocol(myram_ports,PC,word);` instead of
`myram.read(PC,word)`

As in the first solution, this kind of description can be handled by common high-level synthesis tools if they are able to deal with procedures and global signals.

The two solutions presented so far are only two extremes. Other solutions may use a mixture of both methods: one part of the functionality of a general object is distributed among the using processes, whereas the other part is combined into a component.

In the example used so far no object is used from more than one process. This fact makes it easy to deal with the global signals resulting from the translation. In the case where multiple processes access (read and write) a common object, it is necessary to specify the methods of a class that are mutually exclusive, i.e., those that cannot be activated at the same time. From this information some means for arbitration have to be synthesized. Several methods for arbitration are known, e.g.,

- Equal priority
- Unequal priority
- Rotating priority
- Queuing
- Random delay.

An analysis of these methods regarding performance and hardware implementation is given in [Guib89].

7.4.4 Communication

The most important difference between system-level synthesis and high-level synthesis is the synthesis of multiple processes that communicate with each other. All common HLDs support communication between processes through ports. Some languages, additionally, support communication using global (shared) variables or signals. Using these features, the designer specifies a *logical* communication structure of the system. This means that an abstract wire (signal) between two processes defines a path for exchange of information between the processes but not necessarily a physical wire between the processes. Some HDLs provide additional primitives for the activation of communication, e.g., **send, receive**, and predefined protocols for these primitives, e.g., synchronous or asynchronous communication. If the language already supports object oriented description styles, as explained in the previous subsection, communication is just a special application of these features. One can, for example, define a general class **channel** with the member functions **send** and **receive**. Several communications channels can then be declared by building instances of the **channel** class.

The most important tasks for synthesis are:

Mapping the logical communication structure onto a physical communication structure. Two extremes in the design space are single bus communication, on the one hand, and point-to-point connections on the other hand. The selection of the physical communication structure has a large impact on performance and area. A point-to-point connection between processes may be fast, but it is expensive if, for example, the processes are realized on different boards or chips, because of the large number of pins/ ports needed. Currently there is no system that performs this mapping automatically. In the UMIST synthesis system [EdFo87, EdFo88], a simple architecture is selected which supports communication of all processes with each other over a single global bus (see Figure 7.16). Each individual process is synthesized into a so-called node processor that consists of a general communication part to handle bus arbitration, and into a specific part to perform the task of the process. Input and output between the system and the environment is handled by special I/O processors.

Synthesis of communication protocols. If a class is used for the definition of the communication channel, synthesizing a protocol can be accomplished as described above. In this case, both solutions that are described in Section 7.4.3 make sense, too. Taking the first solution results in the distribution of the protocol code among the communicating processes. Using the second solution results in a "communication processor" that controls communication between the processes. Another possibility to provide synchronization for data exchange between processes is control step scheduling (cf. also Section 7.3.2.2).

In the case of predefined language constructs for communication a protocol has to be synthesized according to the semantics of the specific language.

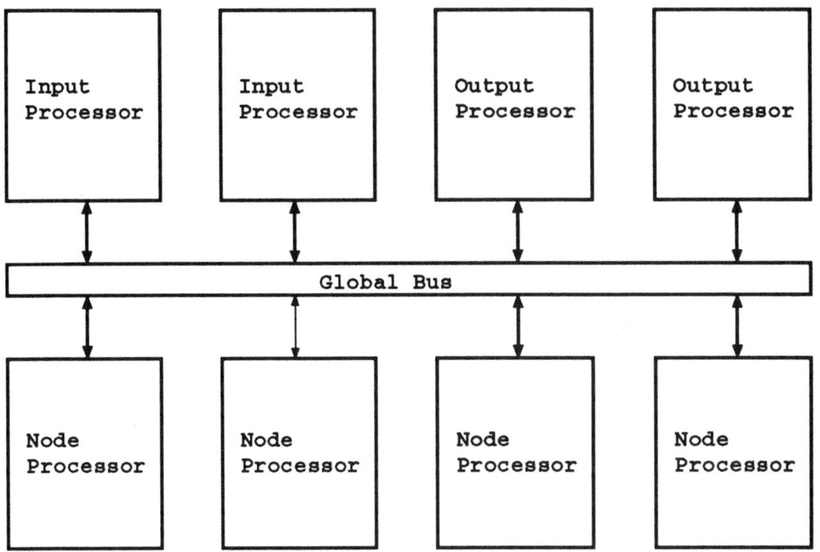

Figure 7.16: Target communication architecture of the UMIST synthesis system.

7.5 Outlook

In this chapter the subject was viewed as an extension of current high-level synthesis technology towards higher levels of abstraction. This provides an attractive conceptual basis for research in the new area of system-level synthesis and opportunities for re-using theoretical developments achieved in the former. Examples are the similarity between scheduling and allocation at the level of operations (high-level synthesis) and at the level of processes or procedures (system-level synthesis).

It is not certain whether this approach to system-level synthesis will be successful from a practical point of view. Despite of the attractive starting point, the biased view of system-level synthesis may result in some potential short-comings.

- System-level synthesis aims for the implementation of abstract behavior through appropriate system structures using fairly fine-grained structural components. However, most current electronic system designs use very complex parts like microprocessors or other standard "off-the-shelf" components. Current synthesis research has not yet solved the combined problems of adequately representing the functionality of complex components and of mapping the behavioral specification of a system onto structures containing these complex parts. Conventional simulation models do not provide the solution because they represent the knowledge about the behavior of the components implicitly only.

- The current approach to system-level synthesis is not tuned for a specific application domain. On the contrary, it aims for a general method like most of the high-level and all of the low-level synthesis methods presented in other chapters of this book. As is explained in Chapter 6 high-level synthesis will be successful only if a large amount of domain-specific knowledge is incorporated into the synthesis procedures. Obviously, this can be expected to be even more true for system-level synthesis.

- System-level synthesis as presented in this chapter is essentially limited to digital electronic hardware. Even if the problems mentioned above could be solved, system-level synthesis would certainly suffer from its dominant focus on hardware issues. Requirements like cost and flexibility dictate that as much as possible of the behavior, i.e., of the algorithms that provide system solutions to certain problems, is implemented by firmware and software. Pure hardware solutions are possible, of course, but they are not cost-effective except for very few exceptional cases which require extreme performance.

These are a few problems associated with the attempt to realize a type of system-level synthesis that is relevant in industrial design beyond the realm of academic research. These problems have to be addressed in the future. The work presented in this chapter is based on standard approaches such as graph algorithms and language compilation including modern techniques like object-orientation. Other approaches might be based on declarative specification of behavior and design knowledge, expert system technology, and formal logic. Some promising attempts exist along these lines, see for instance [BiGS89] and Chapter 8 of this book.

Chapter 8

Formal Methods for Synthesis

Holger Busch, Hans Nusser, Torsten Rössel

8.1 Introduction

As is clear from previous chapters, a number of hardware description languages (HDLs) are used for synthesis. If, however, one tries to argue about designs and their correctness in the strong sense of mathematical rigor and proof, most of these HDLs reveal two common limitations. There is neither a formal definition of their semantics nor a mechanism for safe symbolic reasoning about the objects they describe. This chapter presents a general-purpose formalism, *higher-order mathematical logic*, a technique of safe symbolic reasoning, *computer-aided theorem proving*, and their potential in the areas of hardware specification and verified synthesis.

For reasons of consistency, the notation of one specific *proof system*, namely the LAMBDA system [AHL91], is used throughout the chapter. Logic formulae, rules, and theorems are shown in concrete LAMBDA syntax rather than in abstract mathematical shorthand to emphasize the machine support for the calculus and its underlying language. While LAMBDA was designed as a core tool of a formal design assistant and hence is particularly well suited for this topic, the presented concepts are general and not restricted to this proof system.

Following a motivation for the choice of formalism, the basic notions and features of LAMBDA, its language, and its logic are introduced in Section 8.2. This section also contains a survey of formal hardware description styles and abstraction principles. Two approaches to interactive formal synthesis, called *hierarchical refinement* and *transformational design*, are described in Section 8.3.

Both benefit from ideas to integrate design and verification steps in the framework of formal proof as they have been presented in [FHPZ87, FoMa89, FFFH90, Four90, HaLD90]. Rules of logic are recognized as an appropriate representation for the interim stages of the synthesis process that provides a unified view of design and verification tasks.

Section 8.4 is dedicated to the formal verification of design generating synthesis functions rather than single designs. This allows the creation of verified design instances for a complete range of parameters from a single proof. Section 8.5 contains some conclusions and pointers for further reading in a brief report on other proof systems and their applications to formal hardware verification and synthesis.

8.2 Formal Reasoning about Digital Systems

8.2.1 Motivation and Choice of a Suitable Formalism

Two basic structural objects needed in a formal model of digital hardware are signals and components. Both may be modeled as functions. Doing so means to regard signals as functions from time to some type of signal values, e.g., Boolean, and to think of components as functions on signals. The latter are called *higher-order functions*, as the arguments (signals) they take and the results (again signals) they compute are functions themselves. Under restriction to discrete notions of time, natural numbers may be used to model time. Hence, if s_1 and s_2 are signals, t is a natural number, and C a component producing one output from two input signals, $s_1\,t$, $C(s_1, s_2)$, and $C(s_1, s_2)\,t$ denote the value of s_1 at time t, the output signal of an instance of C having s_1 and s_2 as its inputs, and the value of that output signal at time t, respectively.

Example 8.1 (Boolean Gates) Let *fn* $t \Rightarrow$ *body* denote the function mapping each t to the expression *body* which may or may not depend on t.[1] Let *bop* be an arbitrary Boolean operator on two arguments, such as *and*, *or*, and *xor*. Then

$$fun\ gate(s_1, s_2) = fn\ t \Rightarrow (s_1\,t)\ bop\ (s_2\,t)$$

defines a Boolean gate performing the operation *bop* on pairs of incoming signal values. Note that the gate is assumed to operate delay free at the discrete level of time as its output at time t is determined by its input signal values at time t. Making the operator *bop* just another functional parameter of the component

[1] This is the usual *function* or λ-*abstraction* known from functional programming and the λ-*calculus*. It allows one to write and use functions just as any other expressions without prior definition of an explicit name for them.

results in a definition of a *generic* two-input Boolean gate template[2]

$$fun\ gate_2\ bop\,(s_1, s_2) = fn\ t \Rightarrow (s_1\, t)\ bop\,(s_2\, t)$$

that is easily instantiated to concrete gates for any choice of a suitable operator. It also gives a first hint on how to exploit the parameterization capabilities of higher-order functions in the specification of components and designs.

Components may also be modeled as the *relations* they establish between signals. A natural way to describe a relation is by means of a function R that yields one of the truth values *True* or *False* depending on whether or not the relation holds between its arguments. Functions returning truth values are called *predicates.* The term *higher-order predicate* is used when the arguments of a predicate are allowed to be functions themselves, such as signals, operators, or even other predicates. The body of a predicate often contains some *universal* or *existential quantifiers.* These are used to express that, for example, a relation holds for all instants of time or that there exist objects with certain properties.

Example 8.2 (Multiplexor, Register) Let the formula $\forall x : type.\ P(x)$ denote the statement that property P is satisfied by all x of a given *type.* Then

$$fun\ Mux(sel, i_1, i_2, o) = \forall t : time.\ o\,t = if\ sel\,t\ then\ i_1\,t\ else\ i_2\,t$$

defines a predicate describing the relation between inputs and output of a two-to-one-multiplexor and the predicate

$$fun\ Reg(enbl, d, o) = \forall t : time.\ o(t+1) = if\ enbl\,t\ then\ d\,t\ else\ o\,t$$

describes a register that stores and outputs the data signal value it was most recently enabled to read. Note first that there is no need to specify an initially stored value for the register and that its output (apart from being constant before) is only determined after the first occurrence of *true* on the *enbl* line. While the *sel* and *enbl* signals are required to carry Boolean values to select one of two clauses in an if-then-else expression, there is no way to infer a required type for the other signals. As the fundamental behavior of multiplexors and registers is independent of the type of values they select and store, the signal parameters i_1, i_2, and d are merely required to be of the same but arbitrary type as the corresponding outputs, involving the type of carried values as a *type variable.*

[2]Functions in this chapter take exactly one argument. In $fun\ f(x,y) = body$, this is the pair (x, y) whereas $fun\ f\ x\ y = body$ is merely a shorthand for $fun\ f\ x = fn\ y \Rightarrow body$.

Types containing type variables are called *polymorphic types.* Type systems with polymorphism are comfortable, as they combine the benefits of a strong typing discipline with the capability to collapse many structurally identical descriptions into a single one using type variables. A suitable formalism must finally not restrict the objects one can describe in it to a specific set of predefined types. Beyond the capability to compose existing types into more complex ones there are many situations where it is convenient to introduce new, user-defined data types. The operation codes of an arithmetic logic unit (ALU), the instruction set of a microprocessor, and the abstract states of a finite automaton are all examples of objects that are easier described as an enumeration type of mnemonics than in any numeric encoding. Also concepts and operations involving lists, stacks, trees, or networks of objects are easier to express and understand in terms of their underlying abstract data types than in terms of any data structure restricted by a particular choice of implementation medium.

When the focus is moved from descriptions of structure, i.e., from the target domain of all synthesis methods covered in this book, towards descriptions of behavior which form the top-level input to the overall synthesis process, more of the expressive power in the formalism developed so far is recognized and appreciated.

Example 8.3 (Modulo-Counter) The following predicate is a behavioral description of a *modulo-m*-counter with reset and count input signals *rst* and *cnt* whose output values on *o* have to be subjected to an interpretation as numbers through function *ip*.

$$
\begin{aligned}
&\mathit{fun}\ MC(m, rst, cnt, o, ip) = \\
&\quad \forall t : \mathit{time}.\ ip(o(t+1)) = \mathit{if}\ rst\, t\ \mathit{then}\ 0 \\
&\qquad\qquad\qquad\qquad\qquad\qquad \mathit{else}\ \mathit{if}\ cnt\, t\ \mathit{then}\ (ip(o\, t) + 1) \bmod m \\
&\qquad\qquad\qquad\qquad\qquad\qquad\qquad\qquad\qquad \mathit{else}\ ip(o\, t)
\end{aligned}
$$

For any specific *m* this behavioral description covers the full range of possible designs, as it neither restricts the encoding of numbers (any suitable pair of encoding for the values on *o* and corresponding interpretation function *ip* will do) nor the structure of an implementation for a particular choice of encoding.

The above considerations suggest that a logic of higher-order functions and predicates on a type system with user-definable data types and polymorphism is a suitable choice of formalism for reasoning about digital systems. In more detail, this choice has also been advocated in [Gord86, Joyc91]. An introduction to a specific proof system built on such a logic is given in the following section.

8.2.2 A Proof System for Higher-Order Logic

The LAMBDA proof system [AHL91] was designed as the core tool of a formal design assistant and provides the necessary mechanical support for safe symbolic reasoning in a logic of higher-order functions and predicates. Its type system and syntax closely resemble those of the functional programming language Standard-ML [MiTH90]. As a consequence, every object in the logic has a unique, possibly polymorphic type, and functions defined in the logic can easily be executed and tested by means of any Standard-ML compiler.[3]

Types and Functions

Data types, whether new or predefined, are created and introduced in the system by *data type declarations*. A data type declaration defines one or several *constructors* (separated by "|") that are either constants or functions taking a tuple of arguments to yield an element of the type. The type of *Boolean values* typically used to represent a single bit of data is predefined as

```
datatype bool = true | false;
```

Data type declarations may be *recursive*.[4] The *natural numbers* for example are recursively defined to be either 0 or the successor S of another natural number by the declaration

```
datatype natural = 0 | S of natural;
```

Hence, the numbers 1, 2, and 3 are internally known to the system as $S\,0$, $S(S\,0)$, and $S(S(S\,0))$. This way of introducing the natural numbers, also known as *Peano style*, allows a formal, recursive definition of addition, subtraction, and other basic operations on them, which may then be used in proofs involving arithmetic transformations.[5]

A data type declaration may also introduce a polymorphic type parameterized by one or more type variables. The type family of *lists* containing elements of the same arbitrary type, a list being either the empty list *nil* or built up by *insertion* of an appropriately typed element in front of another list, is a popular example of a polymorphic (and again recursive) type. Its declaration in LAMBDA reads

[3]The LAMBDA system comes with a Standard-ML compiler, as it is entirely implemented in ML and uses ML function calls as its basic command language.

[4]At least within certain restrictions which mainly prevent that empty types and Russel's antinomy can be dragged in and corrupt the logic under the disguise of recursive data type declarations.

[5]Being used to pocket calculators and computers, it takes a while to recognize the reasons for such a complicated approach to a seemingly obvious thing like arithmetic on natural numbers. A calculator, however, does not *prove* that $1 + 2 = 3$, it merely displays the result of a computation.

```
datatype 'eltyp list = nil | 'eltyp :: 'eltyp list;
```

wherein "::" denotes an infix constructor for front insertion of list elements and the parameter *'eltyp* is a type variable[6] representing an arbitrary type of list elements. The shorthand $[a, b, \ldots]$ is used for $a :: b :: \ldots :: nil$.

The *type constructors* "$*$" and "$\longrightarrow$" are used to build *Cartesian product* and *function types.* The Cartesian product of N types, i.e., the type of N-tuples built from elements of type t_1 to t_N, is denoted by $t_1 * \ldots * t_N$. The type of a function from some domain type *dom* to some range type *rng* is denoted by $dom \longrightarrow rng$. Hence, the function *gate* defined in Example 8.1 is of type

```
gate : ((time -> bool) * (time -> bool)) -> (time -> bool)
```

For clearness and brevity, *type abbreviations* which can involve parameters as well are introduced. The following type abbreviations, which are used throughout the rest of this chapter, clarify the ideas of denoting time by natural numbers and of regarding signals as functions from time to some type *'svtyp* of signal values.

```
type time = natural; type 'svtyp signal = time -> 'svtyp;
```

Given these, the type of the above function *gate* nicely rewrites to the following wherein the type variable *'svtyp* has been instantiated to the type *bool.*

```
gate : (bool signal * bool signal) -> bool signal
```

Functions are created and introduced in the system by *function declarations.* A function declaration is given in one or several clauses (again separated by "|"). When a function is applied to an argument an appropriate clause is selected by *pattern matching.* Functions are allowed to be partially specified. Opposed to ML, even partially specified functions are assumed to be *total,* i.e., to perform a terminating computation of a result for any argument in their domain, and clauses must not overlap to make the pattern matching unique independently of the ordering of clauses. Applying a function to an argument not matching any of the specified clauses results in an irreducible term of the logic. Functions on pairs may use infix operator notation if appropriate. The Boolean functions *and* and *or* for example are denoted by the infix operators "&&" and "||" and defined by case splitting on the second argument.[7]

```
fun a && true = a    | a && false = false;
fun a || true = true | a || false = a;
```

[6] As in Standard-ML, a starting prime is used to distinguish type variables from other identifiers.

[7] Of course any other nonoverlapping case split works as well.

Function declarations may be recursive.[8] Addition and subtraction on natural numbers, for example, are recursively defined by case splitting on the second argument.

```
fun m + 0 = m |   m   + (S n) = S(m + n);
fun m - 0 = m | (S m) - (S n) = m - n;
```

Given this partial specification of subtraction on natural numbers (none of the clauses matches expressions $m - n$ with $m < n$), the expression "$2 - 3$" can merely be reduced to "$0 - 1$" which is then handled as an irreducible term of unknown value. As a final example, the polymorphic function that appends two lists of the same, arbitrary element type, is denoted by the infix operator "@" and recursively defined by case splitting on the first argument.

```
fun        []        @ list = list
 | (head :: tail) @ list = head :: (tail @ list);
```

Formulae and Predicates

There is a special type *om*[9] for the truth values and formulae of the logic to keep them clearly separated from Boolean values and expressions. A LAMBDA formula may be any of

- the *constants* `TRUE` and `FALSE`, or any *variable* `P`, `Q`, `R`, ..., of type `om`;
- an *equality* `exprA == exprB` of two expressions having the same type; if the expressions are themselves formulae, equality is a synonym for *bi-implication*;
- a *conjunction* `formA /\ formB` or *disjunction* `formA \/ formB` of two formulae;
- an *implication* `formA ->> formB` or the *negation* `NOT form` of a formula;
- a *universally quantified* formula `forall x. form(x)` or an *existentially quantified* formula `exists x. form(x)`.

A *predicate* then is a function with result type *om*. Applying a predicate to a set of suitable arguments yields a formula.

[8] At least within certain restrictions which prevent violation of the totality assumption by recursive functions that cannot be guaranteed to terminate.

[9] From the Greek Ω.

Sequents, Rules, and Theorems

In LAMBDA, *sequents* are the means to express statements, e.g., a proposition to be proven, in the logic.[10] A sequent consists of a list of *hypotheses* and an *assertion*, all of which are formulae, combined by the *entailment operator* "⊢" in the form

```
hypotheses |- assertion
```

The variables in a sequent are either *bound* by some universal or existential quantifier or *free*, i.e., *not* bound by any quantifier. A sequent then states that the hypotheses *entail* the assertion, which means that after assigning any values to its free variables the assertion will evaluate to TRUE whenever all hypotheses evaluate to TRUE. As there may be any number of implications in the hypotheses and assertion themselves, entailment is best considered as the implication of particular interest in the sequent. A sequent is said *to hold*, if the entailment it claims is provably correct.

Example 8.4 (Sequent) Consider a predicate *Circuit* defining a family of circuits with inputs *i* and outputs *o* spanned up by variations on some parameter data *par*. Let *AccIn* be a predicate characterizing a subset of acceptable inputs, and let *Behavior* be a predicate describing various *aspects* of desired I/O-behavior. The proposition that any circuit from the family given any acceptable input exhibits all aspects of desired behavior may then be formulated as the sequent

```
Circuit(par,i,o), AccIn(i) |- forall asp. Behavior(asp,i,o)
```

wherein *asp* is a bound variable while *par*, *i*, and *o* are all free.

The objects actually manipulated by the proof system are called *rules*. A rule is built from a list of *premises* and a *conclusion*, all of which are sequents, in the form

```
premiseN
...
premise1
----------
conclusion
```

A rule states that if all its premises hold, so does its conclusion. In other words, the conclusion of a rule can be *deducted* from its premises.[11] When read from

[10] This style of implementing a calculus of higher-order logic is hence also known as *sequent calculus* as opposed to the alternative approach called *natural deduction calculus*.

[11] In a sense, the relationship between premises and conclusion of a rule is just yet another level of implication.

the bottom up, a rule may also be regarded as part of a cookbook for proving propositions that match the structure of its conclusion. The recipe tells how to achieve the proof goal by replacing it with simpler subgoals, namely the premises of the rule, and proving these. A rule with an empty list of premises is called a *theorem*. In the cookbook sense, theorems are the most convenient recipes, because once they are applicable there are no subgoals left at all.

Axioms, Derived Rules, and Soundness

Obviously, combining arbitrary hypotheses and an assertion into a sequent rarely results in a sequent that holds. Likewise, a rule built from an arbitrary list of premises and a conclusion is most often *unsound*, i.e., it allows to "prove theorems" that are neither supposed nor intended to be theorems of higher-order logic. Hence, starting from a rule base that is sound, and providing a restricted toolbox for the derivation of new rules that is guaranteed to maintain *soundness*, is a key issue for every proof system. LAMBDA is built on a fairly small set of *axioms* or *basic rules*[12] that define the logic. There are axioms on

- properties of entailment, e.g., reflexivity, transitivity, and monotonicity;
- properties of `TRUE` and `FALSE`, and of the propositional operators `/\`, `\/`, and `->>`;
- properties of equality, e.g., the important substitution principle that allows to replace terms by other ones of equal value;
- properties of the quantifiers `forall` and `exists`;
- properties of function expressions and a few other elements of the language not covered here.

All other rules in the LAMBDA libraries and all rules that a user wants to add have been, or have to be *derived* from these axioms or from previously proven rules. Elementary rules on data types and functions introduced by corresponding declarations form an exception from this principle, as they can automatically be generated by the system in a way that preserves soundness. A seemingly useless mechanism to generate new, sound rules is to compose a single premise and an identical conclusion into a *trivial rule* or *tautology* of the form

[12] More precisely, these are *rule schemata* that may be instantiated to any number of rules by assigning appropriately typed values, e.g., formulae or lists of formulae, to the schema variables.

```
sequent
-------
sequent
```

For most proofs, however, tautologies formed from the sequent to be established as a theorem are used as a starting point. Working from there, the initial premise may iteratively be broken into subgoals and simplified until all premises can finally be discharged, because they reduce to axioms or previously proven theorems.[13]

The central mechanism to infer new rules is applying rules to other sound rules. The procedures involved in rule application are covered in the next paragraph. An efficient tool and key concept to proof automation is *rewriting* with a set of *equations*. An equation is a theorem whose assertion consists of a single equality formula, i.e., a theorem of the form

```
------------------
H |- exprA == exprB
```

that may be used to replace occurrences of an expression in a sequent with another expression of equal value. Rule applications and rewrites may be composed into *tactics*. Functions analyzing the structure of a current goal together with operators for sequencing, repetition, and conditional branching of tactics may be used to encode heuristic proof strategies or even proof procedures for specific classes of propositions.

Applying Rules and Tactics: Unification and Resolution

For a rule to be applicable in the backwards direction, its conclusion has to match the premise of the rule to which it will be applied. The free variables in both of these sequents, the conclusion of the first and the premise of the other rule, may need to be restricted to specific expressions in order to achieve the match. *Unification* is the process that takes two sequents and tries to find values, called *bindings*, for the *flexible* variables in them, such that the sequents become identical. Free variables for which no bindings are made can be turned *non-flexible* or *rigid*. Usually, in the proof of a theorem the variables present in the initial goal are not supposed to be modified through bindings, as this leads to a different theorem.

In general, there may be several ways to unify two sequents with different sets of bindings. These are then delivered by the prover as a *stream* of possible unifications. A user may either *guide* the prover to select an intended unification directly or manually *backtrack* in the produced stream until that

[13] This proof style is usually known as *goal-oriented* or *backward proof*.

unification is reached. *Resolution* is a mechanism to automate the search for suitable unifications. It is particularly valuable when sequences of rule applications only succeeding for certain combinations of unifications are composed into tactics. Whenever a rule in the sequence fails to apply, resolution automatically backtracks in the streams of previous unifications until a successful combination is found or all combinations, and hence the complete tactic, fail.

Example 8.5 (Resolution) Consider the task to prove the following sequent with a backward proof starting from the corresponding tautology.

```
H |- a + b + c == c + b + a
```

Since the addition operator associates to the left, the proof requires a change in associativity on one side of the equality combined with application of a commutativity law for two of the additions. After the expressions on both sides of the equality have been made identical, reflexivity of the equality operator is finally needed to complete the proof. The corresponding rules from the LAMBDA library involved in this proof together with their names, briefly memorizing the purpose of each, are given below. $P\#(expr)$ denotes any correctly typed, *syntactic context*[14] of an expression matching *expr*. For instance, $P\#(x+y)$ may be unified with any formula containing a sum of two terms.

```
addAssoc:             addComm:          eqRefl:

H |- P#(x+(y+z))      H |- P#(y+x)
----------------      ------------      -----------
H |- P#((x+y)+z)      H |- P#(x+y)      H |- x == x
```

The proof sketched above consists of successive applications of these rules which can, in principle, be performed step by step. Obviously, however, there are quite a number of ways to unify the conclusions of the associativity and commutativity rules with the sequent under consideration, and the prover needs to be guided to select the intended unifications. Using instead the combined tactic

```
sequentialUnifyTac [addAssoc, addComm, eqRefl]
```

which sequentially applies the three rules by unification, and exploiting the resolution mechanism, the goal can be proven in a single stroke. Each time the

[14] In LAMBDA, function variables denoting contexts belong to a special set of meta-types called *syntactic functions*. The "#" is used to distinguish application of a syntactic function from applications of the proper, ML-like functions seen so far. It is only the latter, that can be reasoned about in the logic. Context variables are merely used in the process of unification, by which they may be bound to specific syntactic functions. For that reason, LAMBDA is also said to perform *2nd order resolution*.

tactic fails to apply the last rule, i.e., each time the first two rules are applied in a way that does not make both sides of the equality identical, resolution backtracks in the stream of possible unifications. As there are suitable combinations of unifiers for the first two rules to make the last one applicable, the tactic finally succeeds.

A detailed description of resolution, unification algorithms, and their applications is found in [Paul86, Huet75].

8.2.3 Functional and Relational Description Styles

In this section, the consideration of component and design description styles started in Section 8.2.1 is resumed in more detail. The type abbreviations *time* and *'svtyp signal* given there are used here again.

Functional Models

The following declarations may be read as functional descriptions of elementary Boolean gates: an inverter, a two-input and-gate, and a two-input or-gate.

```
fun bInv (a:bool signal) (t:time) = not(a t);
fun bAnd (a:bool signal, b:bool signal) (t:time) = a t && b t;
fun bOr  (a:bool signal, b:bool signal) (t:time) = a t || b t;
```

The types of these functions, with parentheses used to emphasize that the type constructor "→" associates to the right, are

```
bInv      : bool signal -> (time -> bool)
bAnd, bOr : (bool signal * bool signal) -> (time -> bool)
```

Hence, when partially applied to a Boolean signal or signal pair, i.e., not given the time argument, these functions return a result of type *time* → *bool*, which is the same type as *bool signal*.

This model of components considers time only with the granularity of clock ticks of a synchronous circuit and assumes combinational logic to consume zero time. More abstractly speaking, these components transform sequences of input values into sequences of computed output values, and for combinational components corresponding input and output values have the same sequence index. The idea is further illustrated by the following declaration of a one unit delay element for Boolean signals which after starting with an arbitrarily selected value of *true* continues to output its input signal with a time shift of one clock tick.

```
fun bDel (a:bool signal) (0:time) = true
  | bDel  a              (t+1)    = a t;
```

Composing components corresponds to function composition. A multiplexor for instance may be composed from the above gates as follows.

```
fun bMux (a:bool signal, b:bool signal, sel:bool signal)
  = bOr (bAnd (a, sel), bAnd (b, bInv sel));
```

Here, applying functions representing components partially eliminates the need for any explicit reference to time. Furthermore, internal wires of the composite structure are automatically hidden: there is, for instance, no way to refer to the outputs of the embedded and-gates. Finally, the directionality in a functional model has to be noted: signal typed arguments and results of functions correspond exactly to input and output ports of components. Hence, function composition automatically obeys the common design rule "do not connect outputs together."

Recursion is the means to describe feedback. The following declaration describes a *parity checker* having an output value of *true* at time t iff the number of *true* values at input port a up to and including time t is even.

```
fun bParity (a:bool signal) (t:time)
  = let val d = bDel (bParity a) in bMux (bInv d, d, a) t end;
```

In contrast to simple composition, declarations with feedback need an explicit reference to time to avoid nonterminating recursion.[15] This does not prevent connections resulting in zero-delay feedback loops, usually a source of serious inconsistencies, which is a limitation of the model.

The above examples suggest the functional style as a natural way to express a designer's mental model of components formally. In addition, using languages like ML makes functional descriptions directly executable, which is a valuable property. Perfect hiding of internal wires in composite structures and directionality, however, may turn into drawbacks when access to internal signals or bidirectional behavior are required.

Relational Models

Using a relational style, the behavior of components is modeled by predicates on signals. A piece of hardware is regarded as a correct implementation of a component iff it admits those and only those combinations of signals at its ports for which the specifying predicate evaluates to `TRUE`. As every function f can be described by a corresponding relation F, namely the graph of f, relational models obviously cover a broader scope of possible descriptions than functional

[15] If *bParity* is defined such that *bParity a* is no longer a partial application only, the semantics of functions in LAMBDA or ML leads to infinitely many attempts for a complete evaluation.

ones do. To allow for a direct comparison, the following gives redeclarations of the above components using predicates.

```
val bINV(i,o)      = forall t:time. o t == not(i t);
val bAND(i1,i2,o) = forall t:time. o t == i1 t && i2 t;
val bOR (i1,i2,o) = forall t:time. o t == i1 t || i2 t;

val bDEL(i,o) = o 0 == true /\ forall t:time. o (t+1) == i t;
```

The signal names used above suggest which of the arguments are regarded as input and output ports of the described components. Given this interpretation, the predicates characterize exactly the same relation between input and output signals that the above functions established between their arguments and results. Formally speaking, all of the following sequents hold (with an arbitrary list of hypotheses H), or equivalently, are conclusions of provable theorems.

```
H |- bINV(a,bInv(a))          H |- bAND(a,b,bAnd(a,b))
H |- bOr(a,b,bOr(a,b))        H |- bDEL(a,bDel a)
```

There is, however, no formal distinction between inputs and outputs and hence no directionality in the relational model. Composing components corresponds to predicate conjunction, and internal wires are hidden explicitly by existential quantification. For instance, a redeclaration of the multiplexor is given by

```
val bMUX(i1,i2,isel,o) = exists iselbar, w1, w2.
      bAND(i1,isel,w1)    /\ bAND(i2,iselbar,w2) /\
      bINV(isel,iselbar) /\ bOR(w1,w2,o);
```

No additional means are needed to describe feedback. Having the names of internal wires at hand is sufficient, again unfortunately not preventing zero delay feedback loops to be built. Hence, a relational model of the parity checker reads

```
val bPARITY(i,o) = exists l, lbar.
      bDEL(o,l) /\ bINV(l,lbar) /\ bMUX(lbar,l,i,o);
```

Almost unrecognized, hierarchy at this point occurs. While the multiplexor itself is a composite object, it appears to the parity checker as a single unit. The composite parity checker may then be embedded as a single unit into the next, larger description. This procedure creates as many levels of hierarchy and stepwise flattening or unfolding as needed for a structured handling and understanding of complex systems.

Relational models are well suited for formal specification, synthesis, and verification, and most of what follows in this chapter is based on them. While unfortunately one cannot easily animate them as one can animate functional models by mere execution, they provide a unified descriptive style of both behavioral and structural aspects of digital systems. Their strength is to allow focusing on those properties of a system that are essential from any specific point of view. One can precisely specify the effect of a circuit sorting a sequence of input values by a certain criterion without any reference to a sorting algorithm. This gives a degree of freedom to implementors that functional specifications can not provide, while it meets, as well, the demand of most users asking for "*quick sorting*" rather than "*QuickSort*"[16] chips.

8.2.4 Formalizing Abstractions of Data and Time

The relational description style allows one to specify components and systems without detailed reference to the internal structures and procedures they use to exhibit their specific behavior. These features, also known as *structural* and *behavioral abstraction*, are one of the keys of managing complex design tasks. There is another pair of important abstraction principles, which are related to data and time. These allow initial specification of design goals in intuitive terms such as natural numbers, names of operations, or instruction cycles. Only step-by-step, as required by implementation activities, more details on data encoding and timing in cycles of a real clock are added.

If the same level of abstraction is used in describing the set of objects *obj* relevant to the specification and an implementation of a design, one usually tries to verify the sequent

```
Imp(obj) |- Spec(obj)
```

as a correctness proof for the implementation. When the objects in the specification are more abstract than those in the implementation, one has to resort to a slightly different proof goal which reads

```
obj == Abstr(rep), Imp(rep) |- Spec(obj)
```

Here, *Abstr* is the mapping from actual representations used in the implementation to the abstract objects used in the specification.

An abstraction function *listToNat*, interpreting lists of Boolean values as binary representation of natural numbers, is an example of a *data abstraction*. An implementation on Boolean lists may then be verified against a specification on natural numbers by proving the sequent

```
(i,o) == (listToNat(x),listToNat(y)), Imp(x,y) |- Spec(i,o)
```

[16] *QuickSort* is the popular, fast sequential sorting algorithm having an average time complexity of order $n \log(n)$ in the number n of elements to be sorted.

A typical *temporal abstraction* occurs, when a given specification is only dependent on a sample of a signal at every n^{th} cycle of a real clock. With the definition

```
fun Sample n x t = x(n * t)
```

a proof goal involving temporal abstraction may then read

```
y == Sample 8 x, Imp(x) |- Spec(y)
```

A more detailed illustration of formal abstraction principles is found in [Herb89, Melh89].

8.2.5 Generic Hardware Models

The kinds of parameters occurring in the hardware descriptions so far have been mostly restricted to signals and encoding or interpretation functions. Composite systems are often more efficiently and concisely described by models involving additional kinds of parameters. These may also simplify formal reasoning by providing an abstract view of a system freed from inessential details. For instance, a retiming operation, moving registers between inputs and outputs of synchronous components, is independent of functionality [LeSa83]. Parameterization supports powerful proof principles, such as induction. An induction proof is valid for all parameter values. Hence, proof complexity does not necessarily grow with the complexity of hardware components in well-structured problems.

First-order parameters may be instantiated to constant numbers or constants of other types. Examples of these are word lengths and initialization values. The expressiveness of first-order logic, however, is not sufficient to parameterize functionality and behavior, because only higher-order logic allows quantification over functions and predicates.[17]

Using higher-order parameters allows for formulation and proof of very general rules. If distinguished classes of frequently occurring hardware components can be identified, formal specifications and proof results may be reused in many applications. During the resolution process, higher-order variables in generalized rules may then be bound to complex expressions, e.g., complete hardware models. The benefits of higher-order parameters have been demonstrated in functional programming [Back78, FiHa86]. Functional techniques allow a direct manipulation of formal models, which simplifies design and optimization of functional programs. These techniques have also successfully been applied to hardware design [John86, Shee86]. The first approach addresses the derivation of synchronous hardware with separate control, the second one transformational

[17]Note that free variables in logical formulae are implicitly all-quantified.

design of regular circuits. The advantage of generic specifications has finally been recognized in interactive theorem proving [Busc90a, Joyc89, Wind90]. All of these approaches exploit regularity and analogies in proof problems.

The composition of relational models using existential quantification to hide internal wires has been explained in Section 8.2.3. A basic generic model for the composition of two hardware components is given by the function

```
fun com F G x y = exists z. F x z /\ G z y;
```

Corresponding to Figure 8.1, this combinator joins components with one port on either side, hiding their internal connection. This particular style is chosen, because functions provide the concept of partial application. For instance, the formal parameters F and G are partially applied in the composite expression, since the body of the combinator *com* shows a full application of these component parameters resulting in an expression of the formula type *om*.

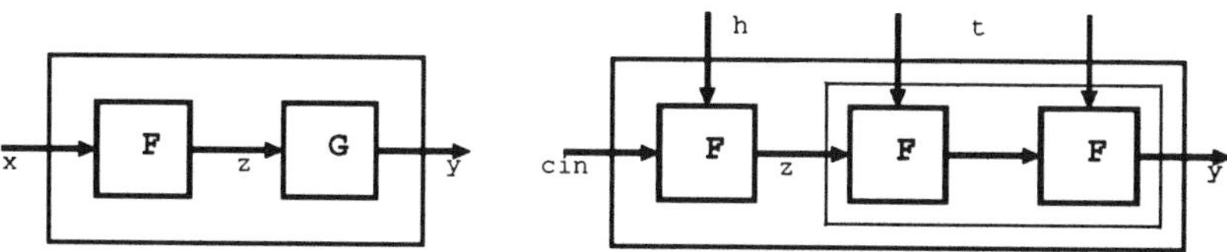

Figure 8.1: Composition and abstract cascade.

All component models observing the type restrictions can be combined. Therefore, actual parameters may even be compositions themselves as in

```
com (com F1 G1) (com F2 G2) x y
```

and as the types of the combinator arguments are polymorphic, the ports may as well be compound objects, such as tuples. Recursive combinator definitions describe structures with parameterizable dimensions. A cascade of components for example, as illustrated in Figure 8.1, is defined by the following function.

```
fun cfold 0     F (cin,[])   y =  y == cin
  | cfold (S n) F (cin,h::t) y =
      exists z. F (cin,h) z /\ cfold n F (z,t) y;
```

The first argument specifies the length of the cascade. The second parameter denotes the basic component that repeatedly occurs in the cascade. Given the following predicate as specification of an adder on natural numbers,

```
fun ADD (x1,x2) y = forall t:time. y t == x1 t + x2 t;
```

an adder cascade may be constructed with the cascade combinator:

```
cfold n ADD x y
```

This cascade expression may again be plugged into higher-order structures, as it defines a binary relation between a pair of input signals and a result port. The following term for instance, denotes a cascade of m cascades of n adders:

```
cfold m (cfold n ADD) x y
```

While components with input and output ports can be modeled as binary relations by collecting each of these port groups into a tuple or list, components having only outputs, such as constant generators, have to be wrapped in appropriate other component models, to fit the binary relation style. The first of the following functions transforms a value into a constant function of time which always returns that given value. The second one is a description of a constant signal generator wrapped in a neutral component that just passes an unchanged input through to its output.

```
fun const c (t:time) =  c;
fun cfst c x (y1,y2) =  y1 == const c /\ y2 == x;
```

Using these components, an initialized adder cascade may be specified by the expression

```
com (cfst 0) (cfold n ADD)
```

On generic descriptions such as these, many well reusable rules may be formulated and proven. In some cases, proving these generic rules is more difficult than proving rules on specific models. Many of these proofs, however, are even simpler to perform, because abstract variables are easier to handle than complex models containing irrelevant information.

8.3 Interactive, Formal Synthesis

As pointed out in Chapter 1, synthesis is a combined process of implementation, refinement, and optimization activities. It leads from a behavioral specification to a structural description of a digital system that satisfies the specified requirements in terms of functionality, performance, and production costs.[18] Interactive, *formal synthesis* fits well into this scheme, although its underlying philosophy is different from that of most synthesis techniques presented so far. The main paradigm of the formal approach is that a proof of correctness for the design is produced simultaneously with the design itself. Basically, as a partial result or side effect of the synthesis process, a sequent of the form

[18] There are of course many other aspects in a system specification, such as testability, reliability, and constraints on area, volume, or power consumption, which are not covered here.

```
synthesized_system, environment |- behavioral_specification
```

is established as a theorem. It states that, at least within the limits of the abstract model,[19] the synthesized system exhibits the required behavior under some explicit constraints on the environment. Thus a proof in the rigorous sense of mathematics replaces an enormous part of the simulation effort that is otherwise needed to achieve a comparable level of security and trust in the correctness of a design.

During interactive, formal synthesis, all the information on and ideas behind the decisions that lead to a specific design are available to ease the construction of a correctness proof for that design. This may result in a significant reduction of overall design costs when compared with a post-hoc verification and redesign style of synthesis. Whereas other synthesis techniques are specialized towards certain abstraction levels, proof-based, formal synthesis is in principle a very general approach. Although in particular higher-order logic allows handling of a wide range of abstractions, down to the logic or even circuit level of hardware description, the system, algorithmic and register-transfer levels are regarded as the proper application domain of formal synthesis techniques. The presented methods are neither intended nor capable to compete in terms of computational efficiency with existing powerful, automatic tools at the lower levels of abstraction, e.g., for logic-level synthesis and verification. It is rather their intelligent combination with these supplementary tools that yields the optimum benefits.

8.3.1 The Hierarchical Refinement Approach

The basis of formal synthesis is an embedding of the design process into a logic calculus. A proof system may then be used to keep track of the consequences of each synthesis step and to establish a "correctness by construction" design style. Such an embedding and a first formal synthesis strategy using the LAMBDA logic and proof system is demonstrated in the following.

Each state in the process of designing a system may be represented by a sound rule of logic. The initial state corresponds to a tautology of the form

```
pending_design_steps |- specification
------------------------------------
pending_design_steps |- specification
```

which simply states that "if the design steps performed in the future yield a system that entails the specification, that is what they do." Step-by-step, as the

[19] Although a trivial observation, it has to be pointed out that there is no magic in formal verification or mathematical proof. Like every other model-based validation technique it depends on the accuracy with which the model used corresponds to reality.

synthesis proceeds, a structural description, e.g., a netlist, is constructed from components or modules in a library. Each major design step corresponds to the application of a tactic, i.e., a combination of rules and rewrites, that captures the conditions and logical consequences of inserting or reusing a component in the netlist. Intermediary design steps may transform the design goal in a way that prepares it for the next intended insertion or reuse of a component. Thus, the initial tautology is successively modified. Each rule representing an intermediate design state contains the information on the three essential aspects of that design state and their logical connection: the specification, the partial design created so far, and the tasks that remain to be solved.

```
partial_design, future_steps |- remaining_tasks
-----------------------------------------------
partial_design, future_steps |- specification
```

These rules read as "if the partial design together with any results from the design steps to come entail the remaining tasks, they even entail the full specification." Applying a tactic to introduce a new component *Comp* derives a new state representing rule of the form

```
Comp, partial_design, further_steps |- reduced_tasks
----------------------------------------------------
Comp, partial_design, further_steps |- specification
```

This reflects both, the enhanced netlist and the reduction of the remaining tasks. At some point in the synthesis process, the remaining tasks are often regarded as constraints on the behavior of the environment of the system rather than tasks to be solved by the system itself. The pattern of a corresponding design state representing rule is then best given as

```
system_structure, environment  |- constraints
---------------------------------------------
system_structure, environment  |- specification
```

If, for instance, the synthesized system is a data path composed from arithmetic components, registers, and multiplexors, the constraints may consist of a list of equations (a table), specifying the control requirements for that data path to perform a desired computation. These may then be passed to a controller synthesis tool which, assuming it works correctly, comes up with a controller making the sequent

```
controller |- constraints
```

hold. Including that controller in the system's environment then allows derivation of a correctness theorem of the initially given form representing the final design state.

A Graphical Interface to Formal Synthesis

In principle, all the operations that produce the sequence of sound rules representing design state can be performed in a command-based, textual mode of interaction with the proof system. Operating such a system from its very ground level, however, usually requires a detailed knowledge and understanding of its rule base, available tactics, and successful proof strategies, and a vast amount of typing. Furthermore, designers rarely think of the synthesis process as a sequence of rule derivations, but rather in terms of evolving schematics or block diagrams and the selection of components and functions used to generate them. Hence, designers appreciate the value of graphical interfaces, as they are widely offered by electronic design automation tools, providing a menu-driven access to the design process and presenting its results in familiar terms.

The feasibility of a graphical interface to the formal synthesis process that mostly hides the mechanics of the underlying proof system from the user, is demonstrated by a prototype shown in Figure 8.2.

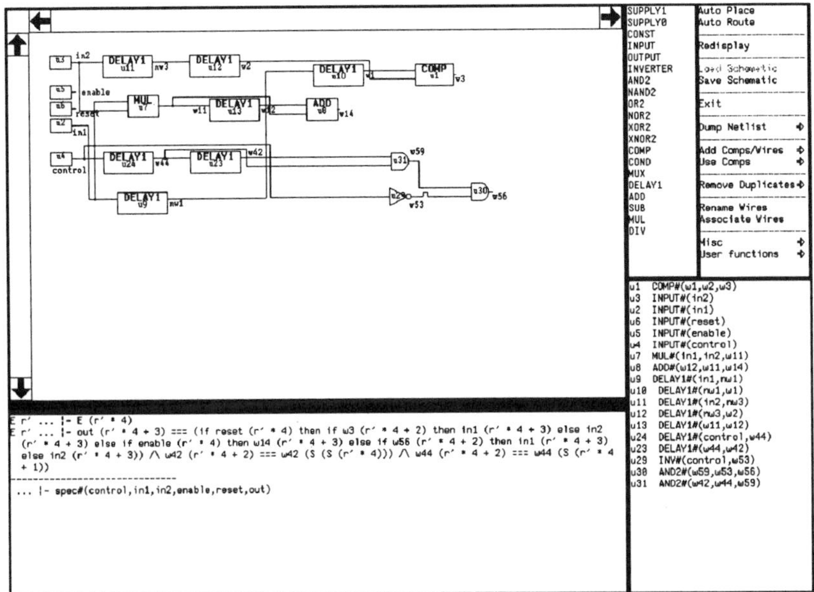

Figure 8.2: Prototype graphical interface based on LAMBDA.

The observable part of the interface splits into five main windows. The upper left window displays a schematic editor that holds a graphical representation of the system structure in the current design state. Through the schematic editing window, a user may drive the proof-based synthesis process as well as view the structural consequences of the steps he/she initiated. The displayed structure is extracted from the design state representing rule and is updated after each change made to that rule by the prover. The same holds for the textual format representation of the current netlist given in the lower right window. In the lower left window, the current design state is displayed in the form of its corresponding rule, focusing on the remaining tasks to be solved, as a basis for decisions on the next synthesis steps.

Finally, there are two main menus in the upper right part of the interface; one for components or modules and one for operations. The entries and submenus offered in both are mostly passed to the interface as parameters. This allows customization of the system for use with various component libraries, and provides specialized sets of typical design operations for particular application domains. Components may range from concrete to very abstract ones, i.e., from cells in a library and modules provided by some generator to mere black boxes with complex behavioral parameters. Many synthesis steps, such as scheduling or schedule-dependent insertion of registers and multiplexors into a data path, can be encoded in the form of tactics. Operations from the menu may then invoke the underlying prover to apply these tactics to a current design goal.

Component Libraries

A user may select component libraries and enhance them with his/her own customized definitions, according to the abstraction level of the specification and the required functionality of the system to be synthesized. Each component is associated with one or several tactics. These tactics hold the know-how concerning which contributions to the completion of a design and which reductions in the remaining synthesis tasks can be achieved with that particular component. For instance, the core rule *MulIntro* within an introduction tactic for a multiplier looks like the following.

```
Mul(i1,i2,o), H |- P#(o(t+1)) /\ i1 t == x /\ i2 t == y
--------------------------------------------------------
Mul(i1,i2,o), H |- P#(x*y)
```

$P\#(x * y)$ again denotes any syntactic context[20] of an expression matching the pattern $x * y$, i.e., it may be unified with any formula containing a product of two terms. The *MulIntro* rule is derived from the behavioral specification of a multiplier *Mul* given by the predicate declaration

[20] See Example 8.5.

```
fun Mul(i1,i2,o) = forall t. o(t+1) == (i1 t) * (i2 t)
```

which states that *Mul* outputs the product of two input values with a delay of one time unit. The rule *MulIntro* can be used to replace a product $x * y$ in any context with the output of a multiplier *Mul* at time $t+1$, given that multiplier is (inserted or present) among the hypotheses (i.e. part of the netlist) and the values of x and y are bound to inputs of the multiplier at time t.

Applying *MulIntro* to a rule *ids* that is related to an intermediate design state with a multiplication among its remaining tasks by resolution yields a new rule *ids'* representing a new design state. The (first) premise of *ids* is unified with the conclusion of *MulIntro* and then replaced by the premise of *MulIntro*, with the variables x, y and P set according to the bindings that resulted from the unification. Depending on the designer's decision, this may lead to insertion of a new or reuse of an already present multiplier in the hypothesis or netlist. Given for instance, a rule *ids* of the form

```
Partial_Netlist |- d == (a*b + c)/2 /\ ...
------------------------------------------
Partial_Netlist |- specification
```

and a decision not to reuse any (if at all present) multiplier from the partial netlist, applying *MulIntro* may derive a new rule *ids'* of the form

```
Mul(w1,w2,w3), Partial_Netlist
  |- d == (w3(t+1) + c)/2 /\ w1 t == a /\ w2 t == b /\ ...
-----------------------------------------------------------
Mul(w1,w2,w3), Partial_Netlist |- specification
```

with the x, y, and context variable P from *MulIntro* bound to a, b, and the syntactic function $P\#(z) = (d == (z+c)/2 \wedge \ldots)$. The newly introduced, free time variable t may then be related to times of other events in a process of scheduling. Going from *ids* to *ids'*, the task of multiplying two values has been simplified to the task of presenting these values at certain ports and times.

When there is more than one occurrence of an operation in the design goal or more than one component to perform it in the netlist, a designer may guide the prover in the assignment of operations to components by simply pointing to corresponding pairs in the schematic editor and rule windows. How this allows for an interactive exploration of the design space is covered in the next paragraph.

Allocation and Scheduling

As elaborated in Chapter 6, high-level synthesis tools perform the allocation of components, assignment of operations to resources, and scheduling of operations to control steps mostly automatically under some user-defined constraints.

In the framework of proof-based, formal synthesis, the degrees of freedom in these activities can be explored interactively, yet safely, until a satisfying trade-off between performance and resource costs is found.

The following sample design goal, which may result from a specification of a simple digital filter, requires a linear transformation of four input values i_1 to i_4 with factors a, b, c, and d to be computed as an output value o.

```
H |- o == a*i1 + b*i2 + c*i3 + d*i4
-----------------------------------
H |- specification
```

Depending on resource constraints and performance requirements, various combinations of adders and multipliers[21] and various assignments and schedules for the operations may be chosen. If the cost of additional control is assumed to be small in contrast to the costs of adding and multiplying, using only one adder and multiplier each leads to minimal chip area and hence minimizes costs. To synthesize a design of this form, a single multiplier is first introduced by applying its associated tactic, and the prover is guided to iteratively reuse it for all occurring multiplications. The resulting rule, with a first component in the hypotheses lists and some new tasks in the assertion of its premise, reads as follows.

```
Mul(w1,w2,w3), H
  |- o == w3(t3+1) + w3(t2+1) + w3(t1+1) + w3(t+1)
  /\ w1 t3 == a /\ w2 t3 == i1 /\ w1 t2 == b /\ w2 t2 == i2
  /\ w1 t1 == c /\ w2 t1 == i3 /\ w1 t  == d /\ w2 t  == i4
-------------------------------------------------------------
Mul(w1,w2,w3), H |- specification
```

All the product terms have been replaced by values on the output wire w_3 of the multiplier at various times, which have to be determined in the later scheduling process. Eight additional equations state the input requirements for the multiplier to compute the desired results. Repeating this procedure with an adder in place of the multiplier leads to the replacement of sums with values on the output wire w_6 of the adder.

```
Add(w4,w5,w6), Mul(w1,w2,w3), H |- o == w6(t6+1)
  /\ w1 t5 == d /\ w2 t5 == i4 /\ w1 t4 == c /\ w2 t4 == i3
  /\ w1 t3 == b /\ w2 t3 == i2 /\ w1 t2 == a /\ w2 t2 == i1
  /\ w4 t1 == w3(t4+1) /\ w5 t1 == w3(t5+1)
  /\ w4 t  == w3(t2+1) /\ w5 t  == w3(t3+1)
  /\ w4 t6 == w6(t +1) /\ w5 t6 == w6(t1+1)
```

[21]Of course, a single type of ALU may do the job as well.

```
-----------------------------------------------------------------
Add(w4,w5,w6), Mul(w1,w2,w3), H |- specification
```

At this stage, allocation, and, as there is only one resource of each type also assignment, are completed. A scheduling process may now assign control steps to the time variables in a way that guarantees values to be provided as outputs before they are consumed as inputs to another computation.[22] Applying, for example, an ASAP scheduling tactic from the operations menu results in the following rule.

```
Add(w4,w5,w6), Mul(w1,w2,w3), H |- o == w6(t+5)
  /\ w1(t+3) == d /\ w2(t+3) == i4
  /\ w1(t+2) == c /\ w2(t+2) == i3
  /\ w1(t+1) == b /\ w2(t+1) == i2
  /\ w1 t    == a /\ w2 t    == i1
  /\ w4(t+4) == w6(t+4) /\ w5(t+4) == w3(t+4)
  /\ w4(t+3) == w6(t+3) /\ w5(t+3) == w3(t+3)
  /\ w4(t+2) == w3(t+1) /\ w5(t+2) == w3(t+2)
---------------------------------------------
Add(w4,w5,w6), Mul(w1,w2,w3), H |- specification
```

This leaves t as the only free time variable. In the next step, constant generating components for a, b, c, and d are introduced. The final task is the insertion of delay elements and multiplexors to achieve the correct data flow. This can again be solved automatically by applying another tactic from the operations menu. The final design state is represented by the rule

```
Complete_Netlist |- o == w6(t+5)
  /\ w2 t    == i1    /\ w2(t+1) == i2
  /\ w2(t+2) == i3    /\ w2(t+3) == i4
  /\ c2 t    == false /\ c3 t    == true
  /\ c2(t+1) == false /\ c3(t+1) == false
  /\ c2(t+2) == true  /\ c3(t+2) == false
  /\ c2(t+3) == true  /\ c3(t+3) == true   /\ c3(t+4) == true
-----------------------------------------------------------------
Complete_Netlist |- specification
```

indicating that the required computational result is available at port w_6 after 5 control steps, given the input values i_1 to i_4 are sequentially presented at port w_2 and the multiplexors are controlled according to the equations on their selector lines c_2 and c_3. The schematic representation of the synthesis result

[22] This constraint is also known as *causality principle*. In high-level synthesis it is usually captured as the partial precedence ordering extracted from a dataflow graph.

is shown in Figure 8.3. It is worth noting how the delay- and multiplexor-introducing tactic minimized the control effort: three of the multiplexors share the same control line c_3.

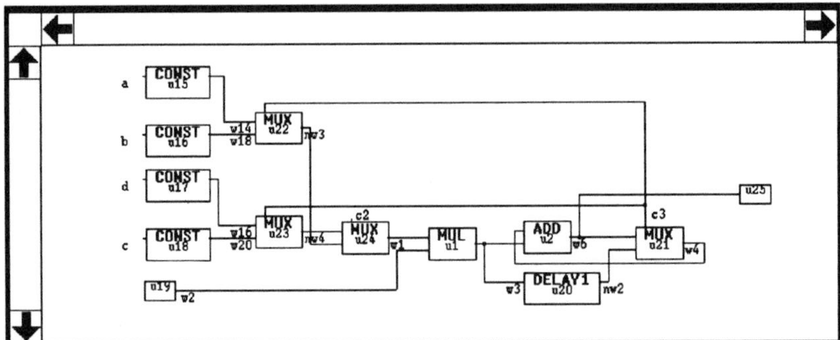

Figure 8.3: Minimum area netlist with one adder and multiplier.

In an attempt to synthesize for maximum performance, the prover may as well be guided to introduce a new multiplier for each product to compute and to spend an extra adder too.[23] Thus, the number of control steps needed to perform the computation can be reduced to 3. A corresponding schematic is shown in Figure 8.4.

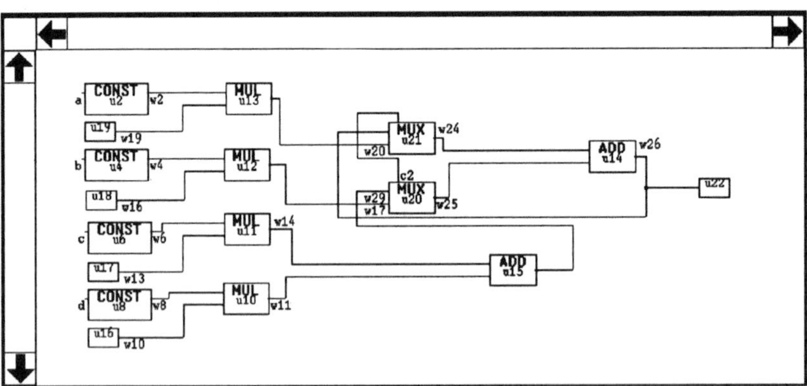

Figure 8.4: Maximum performance netlist with 4 multipliers and 2 adders.

There is also a third feasible design using 2 multipliers and adders each to

[23] Spending more than two adders is of no advantage, since there are never more than two sums ready for computation at the same time.

perform the required computation within 4 control steps, thus representing a compromise between area and performance. All of these variants in the design space have safely (without risk of corrupting correctness) been explored in the interactive, formal style of synthesis presented here.

Hierarchical Design

As mentioned, component libraries may also contain abstract modules or "black boxes" which are the means to introduce hierarchy into the synthesis process. The behavior of a black box is usually determined by a number of parameters which may be functions or predicates of any complexity. This again illustrates the benefits of a higher-order calculus for the purpose of formal synthesis. Each instance of an abstract component with a particular choice of parameters may be regarded as a smaller subsystem to be synthesized separately. The structural implementation of such a subsystem may then again rely on some other black boxes and this may continue over any appropriate number of hierarchy levels. Using this hierarchical structure, designs comprising several thousand components have been synthesized in the interactive, formal style. When a more flattened view of the overall system is needed, the structural details of any black box can of course be expanded into the next higher level of hierarchy.

The following is an example of a simple abstract component computing any function f of two input signal values on x and y and delivering the result at port z after a delay of n time units.[24]

```
Fun2(f,n,x,y,z) = forall t. z(t+n) == f(x t,y t)
```

In the underlying proof system, abstract components, such as *Fun2*, are represented and handled the same way more concrete components are. There is a graphical symbol as well as one or several tactics associated with each of them. The core rule of an introduction and reuse tactic for *Fun2* reads

```
Fun2(f,n,x,y,z), H |- P#(z(t+n)) /\ x t == a /\ y t == b
---------------------------------------------------------
Fun2(f,n,x,y,z), H |- P#(f(a,b))
```

which means, that any context of a function application matching the pattern $f(a,b)$ may be replaced with the same context of the output port z of an appropriate instance of *Fun2* in the netlist at some time $t+n$, given that values a and b are presented to the input ports of that *Fun2* at time t.

[24] For $n > 1$, this describes the behavior of a pipelined component, as inputs are accepted and an output is produced at every time step.

Framework Considerations

The presented synthesis approach and tool are not a fixed, closed system. They are rather understood as an open framework which provides general, safe mechanisms to symbolically manipulate the behavior and structure of digital systems. Within this framework, a designer may use or define a customized selection of data types, functions, components and operations, as needed for a particular application domain.

When setting up a component library, requirements from back end tools must be taken into account. The final result of a formal synthesis process, for example, a netlist on register-transfer level, has to be, at least after some conversion of format, suitable for further processing by subsequent tools, e.g., for logic-level synthesis and technology dependent library mapping. This usually requires a certain compatibility in the basic, concrete components used. Making more synthesis algorithms available in the form of tactics accessible via the operations menu may increase the degree of automation while the possibility of interactive synthesis steps is maintained. Thus, a beneficial partitioning into automated, routine procedures, and creative, human interaction with the system exploiting a designer's experience, may be achieved.

8.3.2 Transformational Hardware Design

In the hierarchical refinement approach to formal synthesis, the subcomponents of a system are considered strictly separated at the lower levels of the hierarchy. Some important and commonly applied design techniques, however, restructure the decomposition of a system model, and hence are not covered by a purely hierarchical synthesis style. By interleaving subcomponents of higher-level modules, for instance, more regular structures are generated. Optimizations leading to efficient pipeline structures with shortened delay paths cannot be kept local either, since they affect and refer to the overall structure as well. A formal counterpart to these design techniques must therefore allow access to system components at different positions in a global structure. Yet it has to solve the task of keeping the complete model with all its implementation details manageable for larger than trivial systems, too.

The transformational approach presented here is a solution to these problems. It organizes design techniques with global effects into sequences of manageable, formal transformations, such that only small, incremental changes between successive stages of the synthesis process have to be proven correct. The transformation rules used abstract from irrelevant details in the models, thus reducing the complexity of formulae to be handled in proofs. Higher-order variables are exploited to keep the transformation rules themselves much more concise than the rules they are applied to. They are also used to achieve most general versions of transformation rules which may be reused in various instances of the same basic problem.

Transformational synthesis starts from the same kind of tautology as the refinement approach.

```
|- specification
----------------
|- specification
```

In contrast to that approach, the evolving system description is kept completely in a single assertion during the synthesis process rather than separating a growing system structure from some diminishing remaining tasks. Individual transformation rules may then be focused to specific subparts of the system model. This provides an alternate way of modularization and significantly eases the handling of transformation rules and corresponding proofs.

Transformations may need preconditions to be applicable. If fulfillment of these preconditions cannot explicitly be derived from the information contained in the current design state, they are introduced as new hypotheses. In the refinement approach, the same mechanism was used to introduce new components into partial netlists. Here, transformation rules in the form of conditional equations such as

```
--------------------------------------
condition |- design == modified_design
```

are used, which allow rewriting an intermediate design or subcomponent under some condition. Thus, from a design state representing rule

```
          |- current_design
---------------------------
conditions |- specification
```

a new one, expected to be closer to the ultimate design goal, of the form

```
  |- modified_design
------------------------------------------
new_condition, conditions |- specification
```

may be derived. Alternatively, an implicative transformation rule of the form

```
------------------------------------
condition, modified_design |- design
```

may be applied to substitute its hypotheses for the assertion of a current design state. While equational transformations may be applied to any subexpression independent of its environment, this does not generally hold for implicative transformations. For instance, the sequent

```
x ->> y |- P#(x) ->> P#(y)
```

is not provable for an arbitrary context P. Establishing equational transformation rules, however, requires more effort since both possible directions of replacing expressions have to be proven correct. Besides, a transformation may lead to specialization. In many examples, an abstract specification merely describing some system properties is refined to an implementation exhibiting a specific behavior among several acceptable ones. In these cases, the relation between implementation and specification is necessarily implicative.

Transformation of Generic Structures

Equational and implicative transformations of generic models in the flavor of Section 8.2.5 are now described in more detail. A first theorem states the associativity of component combination.

```
------------------------------------
|- com (com F G) H == com F (com G H)
```

At the top level of a structural hierarchy, each composite system may be described by an expression of the form *com F G x y*, where F and G may themselves be composite. The associativity theorem for the *com*-combinator is an example of a *reduced equation*, in which the signal arguments have been omitted on both sides of the equation. A reduced equation may be applied directly to subcomponents of composite models.

`|- F2 == F2'` rewrites
`|- com F1 (com F2  F3) x y` to `|- com F1 (com F2' F3) x y`

All variables in these formulae denote specific expressions, whereas the variables in the above associativity theorem are implicitly all-quantified, and can be bound to arbitrary expressions. The component *F2* may be replaced without considering its environment. Therefore, the combinator definitions need not be expanded, which is important to keep the circuit description concise. The same transformation may be achieved by an implication.

`F2' a b |- F2 a b` transforms
`|- com F1 (com F2  F3) x y` to `|- com F1 (com F2' F3) x y`

Generic rules for combinator-based, composite models allow to apply implicative transformations without expansion of combinator definitions, too. These rules use symbolic contexts to exchange subcomponents. After adapting the associativity of a design model to the generic rules, implicative transformations can be performed as easily as equational ones.

Refinement and Constraints

Constraints play an important role in top-down design. They define the circumstances under which a refined structure meets a higher-level specification. For instance, set-up and hold times influence the synchronous behavior of clocked components. Considering data refinement, the range of acceptable values for an integer variable must be restricted according to the finite length of the bit string chosen to represent it.

Post-hoc verification approaches extract the behavior of components from an implementation structure in a bottom-up style, including the derivation of global constraints [Herb89, WaNe89, Weis87]. In the context of synthesis, one is rather interested in how constraints may already be considered during the design process. While this usually complicates refinement operations, it guarantees a system implementation to operate correctly upon its first completion. Refining a realistic system is far too complex a task without modularization. The transformational approach is appropriate for modular refinement of subcomponents and constraint handling.

When building composite structures by means of distinct implementation transformations, one typically incurs internal constraints on the interface signals between subcomponents. As systems are preferably treated as black boxes, internal constraints cannot be checked to hold once the system has been implemented. For this reason, all internal constraints are propagated to the borders of these black boxes. Alternatively, static restrictions on design parameters are sometimes derived that guarantee internal signal constraints are satisfied. These tasks, too, can be performed by transformations.

The following definitions and rules help to handle refinement and constraints in the transformational style. Signal constraints are managed by introducing specific pseudo-components which merely test passing streams of data according to some restriction parameters. For instance, the pseudo-component defined by

```
fun sR rst x y = forall t. rst(x t) /\ x == y;
```

is parameterized with a restriction predicate *rst* that it applies to all elements in a stream of incoming signal values on x while they get passed through to port y. It may be used as an argument of combinators, as it is defined as a binary relation. A constraint on the input signal of an arbitrary component may then be introduced by the following rule.

```
---------------------------
com (sR rst) F x y |- F x y
```

An adder on natural numbers may be considered as an example for *con-*

straint propagation. A constraint on the output of the adder may be transformed to a constraint on its inputs by the following theorem.

```
-------------------------------------------------------
 w1 < w3 == true, w2 < w3 == true,
com (comb2 (sR (isNat w1), sR (isNat w2))) Add (x1,x2) y
 |- com Add (sR (isNat w3)) (x1,x2) y
```

The combinator *comb2* assigns a pair of components to pairs of signals. When the propagation rule is applied, the *smaller*-relations are introduced as additional hypotheses to the current rule. These conditions no longer refer to signals, but to static parameters. In general, constraint propagation must respect the behavior of the component through which the constraint is propagated. The behavior of the adder is indirectly reflected by the two inequality conditions. With pseudo-components for data conversion,

```
fun snb w x y = forall t. isNat w (x t)      /\
                           y t == natToList w (x t);
fun sbn w x y = forall t. length(x t) == w   /\
                           y t == listToNat(x t);
```

the implementation of an ideal adder on natural numbers may be described as a ripple-chain of full-adders.[25]

```
                ---------------
                A#(w) |- Add x y
```

with `A#(w) := com (comb2(snb w,snb w)) (com(FArow w)(sbn(w+1)))`

Here, the component *FArow* takes two bit strings of length w, and produces a bit string of length $w + 1$. The pseudo-components *snb* implicitly restrict the input-signal values to be smaller than 2^w. The propagation of an output constraint to the inputs of this implementation structure can then be achieved with the next rule.

```
-------------------------------------------------------
 w1 < w == true, w2 < w == true,
com (comb2 (sR (isNat w1), sR (isNat w2))) A#(w) (x1,x2) y
 |- com A#(w) (sR (isNat w)) (x1,x2) y
```

With properties of composition combinator and restriction component, the proof of this rule can be simplified by substituting the abstract adder on natural numbers for the adder on bit strings using the implementation theorem. The

[25] Here, the "#"-notation is used again to distinguish mere syntactical abbreviations from function applications.

same kind of strategy applies for more complicated, composite structures. The advantage is that constraint propagation can be performed at an abstract level, which is usually more convenient than at the implementation level. Further theorems allow removal of dual data converters as they occur when subcomponents are refined separately.

Following this procedure, internal constraints are considered at the very time design decisions are made. An interactive proof system is particularly well-suited for this purpose, as it makes the full mechanics of rule manipulation available to store and keep track of changing constraint information.

Example 8.6 (Refinement of an Adder Cascade) This example is an illustration of the modular refinement procedure for a composite structure. Let a cascade of adders on natural numbers be the initial specification.

```
|- com (cfst 0) (cfold (S n) ADD) x y
```

Assume that the following three implementation theorems are given.

```
----------------        ----------------------
A#(w) |- Add x y        C#(w) x y |- cfst 0 x y
              --------------------------
              com R#(w) T#(w) x y |- id x y
```

with

```
A#(w):= com (comb2 (snb w,snb w)) (com (FArow w) (sbn(w+1)))
C#(w):= com (cfst (mklist w false)) (comb2 (sbn w,id))
R#(w):= sR  (isNat w)
T#(w):= com (snb(w+1)) (com (Trunc w) (sbn w))
```

The first one was already presented above. The second one refines a generator of constant signal values used to initialize the input of the adder cascade. Therein, *id* denotes a binary identity relation. The last one, again using *id*, introduces a module truncating the most significant bit of an incoming bit string. The truncation has no effect on the result of type *natural*, if the input is restricted according to the output word-length. The corresponding restriction is specified by the pseudo-component *R#(w)*.

The abstract structure is first transformed into an equivalent structure of abstract units that directly correspond to the partial implementations listed above.

```
|- com (cfst 0) (cfold (S n) (com ADD id)) x y
```

Specific generic rules then allow application of the implementation theorems to the cascade structure, again without unfolding the combinator definitions.

```
|- com C#(w) (cfold (S n) (com A#(w) (com R#(w) T#(w)))) x y
```

The internal constraints get substituted by new constraints on the inputs of the cascade.

```
r <= w == true, n+1 < 2^(w-r) == true
|- com C#(w)
  (cfold (S n) (com (comb2 (id,R#(r))) (com A#(w) T#(w)))) x y
```

In the substitution proof, the following approximation is used.

```
-------------------------------------------------------
n+1 < 2^(w-r) == true |- (n+1) * (2^r - 1) < 2^w == true
```

In the expanded structure, truncation and pairs of data converters between full-adder rows occur, but these components belong to different cascade stages. After changing the cascade hierarchy, these adjacent, dual data converters can be removed. Further generic transformations finally lead to the adder cascade structure shown in Figure 8.5 and formally represented by the following sequent.

```
r <= w == true, n+1 < 2^(w-r) == true
|- com  (cmap (n+1) (com (sR(isNat r)) (snb w)))
    (com (cfst (mklist w false))
    (com (cfold (n+1) (com (FArow w) (TRUNC w)))
         (sbn w))) x y
```

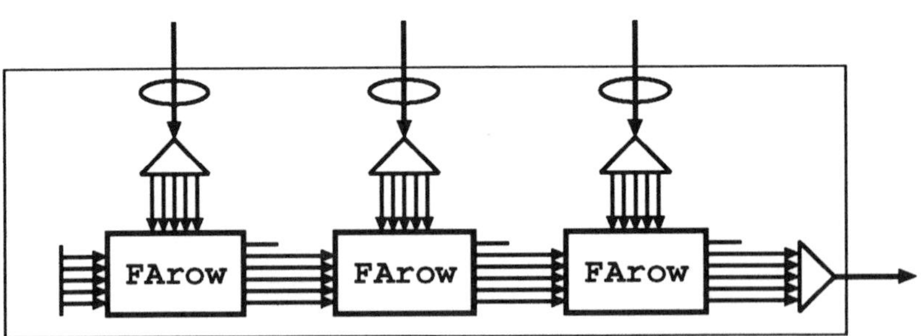

Figure 8.5: Implementation of an adder cascade.

The triangles symbolize data converters, the ellipses signal constraints. The implementation model has pseudo-components at its primary ports only, while the internal structure purely consists of bit-level components. Thus, the kernel can be mapped to hardware components. If the input values do not violate the propagated global constraints, internal overflows are precluded and the cascade is a valid implementation of the summation function. The adder cascade may be combined with partial-product gates yielding a multiplier structure. Again, the internal data converters and signal constraints may be propagated or removed.

8.4 Formally Verified Synthesis Functions

The effort spent during proof-based synthesis of a non-trivial system may be considerable. Therefore, verifying the correctness of some synthesis functions themselves rather than the correctness of each single result they produce may sometimes be more efficient. When applied to specific arguments, verified synthesis functions return designs that are guaranteed to be correct, or can at least be proven to be correct by simple checks on the arguments.

There are basically two approaches to synthesis by means of formally verified generators. The first one is based on generic, formal models of parameterized designs. The second approach uses synthesis functions which process hardware descriptions in the form of list constants. A short tour through various research activities in this field is made in the next section. Using previous specifications, the instantiation of a generic design is illustrated in Section 8.4.2.

8.4.1 Overview of Research Activities

Generic ROMs

The HOL system, a proof checker for higher-order logic [Gord87], has been used to specify and verify generic ROMs [Joyc89], exploiting the observation that the fundamental structure of all ROM instances is identical. Merely the actual position of individual pull-down transistors in a NOR-array needs to be specified, which is done by a "personalization" function. The same generic NOR-array is used for the address decoder and the actual storage of a ROM. A generic correctness theorem states that for any personalization function, the generated ROM always returns the word stored at the addressed location.

In the specification, the personalization function is a parameter describing the expected output of a ROM for well-formed input addresses. In the implementation, this function parameter is used to describe the position of pull-down transistors in the storage part of the ROM. The decoder is defined in terms of a fixed personalization function that is applied to row and column indices of an n-bit decoder. This personalization function corresponds to a verified generator for patterns of pull-down positions in decoders of arbitrary size.

Multiplier Arrays

Synthesis functions for multiplier arrays are described in [Chin90]. Using higher-order logic, a hierarchy of verified synthesis functions is constructed in a bottom-up style. While the lowest-level functions just return half and full adders, the highest-level ones generate carry-save adder arrays.

Components and netlists are represented in a functional form using "meta-functions." Full evaluation of a meta-function application to some arguments just returns the output values of the corresponding component or netlist. Partial evaluation is the key to synthesis. If the evaluation of a meta-function application to specific arguments is appropriately stopped, it yields a functional expression corresponding to a netlist which correctly implements a multiplier with the desired parameters.

Verification of Module Generators

In the context of Cathedral-II [DeMa90] (cf. Chapter 6), the verification of module generators has been studied. These generate specific functional units according to user-defined parameters. Since exhaustive simulation of the units generated for all valid choices of parameters is not feasible, an alternate approach to generator verification is needed. Using the Boyer-Moore theorem prover [BoMo88], the functional behavior of parameterized modules corresponding to families of generated units has been formally verified [VeCD89]. In the case study, the kernel of a parameterized ALU with 16 logic and arithmetic functions was proven correct. Parameters of the module generator comprise word length and optional inclusion of carry-bypass logic.

List Constants

A different strategy for formal synthesis is followed in [BrHu89, BrHu90], where a simple hardware description language is formalized in the first-order logic of the Boyer-Moore theorem prover. In contrast to the previous approaches, hardware descriptions are given in the form of list constants rather than logic formulae. This way, they are regarded as mere data, to which meaning is only attached through some interpreter functions in the formal logic. The gap between conventional hardware descriptions and formal proof techniques is thus reduced.

Interpreters perform checks and evaluation of circuit descriptions. They may also be used as heuristics within circuit generators. The generator functions are comparable to the parameterized hardware modules described above. Again, rather than verifying individual circuit descriptions, generator functions producing such descriptions are proven to be correct.

8.4.2 Generation of Instances

A proof-based design process as described in Section 8.3 can result in a generic implementation to be used as verified synthesis function. Along with a generic implementation comes a list of restrictions on its parameters, defining a space of valid instances. While the proof process transforming an abstract specification into the generic implementation can hardly be automated, automatic generation of valid instances is possible.

The instantiation procedure may be exemplified using the adder cascade from Section 8.3.2. The rule representing the final design state, and relating the generic implementation and its parameter restrictions with the original specification, is equivalent to the following.[26]

```
|- r <= w == true
|- n+1 < 2^(w-r) == true
-------------------------------------------
com  (cmap (n+1) (com (sR(isNat r)) (snb w)))
(com (cfst (mklist w false))
(com (cfold (n+1) (com (FArow w) (Trunc w)))
     (sbn w))) x y
|- com (cfst 0) (cfold (n+1) Add) x y
```

The parameter restrictions appear as assertions of individual premises. By means of an appropriately defined ML function, all flexible variables in this rule may automatically be instantiated to some user-defined values. Selecting for instance the parameter values $r = 4, w = 7$, and $n = 9$, leads to the following specialized rule.

```
|- 4 <= 7 == true
|- 10 < 2^(7-4) == true
------------------------------------------------------
com  (cmap 10 (com (sR(isNat 4)) (snb 7)))
(com (cfst ([false,false,false,false,false,false,false]))
(com (cfold 10 (com (FArow 7) (Trunc 7)))
     (sbn 7))) x y
|- com (cfst 0) (cfold 10 Add) x y
```

The next task is to check whether the choice of parameters meets the restrictions given in the premises. This may be automated by rewriting with an appropriate set of equations. Here, equations for basic arithmetic and orderings on natural numbers are sufficient. The selected internal word length $w = 7$,

[26]Moving hypotheses from conclusion to the assertion of a new premise is a reversible, soundness preserving transformation.

however, is too small for the representation of the sums, and consequently, rewriting fails to prove the second premise.[27]

The search for an optimal solution may be mechanized, e.g., by iteratively incrementing the internal word length until fulfillment of all restrictions is provable. In more complex examples with larger parameter spaces, automatic checking of each solution proposed by some heuristic procedure against the restrictions may become computationally expensive when performed by the prover itself. In these cases, generating and evaluating corresponding test functions in ML helps to significantly increase the performance of the search procedure. In order to achieve maximal safety, the finally determined solution may then again be checked by the prover.

The presented example, although simple, is an illustration of how a theorem prover may be used to automatically derive verified implementations for particular classes of design problems. Optionally, the instantiated structure may be expanded to a flat netlist of primitive components by unrolling the recursive definitions of the combinators. Provided the primitives correspond to cells of some existing library, the way down to layout and silicon is straightforward.

8.5 Outlook

The synthesis of digital systems is a complex process in which human interaction is often both a vital necessity as well as a potential source of design errors. It is the goal of this chapter to demonstrate how higher-order logic and computer-aided theorem proving may be used to establish a safe, formal basis for symbolic reasoning about digital designs. It is aimed at introducing the fundamentals of interactive, formal synthesis in various styles. By means of an appropriate representation for the interim stages of the synthesis process, namely as sound rules of the logic, design and verification steps are integrated in the framework of formal proof. Thus, maximum evidence for the functional correctness of synthesized implementations is achieved.

Besides LAMBDA, there are other proof systems which are applied in the field of hardware specification, design, and synthesis, some of which are referenced in Section 8.4. A brief report on these and some recommendations for further reading are given in the following paragraphs.

The Boyer-Moore Theorem Prover

The Boyer-Moore theorem prover [BoMo88] is a highly-automated proof system for first-order logic without explicit quantification. It is particularly well suited

[27] Keeping r and n fixed, any selection of $w \geq 8$ will do.

for induction proofs over recursive functions and has been mainly applied to pure hardware verification tasks. More synthesis-related work involving the Boyer-Moore prover is found in [BrHu89, BrHu90] and [VeCD89].

The HOL Proof System

The HOL proof system [Gord87] is based on a higher-order logic quite similar to that of LAMBDA. Like the Boyer-Moore prover it has been applied in numerous pure hardware verification projects. A good reference to more synthesis-related work involving HOL is [Chin90].

The Interactive VERITAS$^+$ Environment

The Interactive VERITAS$^+$ Environment (IVE) [DaLH89, HaDL89] uses a higher-order logic with a more sophisticated type system including subtypes and dependent types. It allows for an explorative manipulation of proof trees in a divide and conquer style, separating the division of goals into subgoals from the necessary validation for these goal-splitting activities. The application of IVE to formal synthesis of digital systems is covered in [HaLD90].

For a thorough overview on the field, the interested reader is encouraged to refer to [Miln89, BiSu89, Clae90, Lees90, Subr91].

8.6 Problems for the Reader

1. Define two functions *sum* and *carry* of type $(bool * bool * bool) \rightarrow bool$, computing the sum and carry of three input bits. Hint: start *sum* and *carry* with a definition of the Boolean operator *xor*. Use to define a function $sumL(a, b, c)$ that takes two Boolean lists a and b of equal length together with an initial carry value c and recursively computes their sum. Consider the head of lists to be the least significant bit in the string and drop any final carry value so that the resulting list has the same length as the arguments a and b. What is the type of *sumL*?

2. Write down a component description (a predicate) of an adder for Boolean lists consuming one time unit to perform its operation. Use the function *sumL* from above and try to keep the predicate as short as possible. Hint: use the abstract *Fun2* component from Section 8.3.1.

3. The rule for Boolean induction (case analysis), the axiom on reflexivity of equality, a rule to make implicit all-quantification of free variables explicit, and four equalities on Boolean conjunction read as follows.

```
H |- P#(false)
H |- P#(true)                                  H |- forall x. P#(x)
--------------------   -----------   --------------------
H |- forall b. P#(b)   H |- x == x   H |- P#(x)
```

```
true && b == b    false && b == false
b && true == b    b && false == false
```

Use only these to prove the following sequent in a backwards style, starting from the corresponding tautology.

```
H |- a && b == b && a
```

4. Show that the simple parity checker from Section 8.2.3 really does what the accompanying text claims it to do. Hint: recall the specific behavior of the delay element at time $t = 0$.

5. Give a definition for the abstraction function *listToNat* from Section 8.2.4. The head of a Boolean list may be interpreted as the least or most significant bit of a binary representation of natural number. Which interpretation do you prefer and why?

6. Consider the two-valued ("pressed," "not-pressed") signal delivered from a mouse button at a rate of 1 MHz. Suppose the smallest time unit of interest at the user level to be 1/10 of a second. Try to formalize an abstraction of the signal at the 1 MHz rate to a signal of mouse button events, having values such as "NoEvent," "Click," "DoubleClick," "Pressed," and "Released," at the user level of time, i.e., at a rate of 10 Hz.

7. Write down a predicate defining a two-to-one-multiplexor and figure out the core rule of a corresponding introduction tactic. Hint: take a look at the *MulIntro* rule in Section 8.3.1.

8. Prove the theorem

```
---------------------------
com (sR rst) F x y |-  F x y
```

used in Section 8.3.2 to introduce a constraining pseudo-component *sR*.

9. Prove the associativity theorem for the *com*-combinator.

```
-------------------------------------------
com F (com G H) x y |- com (com F G) H x y
```

10. The following rules are equivalent representations of the same valid inference in the logic.

```
A1 |- B1
A2 |- B2
--------     ------------------------------------
A  |- B      (A1 ->> B1) /\ (A2 ->> B2) |- A ->> B
```

    Show that introducing a new hypothesis P into each premise and the conclusion of the left representation yields a sound rule.

11. Consider the following sequent.

```
a ->> b |- P#(a) ->> P#(b)
```

    Find a context P for which it cannot be proven as a theorem.

12. Show that the following sequent holds for the given context P.

```
forall x,y. Cmp1 x y ->> Cmp2 x y
|- P#(Cmp0,Cmp1,Cmp3,r1,r2) ->> P#(Cmp0,Cmp2,Cmp3,r1,r2)

P#(F1,F2,F3,x,y) := com F1 (com F2 F3) x y
```

Chapter 9

Synthesis Related Aspects of Simulation

Franz J. Rammig

9.1 Introduction

The starting point of a chapter on synthesis related aspects of simulation has to be the assertion that, at least in theory, as any other verification method simulation becomes obsolete by synthesis, because one intent of using synthesis is to produce designs that are correct by construction. So what can simulation be good for in such a context?

First of all, one has to make the distinction between *fully automatic* and *interactive* synthesis. No doubt, some kind of verification is needed for interactive synthesis, to check the manipulations the user imposes on the design. In Chapter 8 formal methods for this purpose are discussed. In contrast to those "analytical methods," in this chapter "experimental methods" are explained. But, one has to keep in mind that simulation is not *in*formal at all. Both approaches, formal verification and simulation, have benefits as well as restrictions and both methods are heavily dependent on intelligent usage. Thus, in this chapter one main emphasis is on how to use simulation properly.

Under the assumption of fully automatic synthesis, one may doubt whether there is any need for verification, especially for simulation, at all. However, using synthesis does not affect the correctness of an initial specification. Since specification is the first formal document, it can only be proven to be consistent and in accordance with certain fixed or generic rules. By no means can it be verified analytically whether or not the initial specification matches the designer's intent. Because synthesis even in the ideal case can guarantee nothing more than an implementation which is correct in relation to a given specification, the term *correctness by construction* has a meaning that reduces correctness to correctness modulo specification. It can be concluded, therefore, that independent of the synthesis strategy (fully automatic or interactive) and independent of the availability of analytical (formal) methods, the initial specification has to be verified by simulation. Using simulation, intelligent experiments have to be carried out to ensure that the specification is in accordance with the designer's intention.

In this chapter, the term "simulation" is used slightly differently from the traditional usage. Simulation, in general, is used to validate that design steps have been carried out correctly. Consequently, simulation systems in the past have been optimized for lower levels of abstraction (logic level or circuit level). When used in the context of synthesis, simulators are needed at higher levels, both for checking adequateness and performance. Most important, however, in the case of synthesis the designer has to be supported validating his or her initial specification. This implies that planning, performing, and analysis of intelligent experiments have to be supported. This is an aspect missing in most traditional simulation systems.

The question remains about the relevance of simulation at lower levels of abstraction. As already mentioned, it is needed for interactive synthesis, competing—or better cooperating—with analytical methods. In the case of fully automatic synthesis it is, in principle, obsolete. However, designers tend to distrust synthesis systems. Of course, like any other software, synthesis systems are not free of "bugs." Therefore, low-level simulation offers an opportunity to crosscheck the synthesis results. Another argument for low-level simulation can be seen in *synthesis for simulation*, when a hardware specification, unknown to be correct, may quickly be translated into a low-level implementation and efficiently be simulated at this level of abstraction. The behavior at the lower level, then, has to be retranslated to the higher level of abstraction. In the extreme case, the low-level simulation may be performed on a so-called *simulation engine* (hardware accelerator), which has a multiprocessor architecture and whose instruction set is optimized for the execution of logic-level simulation, i.e., synthesis can be used as a compiler frontend for the respective simulation engine. This method has its analogy in software design, where it has been observed that low-level instruction sets allow a more efficient execution of high-level programs (after compilation) than architectures that support the high-level languages directly. Therefore, experiments (tests) with high-level programs are carried out after compilation at the lower level, and only the diagnostics are retranslated to the higher level of abstraction.

The third argument for simulation of synthesized designs at lower levels of abstraction derives from the fact that the transformation from high- to low-level descriptions provided by synthesis is by no means unique. Synthesis results can vary greatly in low-level characteristics, such as delays and power consumption, which are not or only incompletely specified in the input description. Low-level simulation can be used in this situation to evaluate the quality of synthesized implementations.

To conclude, simulation is needed within the context of synthesis, both for automatic and interactive synthesis. There are good reasons to consider simulation at high and low levels of abstraction. Simulation systems, therefore, have to support the planning, execution, and analysis of intelligent experiments.

9.2 Multi-Level Modeling

9.2.1 Levels of Abstraction

As described in Chapter 1, a widely accepted scheme of levels of abstraction and domains of description is used in the area of digital system design. In the context of simulation, the behavioral domain—especially the modeling of time and data—is of main interest.[1]

At the *circuit level* behavior of electronic circuits built from resistors, capacitors, etc., is modeled over the time axis. This is achieved by a system of differential equations, i.e., both the time axis and the observable data (currents, voltages) are represented by a continuous value domain. There are some simulators at this level of abstraction, with SPICE [Nage87] in numerous versions being the most widely used.

The *logic level* has a long tradition in digital system design. This level has solid mathematical foundation in Boolean algebra. However, with Boolean algebra only the timeless behavior can be modeled. Some additional concepts have to be considered in order to cover time as well. In the ideal case the value domain[2] is restricted to Boolean values 0 and 1 while the time domain remains continuous, but the problem of uncertain values necessitates the introduction of additional values. As a consequence, in most cases the underlying algebra is no longer Boolean. The operators, however, are always Boolean (logic) operators. Because of the long tradition of using the logic level for simulation purposes, numerous simulators are available for this level of abstraction. They differ in

[1] A level of abstraction, which plays a major role in simulation, located between circuit and logic level, is the switch level, where circuits are modeled as networks of ideal switches and capacitors. As this level plays no role in synthesis, it is not further discussed here, and is not included into the representation of the Y-chart of Chapter 1.

[2] The term "domain" is used here in another sense than in the context of the Y-chart where it denotes the view or aspects of a design object one is interested in. Here, domain refers to a range of values.

their definition of value domains and timing models. HILO [HILO91], CADAT [CADA91], DISIM [DISI91] are some examples of commercially available simulators.

At the *register-transfer level* a specific mode of operation is assumed. Components continuously observe specific, associated conditions. Whenever the condition associated with a component becomes true, this component performs its specific operation. Operations can be interpreted as a transfer of data between registers; the data transferred may be modified during this operation. The value domain at this level of abstraction is given by (uninterpreted) bitstrings, whereas the timing model is that of counting clock ticks. Thus, the time domain has become discrete as well. The register-transfer level is very helpful for well-structured synchronous designs, since it enforces to some extend such a design style. Simulation at this level of abstraction has been mainly studied in academic institutions [Borr81, Chu74, DuDi75, Hart77].

At the *algorithmic level* the reactive point of view of the register-transfer level is inverted to an imperative one. While at the register-transfer level the system is seen from the point of view of the individual components, at the algorithmic level the controller's point of view is taken. In contrast to ordinary algorithmic descriptions, however, concurrency plays an important role in hardware design and, therefore, at this level of abstraction. In contrast to the register-transfer level, where it is precisely specified which conditions trigger an operation, at the algorithmic level this information is not explicitly provided. Only the discrete point of time at which an operation has to be carried out, is identified. The domain of values can be defined in various ways, but is usually restricted to bitstrings with interpretations attached (e.g., two's complement integer, IEEE standard floating point number, ASCII character). The timing model at this level of abstraction is either that of counting clock ticks as at the register-transfer level or a purely causal one. In the latter case, simply a causality structure is assumed as it is known from general programming languages. Only a few commercial simulators are available for this level of abstraction [DACA90, Thom91].

Finally, at the *system level* the entire system is considered as a set of cooperating processors. In this context the term processor is used in a broader sense, namely to denote a subsystem which exports certain services (instructions). Both the value domain and the timing model are purely symbolic at this level of abstraction. Types can be defined freely and associated with arbitrary semantics. The sequence of operations is only defined by causality relations. The system level is supported by only very few commercial simulators. More simulators working at this level of abstraction will evolve as synthesis approaches this level of abstraction.

9.2.2 Modeling Concepts

In the context of synthesis three major modeling concepts have to be considered by a simulation system: imperative models, reactive models, and stimulated equations (strictly functional models).

Imperative models originate from algorithmic programming languages for von Neumann type processors. High-level synthesis in most cases starts from a high-level description in imperative style (cf. Chapter 6). Therefore, the support of this model is essential in the context of this book. While in traditional imperative programming languages a strictly sequential operation is assumed, this concept has to be generalized to concurrent algorithms in the domain of hardware design. This is necessary, because hardware systems are concurrent by nature, i.e., performance is gained by the use of parallelisms. An obvious way for generalizing sequential algorithms to concurrent ones is the approach of communicating sequential processes (CSP) [Hoar78, Hoar85]. Therefore, this model may serve to introduce this concept.

In CSP the entire system is represented as a system of concurrently active processes, where each process is strictly sequential. This concept, perfectly implemented in the programming language OCCAM [May83], can also be identified in VHDL (cf. Chapter 2). In CSP, a single process is represented using the constructs *assignment*, *guarded command*, and *iteration*. The processes are completely independent, i.e., they are not allowed to share any resources other than communication channels. In this aspect, VHDL is less strict, because shared resources are allowed. Any conflicts, however, have to be resolved statically using *resolution functions* in VHDL.

Events and processes. Objects to be specified are described using *events*. Such an event is treated as an atomic action. Typically, an event is the assignment of a value to a variable. The set of all events used to describe an object is called *alphabet* of this description (of this object). A *process* is an arbitrary behavior of an object that can be expressed using the alphabet of this object.

Sequential execution. Let a be an event and P a process. By "**seqbegin** a; P **seqend**" it is denoted that event a has to happen first, before process P can be started. By definition, a is a process, and if P is a process, "**seqbegin** a; P **seqend**" is a process, too.

Recursion. Let P be a process. By "**while** true **do** P" it is denoted that P has to be repeated infinitely. If P is a process, "**while** true **do** P" is a process, too. Let con be a binary variable. By "**while** con = true **do** P" it is denoted that P has to be executed as long as con has the value "true." If P is a process, "**while** con = true **do** P" is a process, too.

Case distinction. Let $P_1, \ldots, P_n$ be processes, *cnt* a variable with domain $\{c_1, \ldots, c_n\}$. By "**case** *cnt* **of** $c_1 : P_1; c_2 : P_2; \ldots; c_n : P_n$ **caseend**" it is denoted that only that P_i has to be executed whose c_i is the current value of *cnt*. If $P_1, \ldots, P_n$ are processes, "**case** *cnt* **of** $c_1 : P_1; c_2 : P_2; \ldots; c_n : P_n$ **caseend**" is a process, too.

Input/output. Let *chan* be a special variable of type **channel**, denoting a communication channel. Let *var* be an arbitrary variable. By *chan!var* an output operation is denoted. The actual value of *var* is transmitted via channel *chan*. The operation is not completed until a concurrently active process (see below) has read this value (rendezvous concept). By *chan?var* an input operation is denoted. The actual value taken from the channel *chan* is assigned to *var*. This operation can be carried out only if the channel is not empty. Initially, a channel is empty and a non-empty channel is emptied by executing an input operation on it. Like assignments, input/output operations are treated as events. By definition, neither an input operation nor an output operation can be executed if no concurrently active additional process exists. It is assumed that a function **test**(*chan*) exists for each channel *chan*. It returns **true** if *chan* is not empty, **false** otherwise. This function can be used only within control expressions and is not treated as an input operation (does not empty a channel and does not block on an empty channel).

Sequential process. A *sequential process* is an arbitrary sequence of the above constructs, and nothing else.

Concurrent process. Each sequential process is also a *concurrent process.* Let P be a sequential process and C a concurrent one. By "**conbegin** P; C **conend**" it is denoted that P and C are executed concurrently, i.e., P and C are initiated at the same point of time and, then, run completely independent as long as they do not communicate via a shared channel. The process "**conbegin** P; C **conend**" terminates when the last one of the contained processes P and C terminates. As mentioned above, processes within a concurrent process are not allowed to share resources other than channels. If P is a sequential process and C is a concurrent one, "**conbegin** P; C **conend**" is a concurrent process, too.

Locally, VHDL has a similar point of view. However, internally all processes are strictly sequential and are treated as looping infinitely, the loop being interrupted in each cycle until events specified in the *sensitivity list* of the process occur. Therefore, VHDL follows a reactive modeling concept.

With the imperative model, a system is observed from the controller's point of view. In this sense, a controller is an object that causes other objects to execute operations in a well defined (partial) order. The reactive model inverts this point of view. In this case, the entire system is observed from the position of the controlled objects. From this (local and partial) point of view the partial ordering of the operations has no meaning. For any specific object, it is relevant only that a certain action has to be executed whenever a certain condition becomes true. Such an action may include a modification of certain conditions, thus implicitly causing other objects to execute operations. The descriptive power of the reactive model is the same as that of imperative modeling. However, the global control concepts of a description are no longer visible. The following simple example may help to illustrate this (VHDL notation):

```
process begin
        operation_0;
        operation_1;
        operation_2;
end process
```

This is equivalent to

```
signal sequence: Natural := 0;
-- auxiliary signal, initialized to 0
process begin
        operation_0;
        sequence <= 1;
        wait until sequence = 0;
end process;
process begin
        operation_1;
        sequence <= 2;
        wait until sequence = 1;
end process;
process begin
        operation_2;
        sequence <= 0;
        wait until sequence = 2;
end process;
```

After the start of the system all three processes are continuously active. Whenever the condition "sequence = i" becomes true the respective process is (re-) started and the two actions "operation_i" and sequence "$\Leftarrow ((i+1) \bmod 3)$" are executed.[3] The sequence in which the processes are written down in the descriptive text is of no influence.

The reactive model seems to be quite natural for hardware descriptions, because it takes a structural point of view with behavior attached, and it is the basic modeling concept at the register-transfer level. At this level of abstraction, the basic operation is the storage of information (possibly after modification) into a destination register under certain conditions. This is nothing else than inverting the point of view of microprogramming, or looking at a microprogram with the controlled object's eyes. The global semantic concept of VHDL is, as such, a reactive one.

What happens if the conditions in a reactive model are always true? In this

[3] Due to technical reasons, VHDL requests a resolution function to be defined for the signal sequence. For reasons of readability, this function is omitted here.

case, the attached operations have to be executed continuously. Restricting reactive models to those for which every condition is always true, does not restrict the modeling power. Since, obviously, logic-level implementations exist of all digital systems described at the RT-level, this may not be too surprising. The same can be shown within the modeling context as well:

A reactive model is represented by a set

$$R = \{\textbf{at } c_i \textbf{ do } t_i := f_i(s_i) | i = 1 : n\}$$

, i.e., a set of "guarded commands" with the meaning that whenever c_i becomes true, the value calculated by function f_i on its arguments s_i is assigned to t_i. It is not requested per se that the destination ranges t_i are disjoint, but it is always possible to rewrite R in such a way that this becomes true. In this case, for each destination object the sequence of its values is defined. On the other hand, each f_i can be modified in such a way that the executability condition c_i becomes an argument of this function:

$$f_i'(s_i, c_i) := \begin{cases} f_i(s_i) & \text{if } c_i = \text{true} \\ t_i & \text{else} \end{cases}$$

But then, the prefix "**at** c_i **do**" becomes obsolete and can be omitted. By the procedure described above a system of equations is obtained. In the stable state all equations contained are in equilibrium (the system of equations is solved). By changing the value of an arbitrary object, the equilibrium is distorted, resulting in an instable global state. As a reaction, the system of equations trys to restabilize (recalculate a solution). Of course, the existence of a global stable state (a solution) is not always guaranteed. In that case, the system will continuously attempt to reach a stable state, without success.

In the above it is assumed that the left-hand side and the right-hand side have different meanings (destination and source). In this case, a system of unidirectional value assignment is modeled. In a system like this a well-defined flux of distortions through the system exists, similar to a wavefront. If there are no different roles, the assignment symbol becomes an equality sign. Of course, the flux of distortions then becomes much more complex. Both models, unidirectional and bidirectional flux of distortions make sense in hardware modeling. This kind of functional modeling is the most natural one at lower levels of abstraction, as well. Unfortunately, it is not at all supported by VHDL.

9.2.3 Modeling of Time

To model the timing of a system is of crucial importance, even though it adds to the complexity of simulation. In addition, the problem exists that in real systems a value at a certain point of time is not only dependent on calculations based upon arguments at exactly the same point of time, but on a history of values over the time axis. In the range of abstractions to be considered in this context time dependencies can be reduced to delays, which are either inertial or inertial free delays. This concept can be introduced quite easily at the logic level and can be generalized to other levels of abstraction, afterwards. The following discussion is based on the assumption that the observable values are modeled by a set of five elements $\{1, 0, p, n, u\}$ where 1 and 0 stand for stable values, n and p for negative or positive edges, and u for an uncertain value [Ramm80a]. This value model is used in this section for simplicity reasons, only; for more advanced models the reader is referred to [Coel89, Dani70, Haye86, KrAn90, LeRa83].

Definition 9.1 (Real continuous signal set, RCS) *Let $[L, R] \subset \mathbb{R}$ be a real interval. A set of differentiable real functions $RCS \subset [L, H]^{\mathbb{R}}$ is called real continuous signal set, $rcs \in RCS$ real continuous signal. L and H stand for the lowest and highest voltage available, for example, in the particular technology.*

This infinite set of observable values has to be mapped onto a set of only five elements.

Definition 9.2 (Five valued continuous signal set, $FV^{\mathbb{R}}$) *Let $H > TH > TL > L$ be reals, $FV := \{0, p, n, 1, u\}$. $FV^{\mathbb{R}}$ is called five valued continuous signal set, $fcs \in FV^{\mathbb{R}}$ is called model of a $rcs \in RCS :\Leftrightarrow fcs := f(rcs)$ with*

$$
\begin{aligned}
\forall t \in \mathbb{R} : \ & f(rcs_t) = 1 :\Leftrightarrow rcs_t \geq TH \\
& f(rcs_t) = 0 :\Leftrightarrow rcs_t \leq TL \\
& f(rcs_t) = p :\Leftrightarrow TH > rcs_t > TL \ \wedge \ drcs_t/dt > 0 \\
& f(rcs_t) = n :\Leftrightarrow TH > rcs_t > TL \ \wedge \ drcs_t/dt < 0 \\
& f(rcs_t) = u :\Leftrightarrow \text{otherwise}
\end{aligned}
$$

To achieve a further abstraction, the time set is restricted to a discrete one:

Definition 9.3 (Five valued discrete signal set FV^T) *Let T be a countable subset of $\mathbb{R}$ with metric and ordering inherited from $\mathbb{R}$. FV^T is called five valued discrete signal set, $fds \in FV^T$ five valued discrete signal.*

Let $\ldots, t_{i-1}, t_i, t_{i+1}, \ldots$ denote adjacent elements (points of time) of T. $fds \in FV^T$ is called model of a $fcs \in FV^{\mathbb{R}} :\Leftrightarrow fds = f(fcs)$ with:

$$
\begin{aligned}
\forall t_i \in T : \forall w \in \{0, p, n, 1\} : & \\
f(fcs_{ti}) &= w :\Leftrightarrow \forall t_{i-1} < t \leq t_i : fcs_t = w \\
&= u \ \text{otherwise}
\end{aligned}
$$

With the above abstraction, signals can be represented by traces [Snep85]. For instance, a 50 MHz clock signal with a transition time of 4 ns for the raising edge and 6 ns for the falling edge is modeled by the sequence[4]

$$(pppp11111nnnnnn00000)^*.$$

It is a basic principle for the modeling of real time behavior to distinguish between purely functional aspects and timing.

Definition 9.4 (Timeless Boolean function) *Let a_t denote the value of a signal at point of time t ("projection at t"), as already used above. $h : (FV^T)^n \rightarrow FV^T$ is called timeless Boolean function*
$:\Leftrightarrow$
$\exists (h' : FV^n \rightarrow FV) : \forall a \in (FV^T)^n : \forall t \in T : (h(a))_t = h'(a_t)$

The purely functional aspects are modeled perfectly by Boolean functions. In addition, timing effects have to be considered:

a) **Delay:** It is assumed that a value calculated by an operation comes to effect only some time period after the causing arguments have been applied. No distortion of the waveform takes place.

b) **Inertia:** Usually, switching elements react only to signals that are stable for a specific minimal period of time.

c) **Edge triggering:** Certain switching elements react on transitions rather than on stable values. Usually, a minimal slope requested for the transition has to be accepted.

d) **Transition time modification:** Switching elements may either decrease the slope of transformed signals, or act as pulse sharpeners.

In the following, a number of elementary time dependent functions are introduced. They can be used to compose realistic models of switching elements.

Definition 9.5 (Transport delay function)
Let $TD := (t_{PLH}, t_{PHL}) \in \mathbb{N}_0^2, t_{PLH} + t_{PHL}$ finite, a transport delay specification, $md := max\{t_{PLH}, t_{PHL}\}$.

The function $tdf_{TD} : FV^T \rightarrow FV^T$ is called transport delay function with respect to TD
$:\Leftrightarrow$
$\exists (tdf' : FV^{md} \rightarrow FV) : \forall a \in FV^T : \forall t \in T :$
$\quad (tdf_{TP}(a))_t = tdf'(a_t, \ldots, a_{t-md})$ *with*

[4] "*" denotes repetition.

$$tdf'(a_t,\ldots,a_{t-md}) = \begin{cases} a_{t-t_{PLH}} & \text{if } a_t \in \{1,p\} \vee (a_t = u \wedge t_{PLH} \geq t_{PHL}) \\ a_{t-t_{PHL}} & \text{otherwise} \end{cases}$$

Transport delay is offered by VHDL, but the distinction between t_{PLH} and t_{PHL} (between positive and negative transitions) is not supported as a primitive.

Definition 9.6 (Inertial delay function)
Let $ID := (t_{PLH}, t_{PHL}, t_0, i_1) \in \mathbb{N}_o^4$, $t_{PLH}+t_{PHL}+i_0+i_1$ *finite,* $i_0 < t_{PHL}, i_1 < t_{PLH}$ *be an inertial delay specification. (i specifies how long a signal has to be stable to be accepted.)*

mid $:= \max\{t_{PHL}, t_{PLH}\} + \max\{i_0, i_1\}$.

The function $idf_{ID} : FV^T \rightarrow FV^T$ *is called* <u>*inertial delay function*</u> *with respect to* ID

$:\Leftrightarrow$

$\exists(idf' : FV^{\text{mid}} \rightarrow FV) : \forall a \in FV^T : \forall t \in T :$
$(idf_{ID}(a))_t = idf'(a_t, \ldots, a_{t-\text{mid}})$ *with*

$$idf'(a_t,\ldots,a_{t-\text{mid}}) = \begin{cases} pr_{t-t_{PLH}}(a) & \text{if } a_t \in \{1,p\} \vee (a_t = u \wedge t_{PHL} \geq t_{PLH}) \\ pr_{t-t_{PHL}}(a) & \text{otherwise} \end{cases}$$

with (for $d \in \{t_{PLH}, t_{PHL}\}$*) :*

$$\begin{aligned} pr_{t-d}(a) = u \Leftrightarrow \quad & i) \; a_{t-d} = u \\ \vee\; & ii) \; a_{t-d} = 0 \wedge \exists t > t' > t-d > t" > t - mid : \\ & \quad a_{t-d} \neq a_{t'} \wedge a_{t-d} \neq a_{t"} \wedge t' - t" \leq i_0 \\ \vee\; & iii) \; a_{t-d} = 1 \wedge \exists t > t' > t-d > t" > t - mid : \\ & \quad a_{t-d} \neq a_{t'} \wedge a_{t-d} \neq a_{t"} \wedge t' - t" \leq i_1 \\ = a_{t-d} \quad & \text{otherwise} \end{aligned}$$

Example. Consider the inertial delay specification $id1 = (3, 4, 3, 2)$. Then, the following transformation can be observed from signal a to signal b by idf_{id1}:

```
a  =  00000p1n00000p11111n0p1111
b  =  ???00000pun00000p11111nup1111
```

Inertial delay is supported by VHDL as a primitive; however, no distinction between different values is made.

Edge-triggering needs a bit more consideration. In a discrete model the slope of a transition is represented by the period a value p or d can be observed. In the case of a period longer than a certain threshold, the slope is assumed to be too low for the edge to be accepted as trigger.

Definition 9.7 (Edge triggered inertial delay function)
Let $ED := (t_{PLH}, t_{PHL}, t_0, i_1, ep, en) \in \mathbb{N}_0^6$, $(t_{PLH}, t_{PHL}, i_0, i_1)$ *be a valid inertial delay specification,* $ep + en$ *finite be an edge triggered inertial delay specification. (By* ex *the longest time-period is denoted during which a signal has the transition value* x *to be accepted as trigger,* 0 *stands for "no edge triggering").*
$med := \max\{t_{PLH}, t_{PHL}\} + \max\{i_0, i_1, ep, en\}$

The function $edf_{ED} : FV^T \to FV^T$ *is called* <u>*edge triggered inertial delay function*</u>

$:\Leftrightarrow$

$\exists(edf' : FV^{\text{med}} \to FV) : \forall a \in FV^T : \forall t \in T :$

$(edf(a))_t = edf'(a_t, \ldots, a_{t-\text{med}})$ *with*

$$edf'(a_t, \ldots, a_{t-\text{med}}) = \begin{cases} pr_{t-t_{PLH}}(a) & \text{if } a_t \in \{1, p\} \vee (a_t = u \wedge t_{PHL} \geq t_{PLH}) \\ pr_{t-t_{PLH}}(a) & \text{otherwise} \end{cases}$$

with (for $d \in \{t_{PLH}, t_{PHL}\}$*):*

$$\begin{array}{rll} pr_{t-d}(a) = u \Leftrightarrow & i) & a_{t-d} = u \\ & \vee\ ii) & a_{t-d} = 0 \wedge \exists t > t' > t - d > t" > t - med : \\ & & a_{t-d} \neq a_{t'} \wedge a_{t-d} \neq a_{t"} \wedge t - t" \leq i_0 \\ & \vee\ iii) & a_{t-d} = 1 \wedge \exists t > t' > t - d > t" > t - med : \\ & & a_{t-d} \neq a_{t'} \wedge a_{t-d} \neq a_{t"} \wedge t - t" \leq i_1 \\ & \vee\ iv) & ep \neq 0 \wedge a_{t-d} = p \wedge \exists t > t' > t - d > t" > t - med : \\ & & \forall t' \geq t^+ \geq t" : a_{t^+} \in \{a_{t-d}, u\} \wedge t' - t" \geq ep \\ & \vee\ v) & en \neq 0 \wedge a_{t-d} = n \wedge \exists t > t' > t - d > t" > t - med : \\ & & \forall t' \geq t^+ \geq t" : a_{t^+} \in \{a_{t-d}, u\} \wedge t' - t" \geq en \\ = 1 \Leftrightarrow & vi) & ep \neq 0 \wedge a_{t-d} = p \wedge \exists t > t - d > t' > t - med : \\ & & a_{t-d+1} = 1 \wedge a_{t'} = 0 \wedge t - d + 1 - t' < ep \\ & & \wedge pr_{t-d+1}(a) \neq u \wedge pr_{t'}(a) \neq u \\ & \vee\ vii) & en \neq 0 \wedge a_{t-d} = n \wedge \exists t > t - d > t' > t - med : \\ & & a_{t-d+1} = 0 \wedge a_{t'} = 1 \wedge t - d + 1 - t' < en \\ & & \wedge pr_{t-d+1}(a) \neq u \wedge pr_{t'}(a) \neq u \\ = 0 & & \text{otherwise} \end{array}$$

Comments. By inertia the delayed value is mapped on u whenever it is not stable long enough (ii, iii). The edge-triggering maps a transition to value 1 at a single point of time at the end of the edge if the slope of the transition is "high enough" (vi, vii). Smooth edges are mapped on u, too (iv, v). All other cases are mapped to 0.

Example. Consider the edge triggered inertial delay specification $ed_1 = (4, 4, 3, 2, 2, 0)$. Then the following transformation occurs from signal a to signal b by edf_{ed1}:

```
a  =  00000p1n00000ppp11111nnn00000pp11111
b  =  ????000000u000000uuu0000000000000010000
```

As the last part of the assembly kit, transport delay functions are defined that effect the "slope" either by increasing or decreasing it.

Definition 9.8 (Smoothing transport delay function)
Let $SD := (t_{PLH}, t_{PHL}, su, sd) \in \mathbb{N}_0^2 \times \mathbb{Z}^2, t_{PLH} + t_{PHL} + |sv| + |sd|$ *finite,* $su < t_{PLH}, sd < t_{PHL}$ *be a smoothing transport delay specification. (The values su and sd are additive constants, with the meaning that a slope represented by a sequence of n symbols x is transformed to one represented by a sequence of* $n + m$ *symbols,* $m \in \{su, sd\}$*, su affecting positive edges, sd negative ones.)*

$msd := \max\{t_{PLH}, t_{PHL}\} + \max\{su, sd\}$

$sdf : FV^T \rightarrow FV^T$ *is called* <u>*smoothing transport delay function*</u>

$:\Leftrightarrow$

$\exists(sdf' : FV^{\text{msd}} \rightarrow FV) : \forall a \in FV^T : \forall t \in T :$

$(sdf(a))_t = sdf'(a_t, \ldots, a_{t-msd})$ *with*

$$sdf'(a_t, \ldots, a_{t-msd}) = \begin{cases} pr_{t-t_{PLH}}(a) & \text{if } a_t \in \{1, p\} \vee (a_t = u \wedge t_{PHL} \geq t_{PLH}) \\ pr_{t-t_{PHL}}(a) & \text{otherwise} \end{cases}$$

with (for $d \in \{t_{PLH}, t_{PHL}\}$*):*

$$\begin{aligned}
pr_{t-d}(a) = 0 \Leftrightarrow \quad & i) \quad a_{t-d} = 0 \wedge sd \leq 0 \\
\vee\ & ii) \quad a_{t-d} = 0 \wedge sd > 0 \wedge \forall t - d - sd \leq t' \leq t - d : a_{t'} \neq n \\
\vee\ & iii) \quad a_{t-d} = n \wedge sd < 0 \wedge \exists t - d \leq t' \leq t - d - sd : \\
& \qquad a_{t'} = 0 \wedge a_{t-d-1} = n \\
= 1 \Leftrightarrow \quad & iv) \quad a_{t-d} = 1 \wedge su \leq 0 \\
\vee\ & v) \quad a_{t-d} = 1 \wedge su > 0 \wedge \forall t - d - su \leq t' \leq t - d : a_{t'} \neq p \\
\vee\ & vi) \quad a_{t-d} = p \wedge su < 0 \wedge \exists t - d \leq t' \leq t - d - su : \\
& \qquad a_{t'} = 1 \wedge a_{t-d-1} = p \\
= p \Leftrightarrow \quad & vii) \quad a_{t-d} = p \wedge su \geq 0 \\
\vee\ & viii) \quad a_{t-d} = p \wedge su < 0 \wedge \forall t - d \leq t' \leq t - d - +su : a_{t'} \neq 1 \\
\vee\ & ix) \quad a_{t-d} = p \wedge su > 0 \wedge \exists t - d \leq t' \leq t - d + su : a_{t'} = p \\
\vee\ & x) \quad a_{t-d} = p \wedge su < 0 \wedge \exists t - d \leq t' \leq t - d - su : \\
& \qquad a_{t'} = 1 \wedge a_{t-d-1} \neq p \\
= n \Leftrightarrow \quad & xi) \quad a_{t-d} = n \wedge sd \geq 0 \\
\vee\ & xii) \quad a_{t-d} = n \wedge sd < 0 \wedge \forall t - d \leq t' \leq t - d - +sd : a_{t'} \neq 0 \\
\vee\ & xiii) \quad a_{t-d} = n \wedge sd > 0 \wedge \exists t - d \leq t' \leq t - d + sd : a_{t'} = n \\
\vee\ & xiv) \quad a_{t-d} = n \wedge sd < 0 \wedge \exists t - d \leq t' \leq t - d - sd : \\
& \qquad a_{t'} = 0 \wedge a_{t-d-1} \neq n \\
= u \quad & \qquad \text{otherwise}
\end{aligned}$$

Example. Consider the smoothing transport delay specification $sd_1 = (2,2,-1,1)$. Then, the following transformation exists from signal a to signal b by sdf_{sd1}:

```
a  =  0000ppp1111nnn0000pp1111nn00p11n00
b  =  ??0000pp11111nnnn000p11111nnn0p11nn0
```

In VHDL, neither edge triggered inertial delay functions nor smoothing transport delay functions are offered as primitives. They may, however, be programmed by a user. The four delay functions defined above may be replaced by other ones if different effects have to be modeled. In any case, they may serve as typical examples of such functions. Using such delay functions as a kit, quite realistic real time switching functions can be defined. As an example, this is achieved by composing a smoothing transport delay function, a timeless Boolean function and n edge triggered inertial delay functions. Other combinations are possible as well, but it seems to be the most natural approach to locate smoothing at the output of a switching element, while its inertia and edge sensing are associated with its inputs.

Definition 9.9 (Real time switching function)
Let $SD := (ot_{PLH}, ot_{PHL}, su, sd)$ be a smoothing transport delay specification, $ED_i := (it_{PLH}, it_{PHL}, t_0, i_1, ep, en)_i, i = 1 : n$ edge triggered inertial delay specifications, sdf_{SD} a smoothing transport delay function, edf_{ED_i} edge triggered inertial delay functions ($i = 1 : n$), and f a timeless Boolean function with n arguments.

$$mdi := \max_{1:n}\{it_{PLH_i}, it_{PHL_i}\}, mdo := \max\{ot_{PLH}, ot_{PHL}\},$$
$$min := \max_{1:n}\{i_{0_i}, i_{1_i}, ep_i, en_i\}, ms := \max\{su, sd\},$$
$$his := mdi + mdo + min + ms$$

$rsf : ((FV)^T)^n \to FV^T$ *is called* *real time switching function*

$:\Leftrightarrow$

$\exists rsf' : ((FV)^n)^{his} \to FV : \forall a \in ((FV)^T)^n : \forall t \in T :$
$(rsf(a))_t = rsf'(a_t, \ldots, a_{t-his})$ *with*

$$\begin{aligned} rsf'_t &= sdf_{SD}\ (f_t, \ldots, f_{t-mdo}) \\ &= sdf_{SD}\ (f(\ edf'_1(a1_t, \ldots, a1_{t-mdi-min}), \ldots, \\ &\qquad\qquad edf'_1(am_t, \ldots, am_{t-mdi-min}) \\ &\qquad), \ldots, \\ &\qquad f(\ edf'_1(a1_{t-mdo}, \ldots, a1_{t-his}), \ldots, \\ &\qquad\qquad edf'_m(am_{t-mdo}, \ldots, am_{t-his}) \\ &\qquad) \\ &\quad) \end{aligned}$$

The approach presented above allows for relative precise modeling of timing effects. It may easily be modified to support logics with another value-set, or to describe other timing effects.

The question arises, whether it makes sense to model the timing behavior in such a precise way. Definitely, it is not necessary at the algorithmic level and at register-transfer level. It can even be omitted at the logic level as long as strictly synchronous designs with a well calculated clocking rate are used. At the algorithmic level, in most cases a causal timing is used, i.e., only the relative ordering of actions is specified. For this purpose, algorithmic constructs or explicit causality structures using preconditions/postconditions are sufficient.

The typical timing model at the register-transfer level is the counting of clock ticks. All other actions are related to these clock ticks. This concept can be modeled very easily. In a strictly synchronous design this relation to clock ticks exists at the logic level as well. However, for performance reasons many designers tend to use less strict synchronization schemes. Hence, whenever interactive synthesis has to be supported by simulation, certain timing tricks used by a designer have to be considered. To handle these timing tricks in analytical methods is very difficult, though attempts are made to cover them, e.g., [SaBo91]. Therefore, time accurate modeling is a domain, in which simulation really can prove its strength. However, accurate timing costs a lot of simulation performance. Clever designers will, therefore, model only as accurately as really necessary, and clever implementors of simulation systems will design adaptive algorithms that model only as accurately as needed for the actual design being simulated.

In this section an abstract model of timing has been introduced. Hardware description languages are intended only to specify the timing parameters. In most cases, the point of view is different from the one used in this section. Timing has been introduced by defining current values as a function of past sequences of values (backward oriented modeling) while in most hardware description languages timing is expressed by scheduling future events (forward oriented modeling). Backward modeling is mathematically easier to describe. It allows for specifying timing simply by functions while forward oriented modeling needs a procedure of scheduling and, possibly, rescheduling of events. CONLAN [Pilo83] is one of the few hardware description languages completely bound to backward oriented modeling. VHDL, on the other hand is mostly forward oriented, but supports backward oriented modeling to a limited extend as well.

9.3 Simulation Techniques

The task of a simulation algorithm comprises the effort to map a modeling concept (or a variety of modeling concepts) onto the architecture of a host computer. Of course, the most efficient simulation is achieved if the architecture of the host computer is identical or at least similar to the modeling concept to be supported.

This is the basic idea of a class of dedicated simulation machines (hardware accelerators) [BDPV88, HaFi85, HaHE90, KoNT90]. Another class of such machines makes use of pipelining to accelerate sequential algorithms.

In most cases, however, a conventional von Neumann computer has to be used as host computer for simulation. This case is discussed in the following. The main problem of mapping onto a strictly sequential machine is the high degree of parallelism usually contained in concepts for hardware implementations.

There are three main techniques to tackle the problem:

- *Streamline Code Simulation* (SCS)
- *Equitemporal Iteration* (EI)
- *Critical Event Simulation* (CES)

These approaches are discussed in the following sections.

9.3.1 Streamline Code Simulation

The basic idea of this approach is to generate code from the hardware description that can be executed directly by the host computer. Therefore, this technique is often called *compiled mode simulation*. Of course, it is possible to generate directly executable codes for any modeling technique, but SCS in its strict sense, without any interpretative and scheduling components, usually is restricted to

- continuous evaluation as modeling concept,
- combinational or strictly synchronous circuits,
- models for which timing information is not important.

The classical application area of SCS is the simulation of combinational circuits at the logic level. This example, therefore, is presented first.

A combinational circuit can be represented as a *directed acyclic graph* (DAG). The nodes of the DAG represent the gates of the circuit, while each connection from a gate output to a gate input is represented by an edge of the DAG (cf. Figure 9.1).

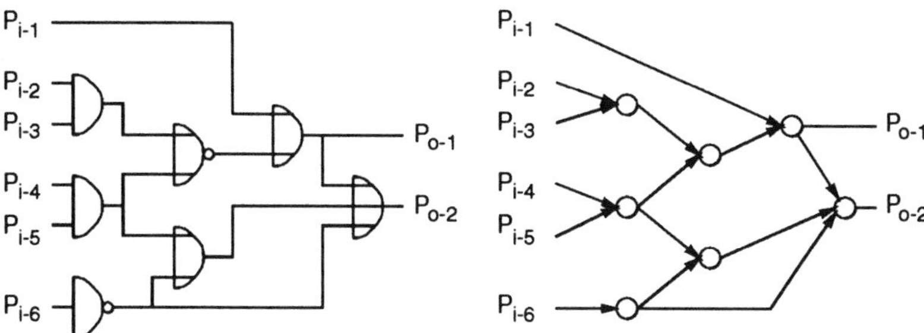

Figure 9.1: Combinational circuit and its representation as a DAG.

It is a straightforward task to semi-order the nodes of a DAG with respect to the length of the longest path from a node to a primary input. This technique is called *levelizing*:

1. Assign to each primary input in_i : $level(in_i) = 0$
2. Assign to each other node n_j : $level(n_j)$
   $= (1 + \max\{level(n_k) | \exists \ edge \ from \ n_k \ to \ n_j\})$

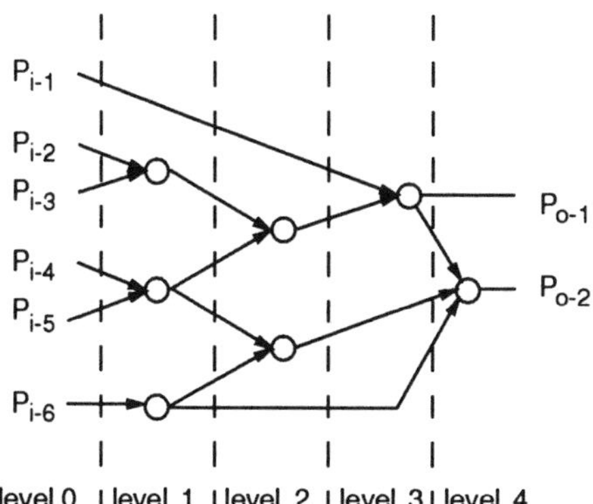

Figure 9.2: Levelized DAG from Figure 9.1.

Then, the levels may be interpreted as follows:

i) There is no influence of a node at level i on a node at level $j, j < i$.

ii) There is no cross-influence between nodes at the same level.

iii) A node at level i may (but is not required to) influence any node at level $k, k > i$.

By levelizing, therefore, a dependency relation is introduced for the circuit. This is a further abstraction of the dependency relation already given by the DAG. It is precise enough to define a save sequence of calculations at the different nodes, i.e., a sequence in which no values are referred to before they have been calculated: it is only necessary to arrange the code in accordance to ascending levels of the represented nodes. The sequence within a level is arbitrary because no interdependence between nodes of the same level exists.

In the case of logic-level simulation, the code needed for a gate is given by only a few instructions of the target computer. The connecting nets are represented as variables (memory locations in the main or virtual memory of the host computer). The example used in Figure 9.1 may be translated into the PASCAL-like code shown in Figure 9.3:

```
var pi_1, pi_2, pi_3, pi_4, pi_5, pi_6: word;
    int_1, int_2, int_3, int_4, int_5: word;
    po_1, po_2: word;
begin
 { level 1 computations }
 int_1  := pi_2 and pi_3;
 int_2  := pi_4 and pi_5;
 int_3  := not pi_6;
 { level 2 computations }
 int_4  := int_1 nor int_2;
 int_5  := int_2 nor int_3;
 { level 3 computations }
 po_1   := pi_1 or int_4;
 { level 4 computations }
 po_2   := po_1 or int_5 or int_3;
end
```

Figure 9.3: Code for example used in Figure 9.1.

The PASCAL-like language used as the target code in this example can easily be replaced by the machine code of an arbitrary processor, replacing declarations by allocating memory locations and assignment statements by executable instructions of the target machine. The example above describes a calculation based on one single input pattern. It may easily be extended to process an arbitrary sequence of input patterns. By introducing simulated shift registers arbitrary sequential circuits can be simulated as well, even asynchronous ones. The latter is of minor interest in the context of synthesis as long as no tricky manual modifications are performed in an interactive synthesis environment. Usually, when SCS is applied, sequential circuits are supported using another and much simpler (but less accurate) approach: It is assumed that no timing information of granularity finer than clock ticks is required. Thus, any sequential circuit may be represented in Huffman normal form.

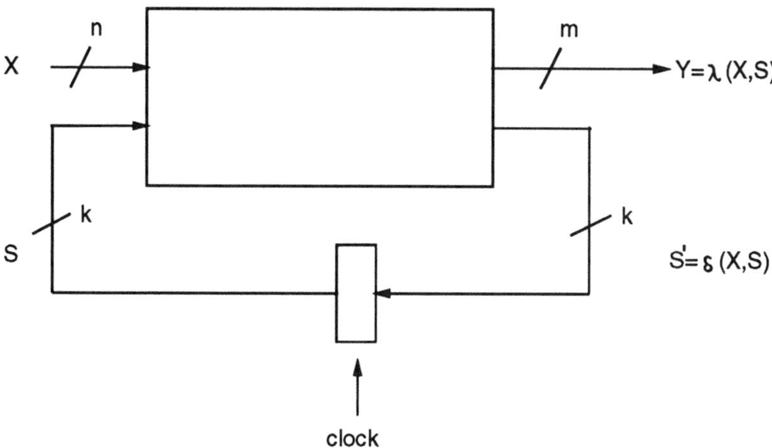

Figure 9.4: Sequential circuit in Huffman normal form.

Two combinational functions have to be calculated:

- $y = \lambda(x, s)$ (Mealy model)
- $s' = \delta(x, s)$.

These functions can be calculated (in a timeless manner) by the basic SCS approach. If one loop iteration of the algorithm is interpreted as one clock cycle, state register on the feedback loop of the circuit do not need to be modeled explicitly. It is sufficient that the assignment to the variables representing the state takes place at the end of each iteration.

SCS is a very efficient approach. It requires more effort in the compilation phase but avoids any overhead caused by interpretation. Therefore, whenever applicable, compiled mode simulation becomes more and more the preferred method. In the context of synthesis, especially high performance register-transfer level simulation can be achieved using this method. The approach presented above to support precise timing, however, causes an enormous overhead. Whenever such precise models are requested (and simulation seems the technique best suited in this area) other techniques are preferable. They are outlined below.

9.3.2 Equitemporal Iteration

As already mentioned, especially those levels of abstraction that are used as input levels to synthesis need to be supported by simulation. For high-level synthesis this is the algorithmic level. SCS is a general technique, but not necessarily the most efficient one at the algorithmic level. A classical, very simple table-driven simulation is *equitemporal iteration* (EI). It is directly applicable to VHDL. As for SCS, an iterative approach is used in EI to calculate the state of the entire system being simulated. After each iteration the global simulated time is increased by a fixed increment. The step size may vary from iteration to iteration, but is always equal for all components of the system to be simulated. (In this context, this technique is referred to as equitemporal.)

The EI-algorithm is very simple and easy to implement. Unfortunately, it is rather inefficient if the percentage of components being executable at a certain point of time is low. This is due to the fact that at every point of time (within the observation resolution) each component has to be visited and checked whether its executability condition is true. At the logic level, typically, the probability for this is about 0.05. The relative efficiency of EI increases with the instability ratio of the system to be simulated. This instability ratio increases either if the resolution of observation time is decreased, or if a larger domain of observable values is used. Consequently, EI is mostly used at the RT-level (coarse time resolution) and the circuit level (continuous domain of observable values).

9.3.3 Critical Event Scheduling

This is the most frequently used simulation technique. The approach is aimed at the task to overcome the efficiency problems of EI by restricting calculations to necessary (i.e., non identity) calculations. CES can be applied to any modeling concept in which the following restrictions are respected:

i) The point of time of the next event is predictable.

ii) If the point of time at which a certain event takes place next, is not predictable, this event does not take place, unless it becomes predictable by another event taking place.

These restrictions are fulfilled by all modeling concepts to be considered in our context, thus leaving CES to be a universal algorithm in the field of synthesis-related simulation.

For CES it is assumed that the system to be simulated is partitioned into components at compile time. Furthermore, a dependency relation has to be introduced. If there is a direct dependency of component A from component B, A is called *influencer* of B and B *influencee* of A. This dependency relation allows the CES-algorithm for the decision which components are influenced by the assignment of a new value to a data object. If the necessary timing information is provided, in CES-algorithm it can also be decided precisely when this has to happen. Only those parts of the system being influenced by an event, and only those points of time an event is scheduled for, need to be considered further. Therefore, in contrary to EI, CES is a local approach.

In Figure 9.5 a skeleton of the algorithm is shown:

```
type event = record component_id, event_time: integer;
                    new_value: word
             end;
var current_event, new_event: event; event_queue: queue of event;
begin
  simulated_time := 0;
  while simulated_time <= final_time & queue_level <> empty do
  begin
    current_event := remove (event_queue);
    current_time := current_event.event_time;
    changed := data [current_event.component_id] <>
                     current_event.new_value;
    if changed then
    begin
      data[current_event.component_id] := current_event.new_value;
      if no_of_influencees[current_event.component_id] <> 0
      then
        for i:= 1 to no_of_influencees[current_event.component_id]
        do
        begin
          component:=
             influencee [current_event.component_id, i];
          if executable (component) then
          begin
            new_event.component_id := component;
            new_event.new_value := action (component);
            new_event.event_time := current_time+
                                    elapses (component);
            insert (event_queue, new_event)
          end
        end {for loop body}
      queue_level := test_level (event_queue)
    end {if changed then begin}
  end {while loop body}
end
```

Figure 9.5: Skeleton of the CES algorithm.

Some comments. In the algorithm shown in Figure 9.5 the existence of an event queue which is always kept sorted in ascending order with respect to the event times is assumed. The operation *remove* deletes the first element from the queue, the operation *insert* inserts an element into the queue, preserving the order of elements, and, finally, the operation *test_level* returns the current number of elements in the queue. The algorithm first sets the initial value of the simulated time (line 6) and, then, starts the main loop (lines 7–34). The algorithm is stopped if either the final time is reached or no more events are left (line 7). The loop body starts by extracting the first element of the event queue (lines 9, 10) and checking whether the new value to be assigned differs from the old one (line 11). The special treatment of this question is called *selective trace*. It can be observed that time is not incremented in fixed steps; rather, the simulated time is changed directly to the point of time of the extracted event. If a change of value occurs and influencees exist, all these influencees have to be processed (lines 13 and 17). If the influencee is executable, an event has to be created and inserted into the queue (lines 24–30).

This algorithm has to be combined with a data structure which describes the structure of the system to be simulated; to generate it, the description of the system is compiled. It provides storage elements for all data objects to be considered (assumed to be arranged in a one-dimensional array, see lines 11 and 15), information on the dependency relation (again assumed to be arranged as arrays, see lines 16, 18, 22), component identification (line 25), component operation (line 26), and component delay time (line 28). The clever organization of these data structures heavily influences the performance of the simulator.

The CES algorithm, too, is relatively simple. It is very efficient in most cases, because no time is wasted for unnecessary calculations. This has to be paid for by the requirement to keep the event queue sorted at all times. Insertions into and deletions from priority queues can be accomplished in $O(\log n)$ time, with n being the number of items in the queue. This is acceptable, especially if one considers that the event queue is reasonably short at any time. This is due to the fact that at most times approximately 95% of a circuit is stable, i.e., only few future events are known.

In situations where maintaining the priority queue tends to become the bottleneck of the implementation, the time-wheel approach can be used. The key idea of this approach is to use a circular data structure with a fixed number n of slots to implement the priority queue. Each slot i contains all known future events to take place at the simulated point of time

$$time = (starttime\ of\ actual\ cycle\ + i) \bmod n,$$

i.e., events to be scheduled are inserted into the proper slot. This is nothing else than a special implementation of a priority queue (based on the address calculation method used in the well-known bucket sort), which is possible whenever the priorities can be mapped to discrete values. If the number of slots is a power of two, the identification of the proper slot can be realized by a simple masking operation.

In the CES-algorithm, the simulated time advances in a cyclic manner by a fixed increment stepping forward to the next slot of the time wheel. Then it is checked whether an event to be executed is contained in this slot. If so, it is processed as described in the above algorithm. If not, the next slot is inspected, etc. The fixed time increment is due to the similarity between CES with time-wheel and EI. However, in contrast to EI, not the entire circuit is inspected at each point of time, but only the few active parts.

CES as a very general and efficient algorithm is ideally suited for multi-level/mixed-level simulation, since it supports the algorithmic level equally well as the register-transfer and logic level. Because mixed-level simulation systems are best suited for synthesis, CES seems to be the adequate implementation technique. In addition, it is an ideal algorithm for the entire bandwidth of VHDL.

9.4 Multi-Level Simulation

As already mentioned, within the context of synthesis, simulators are needed that are suited for different levels of abstraction, e.g., to experiment with the specification of a system or to verify one possible implementation. Classical simulators dedicated to a specific abstraction level are not useful for this purpose. Simulation systems are needed that support at least two different levels of abstraction. In addition, any support for comparison of simulation results obtained at different levels of abstraction, applying equivalent stimuli, is helpful.

The term *multi-level simulation* refers to a simulation system that covers more than one level of abstraction. Since by this definition a multi-level simulation system is not necessarily capable of supporting simulations at various levels *concurrently*, often the more precise term *mixed-level simulation* is used. Two main approaches to multi-level/mixed-level simulation can be distinguished:

- either a *broadband simulator* is offered
- or a set of dedicated simulators is integrated reasonably well into one system (*multi-simulator*).

9.4.1 Broadband Simulator

A language like VHDL requires a flexible simulation system covering various levels of abstraction and various styles of description. A single, monolithic simulator seems to be the most natural approach. The system to be simulated can be treated, as described earlier, without the application of any transformation and without having to deal with distributing simulation tasks to various simulators. No tricky modifications of descriptions in order to overcome restrictions of special simulators are necessary. It can be seen that at least based on CES an algorithm powerful enough to support the entire bandwidth of VHDL can be implemented.

Of course, one has to pay for this advantage. Due to the bandwidth covered by VHDL, a complex simulator has to be implemented, which has to support various modeling concepts and heterogeneous domains. Often, very general algorithms, when applied to special cases, tend to be rather inefficient compared to dedicated solutions. The reason is that dedicated algorithms incorporate the knowledge about specific restrictions while general solutions have to cover and to check for a variety of cases, which do not occur in a specific situation. In the case of VHDL, the performance can be increased by translating large parts of a description, especially the sequential bodies of processes, into directly executable code. Then, only the basic handling of concurrency and timing is performed by an interpretative CES-algorithm of high performance. High performance can be achieved because of its restriction to this special type of events. Today's compiler generation systems support the implementation of this approach.

9.4.2 Multi-Simulators

Despite of the advantages mentioned for broadband simulators, multi-simulators constitute reasonable solutions, as well. A typical situation, in which multi-simulators are useful, arises if a semiconductor manufacturer accepts only models suitable for the proprietary simulator. Another problem is the lack of availability of models, either of standard components or of proprietary parts. Often, a large library exists which was written for a specific simulator. The most important application of multi-simulators is the simulation of a completely heterogeneous design specification. Mixed digital/analog simulation provides a typical example. Systems comprising a mixture of electronic and mechanical parts (mechatronic systems) are of similar nature; such systems begin to play an increasingly important role.

Three major problems need to be solved if existing dedicated simulators have to be coupled:

i) the data exchange between the various simulators,

ii) the synchronization of the simulators involved,

iii) a user interface as transparent as possible with respect to the heterogeneity, i.e., hiding the different simulation approaches from the user.

Compared to the other three problems, data exchange [BeLW91] seems easier to solve. Data have to be exchanged on global connections. These connections are those residing in the intersection of the set of primary inputs of a specific model and the primary output of another. In the ideal (but rare) case, the signal value domains of the primary inputs and outputs involved are identical. Then, no conversion of signal values is necessary at all.

Some other cases are relatively simple to handle: those, where a larger domain of values can be partitioned into classes such that each class can be bijectively mapped into a single value of a smaller domain. In any case, it is always reasonably simple to map a larger domain onto a smaller one, a problem typically to be solved when signals are sent from a model at a lower level of abstraction to one at a higher level of abstraction. The opposite direction is more difficult to handle. In this case, assumptions on details of the driver circuit of the sending component are necessary. These assumptions can be made either globally or locally, based on user specifications. Mapping digital signals to inputs of analog simulators is a typical example for a non-trivial problem of this class.

Synchronization is the central problem of multi-simulation. Since the entire system of cooperating simulators models the system to be simulated, the simulators involved have to be kept at least synchronous enough so that at each point of time when a data exchange happens, this point of time is legal for all simulators involved. This definition gives some degree of freedom concerning the synchronization method used. There are two main approaches:

- the *supervisor approach* and
- the *time-warp method.*

The supervisor approach is an over-synchronizing "pessimistic" solution. It requires each simulator involved to hand over control to the supervisor before the simulator advances its modeled time. The supervisor is then able to identify the simulator that plans to schedule an event in the near future; this simulator is activated by the supervisor. The approach is flexible enough to handle all kinds of simulators. To integrate a CES simulator, this algorithm may contact the supervisor just after having extracted the next event from the event queue. Simulators of class SCS or EI may contact the supervisor just before initiating a new cycle. The supervisor approach is very simple, easy to implement, and requires only minor modifications of the simulators involved. Unfortunately, it over-synchronizes the system as no overtaking between simulators is allowed even if no communication occurs during the overlap period. The most serious drawback is the fact that only one simulator can be active at a time. Therefore, it is rarely applicable to multiprocessor systems.

While the supervisor approach over-synchronizes, the time warping method [JeSo82] under-synchronizes. In this method, following a purely optimistic approach, completely free running simulators are assumed without any synchronization besides the exchange of messages. When such messages arrive at a point of time within the local past of a simulator, they cause this simulator to perform *roll-back*, a costly operation. This situation is discussed in more detail.

Let S_1 and S_2 be two simulators, currently simulating at local virtual points of time (LVT) t_{S1} and t_{S2}, $t_{S1} \neq t_{S2}$. It is assumed that S_1 sends a message m = (data,t_m) to S_2 where *data* denotes the transmitted value and t_m the point of time it has to come into effect. Of course, in any case the following holds:

i) $t_m \geq t_{S1}$.

Concerning t_{S2} there are two possibilities:

ii) $t_m > t_{S2}$ or

iii) $t_m \leq t_{S2}$.

Case ii) is simple. The message is handled like any stimuli event, i.e., it is inserted into the event queue.

Case iii) is much more complex. In this situation, S_2 has to be rolled back to a point of time t_h prior to t_m. This ability is required for any simulator to be integrated into a time-warp controlled multi-simulator system. While S_2 is resimulated starting from t_h, all external events between t_h and t_{S2} have to be reconsidered. Therefore, simulators to be integrated need to store the entire history of incoming events (at least for a period determined by the allowed slack between the simulators involved). The handling of past events is accomplished in three phases. In the first phase, called *roll-back*, the state of S_2 at point of time t_h is restored. S_2 may have sent messages to other simulators during the period between t_h and t_{S2}. All these messages, then, have to be canceled. This is performed during a *cancellation phase* by sending an *anti-message* for each message. These anti-messages are handled similarly to ordinary messages. If a simulator receives an anti-message that exists in its local future, this message is removed only from the event queue. If it is located in the local past, it causes a roll-back as well, with the respective message being removed from the stored incoming events. A single roll-back, therefore, may "ripple" through the simulation system by implicitly causing additional roll-backs. Finally, S_2 can resume its normal mode of operation.

The time-warp method, first introduced in [JeSo82], is a very general method and can be implemented efficiently on multiprocessor systems [Apos89, BaSK91, Jeff85a]. A further enhancement to this method has been achieved by *lazy cancellation* [Jeff85b]. Because the time-warp approach becomes less efficient if too many roll-backs occur, it has rarely been used in heterogeneous simulation systems. An example of a heterogeneous simulation framework for integration of arbitrary simulators is the SiCS system [BeLW91, Niem91]. This framework supports both the supervisor approach and the time-warp mechanism with a controllable slack.

To build a unified user interface that introduces transparency to the heterogeneous structure is the last problem to be solved by a multi-simulator system. Such an interface has to allow for the description of circuits in the proprietary language (including graphical ones) of the individual simulators, but should offer a unified description language (e.g., VHDL) as well. The same is true for stimulation and presentation of results. The entire control of the system has to be presented to the user in a strictly unified way. Embedding a multi-level simulation system into an electronic design automation framework [KlKu91, LeWB91, RaWa91] seems to be the most appropriate approach for this purpose.

Multi-level/mixed-level simulation as required in the context of synthesis is widely understood and available. Both monolithic broadband simulators and multi-simulator systems have their specific benefits. Multi-simulators have been studied for a long time [Merm85]. Recently, open frameworks for simulator coupling have been announced [Niem91].

9.5 Outlook

Only the kernel of a simulation system is presented in the preceding sections. The intent of simulation is to have a host system (a computer) show the same behavior as the system to be simulated, and then to perform useful experiments on this simulated system. Therefore, the quality of simulation is determined by

i) the quality of the model of the system to be simulated,

ii) the quality of the mapping onto the host system (i.e., the quality of the simulator itself),

iii) the quality of the experiment performed,

iv) the quality of result analysis.

Unfortunately, for the traditional simulation systems emphasis has been put mainly on i) and ii). The planning, executing and analysis of experiments have remained widely manual tasks.

In the simplest case, at least the application of stimuli and a certain post-processing of result patterns have to be performed. In graphics-oriented environments a graphical waveform editor is often offered for both purposes. This approach seems to be adequate at the logic level. It is convenient to use and it looks familiar to traditional design engineers. In the context of synthesis, this solution is less adequate. First of all, a synthesis-related simulation system has to cover higher levels of abstraction, levels at which waveforms are no longer a natural way of representation. More abstract stimuli have to be applied (e.g., machine programs for a processor to be synthesized) and much more abstract responses are expected (e.g., results of such programs). If low-level waveforms are of interest, they should be hierarchical so that certain patterns can be identified as the equivalent of abstract traces at the higher level of abstraction.

The most serious drawback of waveform editors is that they produce passive stimuli and passively record responses. In most cases, however, the system to be synthesized and, therefore, the system to be simulated as well, are embedded in a predefined environment. This environment, in general, reacts to outputs of the part to be designed, and as part of the reaction produces specific stimuli. Therefore, it is much more adequate to use a model of the environment instead of passive stimuli generators. Such a model can be written in the same language as that the system to be simulated is described in. Therefore, it can be executed by the same simulator which is used for the system to be exercised. This, of course, is true only if the hardware description language used is powerful enough. Because broadband simulation is assumed, this is the case. In the case of a multi-simulator system at least one simulator in the compound should be powerful enough to model even a complex environment.

VHDL uses exactly this approach of modeling the environment in the same language. This idea is perfectly supported by the modular structure of the language. With the aid of this language feature, it can also be expressed easily which part of a description has to be interpreted as a specification for synthesis and which part models only the environment for simulation purposes.

Replacing passive stimuli by a potentially interactive model of the environment does not at all exclude passive stimuli. They constitute the special case of a predetermined environment without any reaction to the simulated system. Such a special case of an environment can be described by the same language, possibly restricted to a subset.

The modeled environment interacting with the simulated system is the first step towards an environment in which experiments can be performed intelligently that help to answer specific questions about specific features of the system under simulation. If this is achieved, the modeled environment becomes a modeled experimentor that in a goal-oriented manner performs experiments directed by intermediate results and abstract features, extracted from these results. As performing such experiments is a non-trivial task which involves a lot of knowledge in order to be performed properly, a knowledge-based approach seems to be an adequate technique to tackle the problem.

Before describing a potential architecture of such a knowledge-based experimentor, the types of features an experimentor is looking for have to be examined. The easiest form of abstractions from values produced by a simulator are *virtual signals*. They are calculated by an arbitrary function using as arguments signal values or streams of such values, as they are produced by the simulator. Any kind of statistics may serve as an example for these virtual signals. A profound examination of simulation results alone, though only a part of an intelligent experimentor, can become a very complex task to be supported by sophisticated systems such as SIMUEVA [BuLa87, Busc90b]. In this section, a distinction is made between *validation, error correction, performance analysis, tolerance analysis, optimization*, and *stimuli validation*. Validation deals with questions, such as respecting predefined domains, delay ranges, bus restrictions, etc. In the case of error correction, hints are given as to what may be the reason for a detected error condition. Such evaluations are essential if a system like SIMUEVA is integrated into an entire experimentor, because for performance analysis delay paths, slopes of transitions, reaction times, etc., are calculated dynamically from simulation results. Tolerance analysis is similar to validation, but under certain assumptions it is calculated whether or not a proper operation is guaranteed over a range of possible operation conditions. Virtual signals for optimization purposes may include the dynamic power consumption, response times, distribution function of delays, etc. Finally, the applied stimuli can be analyzed with respect to the results obtained. Statistics about algorithmic paths visited during a simulation, coverage of cases at switches, grade of concurrency achieved, etc., are typical data of interest, in this case. These are questions similar to those of interest in software testing. This is not surprising, because validation of an initial specification for synthesis is a task similar to validating a piece of software by testing.

If a powerful result analysis system is available, the remaining step towards an intelligent experimentor can be attacked. This leads to the application of artificial intelligence techniques as some authors have indicated [AdPo86, ElÖZ86, EgRo88, LuAd86, Pfaf91, ShMA85]. An experimentor, in this case, becomes a special expert system using simulation results and possible additional external facts as basic facts, planned experiments as rules. By an inference mechanism, conclusions for controlling the execution of the experiment can be deduced. Using a simulation system with a hardware description language of sufficient power, the expert system itself can be built using this simulation system. Such an approach using DACAPO III [DACA90] as simulation system and hardware description language has been described in [Pfaf91]. Obviously, any hardware description language that allows a reactive description (e.g., VHDL) is suited to describe rules and, therefore, facts as well. The main question arises about the inference mechanism. But if one looks at the RETE algorithm used in OPS5 [Forg81], a similarity to CES with selective trace can be observed. Based on this similarity, a CES based inference mechanism for applications in the area of simulation has been described in [Pfaf90]. It depends on the services offered by the underlying simulation system whether the explanation component usually included in an expert system can be offered as well. In any case, it can be constructed by generating proper virtual signals because all information needed for producing explanations are available from the set of simulation results.

In the system described in [Pfaf91] arbitrary rules can be specified by a user. Thus, specific needs arising in a synthesis context can be supported. This system is built on a relatively complete set of basic rules. This rule base contains:

i) general rules such as value checking, cycle checking, etc.;

ii) analog rules such as transition slope checking;

iii) timing rules such as set-up and hold time checking, minimal pulse width checking, etc.;

iv) breakpoint rules affecting the operation of the controlled simulator;

v) tester-oriented rules, checking whether experiments can be performed by a hardware tester;

vi) abstraction level comparison to compare descriptions at various levels of abstraction;

vii) filtering rules for value sampling;

viii) rules to generate virtual signals such as error-functions, ratios, means, etc.;

ix) logic rules;

x) arithmetic rules.

This basic set of rules already allows formulation and performance of reasonably complex experiments that are used to guide a simulation run in a goal-directed way.

Knowledge-based experimentors provide a good argument for multi-simulator systems. If an experimentor is built on top of a powerful simulator, it can be plugged in very easily into any multi-simulator system; an ordinary simulator cooperating with a simulator-based experimentor automatically forms a multi-simulator system.

It can be expected that emphasis on experimentor guided simulation systems will increase. Then, the gap between analytical (formal) verification methods and simulation will become narrower. The main difference, then, will be that for the inference system of an analytical method (its theorem prover) static facts including symbolic signal values are used while a knowledge-based simulation system continues to be based on dynamic facts (simulation results). The emerging similarity—disregarding this difference—offers the opportunity to build cooperative systems including both approaches and inheriting the strengths of both methods. When tightly coupled with a synthesis system within a design framework [RaWa91] the ideal design environment will be achieved.

9.6 Problems for the Reader

1. Model the "philosophers problem" with CSP: Five philosophers are sitting around a table, each one looping between thinking and eating. For eating, a philosopher needs two forks but there are only five forks available, i.e., each pair of philosophers has to share the fork between them. Model the most simple solution, ignoring any deadlock problems.

2. Specify the operations *insert* and *remove* of the time wheel approach in PASCAL-like pseudo-code. The operation *insert* adds one event to slot i of the event-queue, *remove* removes the next event stored in the time wheel. Assume that the time wheel is modeled by a data structure of

```
type time_wheel_queue = array [0:2**size] of
                        record
                          component_id: integer;
                          new_value: word;
                          event_time: integer
                        end;
```

Chapter 10

Synthesis Related Aspects in Testing

Jean Armaos, Wolfgang Glunz

10.1 Introduction

Three questions that have to be answered during the design and production of any system are:

1. Is what I asked for what I really want ?

2. Is what I designed what I asked for ?

3. Is what was fabricated what I designed ?

The first question deals with the validity of the specification, the second one with correctness of the design (modulo specification), and the third one is concerned with problems in the production line. The third question, which is central to this chapter, arises after the system has been built physically and can be answered by performing *production tests.* The aim of the production test is to find physical faults (not design faults!) in the manufactured system. The first task of production test is to find out whether or not the system is faulty (also called *go/no go test*). If the system contains faults, the second task is to locate the faults (also called *fault diagnosis*) in order to remove the faults or their cause if possible. In this chapter the term *test* is used synonymously with production test. These tests require test patterns to be applied to the test objects. Many approaches have been developed to automate the generation of test patterns for digital circuits. The success of these programs depends not only on the techniques and algorithms applied, but also on the *testability* of the circuit. There are many definitions of what a *testable circuit* is. One very general definition is:

> A testable circuit is a circuit, for which a test program of sufficient quality can be generated utilizing available resources (time, CAD tools, computing time, computer memory, etc.).

Two concepts that play a crucial role in testing are *controllability* and *observability*. Internal nodes need to be controlled in order to apply test patterns to internal circuit elements and they need to be observed to detect faulty responses.

In order to measure the quality of a set of test patterns a fault model is needed. The most common fault model used for digital circuits is the *single stuck-at (s-a) fault* model. The stuck-at fault model assumes that every fault in a circuit changes the functionality of the circuit as if some nodes in the circuit were steadily tied to either logic 0 (s-a-0) or 1 (s-a-1). Moreover this model is usually restricted to the *single* stuck-at fault model which assumes that only one node at a time is tied to 0 or 1. A measure for the quality of a test pattern set is the *fault coverage*. The fault coverage is defined as the ratio between the number of detected stuck-at faults and the number of detectable[1] stuck-at faults. In terms of fault coverage, detection of more than 99% of all single stuck-at faults has become mandatory for high quality circuits. To achieve this goal, the testability of the circuit has to be taken into account already in the design phase. This design philosophy is often referred to as *design for testability* (DFT). Several DFT methods have been developed and are supported by appropriate CAD tools. All of the currently known DFT methods are focused at the logic or RT level. This is no problem as long as the design of the circuit is performed at these levels of abstraction as well. The designer can incorporate special hardware elements (DFT aids) where it is needed. However, when automatic synthesis tools are applied, the designer no longer has influence on the logic-level implementation since the design is accomplished at higher levels of abstraction. The design at the logic level is a result of the synthesis system. This requires that the synthesis tools guarantee testability (*testability by construction*), because it is nearly impossible for the designer to improve testability by modifications at the logic level.

In this chapter first some general testability guidelines are presented, then several DFT techniques are described. It is shown how these techniques can be used in a synthesis environment. The last part of the chapter deals with some synthesis related aspects of automatic test pattern generation.

[1] Some faults may not be detectable due to redundancies, cf. Section 10.4.5.

10.2 General Testability Aspects

As mentioned earlier, the testability of a design must be taken into account during the design phase rather than as an afterthought. In addition to the insertion of additional test hardware as described later, following some basic rules can improve the testability of a design substantially. In a synthesis environment, these rules are best applied by the synthesis system.

Synchronous design. The basic requirement for testability is that the circuit has to operate in a synchronous way. Synchronous design does not imply that only one system clock is used, but more generally means:

> In a synchronous design the state of the circuit changes only if special signals—the clocks—are activated. The state of the circuit does not change if data signals only are changed and all the clocks are in their inactive state.

Besides this general requirement there are more specific requirements regarding the clock system of a circuit:

- Clock signals must not be generated on chip. Otherwise testing at reduced speed is not possible.
- The clock system has to work with non-overlapping clocks. Circuits operating with overlapping clocks are very sensitive to delay variations.
- Data signals may only be combined with clock signals in order to inhibit the clock. The data signals inhibiting a clock signal *clk* must be independent from *clk*. This avoids internal race conditions between clocks and data.
- Two clock signals *clk1* and *clk2* may only be combined if the resulting clock signal can be activated by either activating *clk1* or *clk2* (logical OR-ing).
- When a clock is activated, the internal data must not be transferred over more than one level of storage elements. This forbids, for instance, two adjacent latches to be controlled by the same clock.

Synchronous designs can be tested at any clock frequency up to the maximum feasible one. This property is very important since the test equipment to run the test mostly operates slower than the test object in its typical environment. As a side effect of synchronous design, the system is more tolerant to variations of delay times and therefore easier to design and to verify.

Initialization. It must be possible to bring the circuit into a known state with a "short" sequence of external stimuli (*homing sequence*). The best solution is to provide a global reset signal.

Redundancy. Redundancy must be avoided, because it imposes big problems for automatic test pattern generation (ATPG) (see Sections 10.4.5 and 10.5). If functional redundancy is needed (e.g., for reliability reasons) there has to be a special test mode in which the circuit does not behave redundantly.

Sequential depth and feedback. Another significant problem during automatic test pattern generation is the sequential depth of the circuit. The sequential depth of a circuit is the maximum number of storage elements in series. A high sequential depth results in long test sequences to be found by the ATPG-program which again increases both test time and effort to be spent on test pattern generation. The sequential depth must therefore be kept low or broken up in a special test mode. Sequential feedback loops[2] must also be broken in a special test mode.

Combinational feedback loops. Combinational feedback loops (see Figure 10.1) have to be avoided. These loops result in additional hidden storage

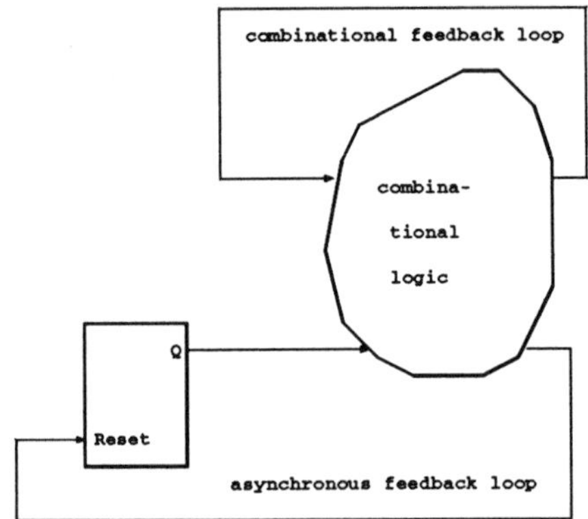

Figure 10.1: Combinational and asynchronous feedback loops.

elements and are therefore hard to test. A special case are asynchronous feedback loops where the data output of a storage element feeds back[3] to the reset/preset input of the same storage element.

Tristate. If tristate busses are used in a design, care has to be taken that in every state of the system exactly one driver actively drives the bus. If there

[2] Feedback loops containing sequential elements.

[3] Directly or through combinational logic.

are states in which this condition cannot be guaranteed, these states must not appear in the test pattern set since the circuit can be destroyed in such a state due to an increased power consumption. This complicates automatic test pattern generation. These states can also prohibit the application of test methods like Scan Path (see Section 10.4.1) or Built-in Self-Test (BIST—see Section 10.4.3) because above condition cannot be guaranteed to hold in these test techniques.

10.3 Test Objects in Synthesis

According to the different levels of abstraction the synthesis process can start from, at each level of abstraction different objects need to be considered for test.

For system-level synthesis, one underlying model is that of communicating processes. In order to apply DFT at this level, the individual processes must be testable independent of one another. This can be accomplished by DFT methods for *modular testing* (see Section 10.4.2), whether the individual processes are distributed among several chips on a board or not. Testability has therefore to be considered during the partitioning into individual processes and the synthesis of the communication network.

The result of high-level synthesis is usually a data path and a controller. The controller can be described as a FSM and is often realized as a PLA or as random logic plus state registers. The data path consists of more complex hardware modules such as multipliers or adders but may also contain simple logic gates. Additionally, the data path includes registers to store local variables. This decomposition into data path and controller can be used to apply DFT. Data path and controller can be separated for modular testing (see Section 10.4.2). For the data path several DFT techniques are available (see Section 10.4). The DFT strategy for the controller depends on the type of its implementation. Whereas for PLA-based controllers BIST techniques are the most appropriate, for random logic based controllers the methods described in Section 10.4.4 are more suitable.

In register-transfer level synthesis the test objects are the different abstract modules (e.g., ALU, multiplier, adder, gates). Care has to be taken that these modules are easily testable per se [Beck85, Beck88].

During logic-level synthesis mainly logic gates are considered. Testability must be considered during logic minimization and state assignment (see Sections 10.4.4 and 10.4.5).

The last step in a synthesis based design process is usually the mapping onto cells of a specific library or modules that can be generated automatically. During this step the DFT cells needed for the application of a specific test strategy have to be incorporated into the design (e.g., scan cells or special test controllers).

10.4 Test Methods for Synthesis

10.4.1 Scan Techniques

Testing sequential logic is very hard if no DFT is provided. The basic problem is that it is much more difficult to control and observe internal nodes in sequential logic than in pure combinational logic. Therefore, methods are required that improve the controllability and observability of sequential logic. Besides many ad hoc techniques such as the insertion of additional test pins, the scan methods described below offer a structured approach to this problem. The basic idea behind the different variations of scan approaches is to connect the storage elements of a circuit to form in a special test mode a serial shift register (*scan chain*). This facilitates control and observation of the internal nodes of sequential logic. For test pattern generation the storage elements included in the scan register can be regarded as directly accessible inputs or outputs of the logic. If all storage elements are included in the scan chain (*full scan*), the problem of test pattern generation for sequential logic is reduced to test pattern generation for combinational logic. A further advantage of scan methods is that it is easy to incorporate them into the circuit automatically. This is specifically interesting for synthesized circuits. Figure 10.2 illustrates the basic principle of scan approaches. In Figure 10.2a the general model of a sequential circuit is shown. The circuit after application of a scan design is shown in Figure 10.2b and the resulting model for test pattern generation in Figure 10.2c.

In the following, different variations and implementations of the basic idea are described:

Level-sensitive scan design. One of the most popular scan methods is the level-sensitive scan design (LSSD) [EiWi77]. LSSD is suited best for latch-based designs. All storage elements are implemented using special *shift register latches* (SRL) (see Figure 10.3). An SRL is composed of two latches—one dual port (*L1*) and one normal latch (*L2*). All these SRLs are connected to one or more scan chains (see Figure 10.3). During normal operation of the circuit two or more non-overlapping clock pulses are applied to the system clock inputs SCK_i.

During testing, first all the latches are tested by activating both scan clocks *SCA* and *SCB*. In this state all the latches are transparent and a data value applied to the *SDI* input is transferred directly through all the latches to the *SDO* output. This test is also known as *flush-test*. After this test the

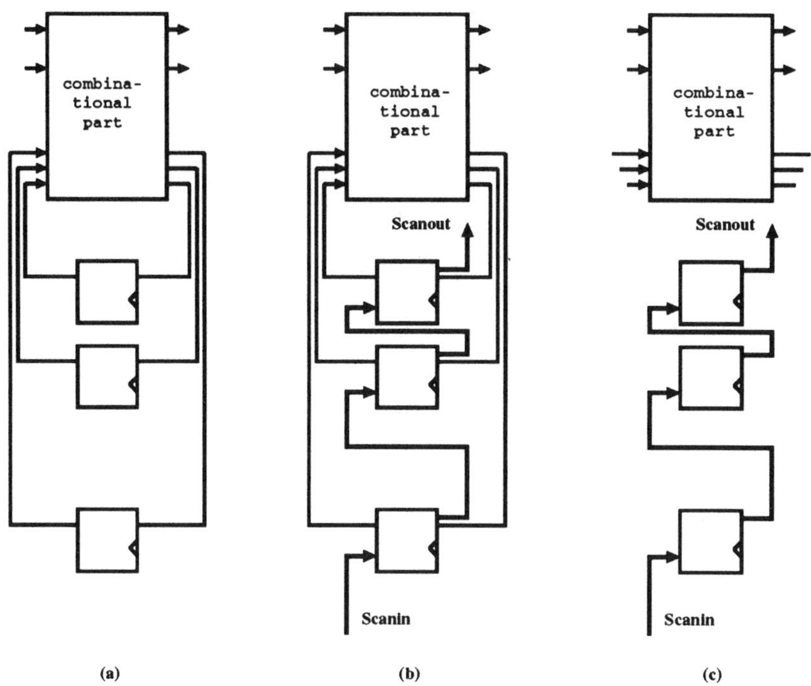

Figure 10.2: Scan design principle.

shift register function is tested by simply shifting in and out a "001100..." pattern. This pattern tests each latch for all possible transitions ($0 \rightarrow 0$, $0 \rightarrow 1$, $1 \rightarrow 1$, and $1 \rightarrow 0$). Then the following procedure is repeated for each pattern generated for testing of the combinational logic.

1. Shift in the values of the test pattern at the SDI (serial data in) port. This is achieved by alternately clocking SCA and SCB.

2. Set the primary inputs to the corresponding values of the test pattern.

3. Check the response of the circuit at the primary outputs.

4. Apply one clock pulse at the system clocks SCK_i. This is necessary in order to load the shift register with the responses of the combinational parts of the circuit.

5. Shift out the values of the internal nodes by alternately clocking SCB and SCA.

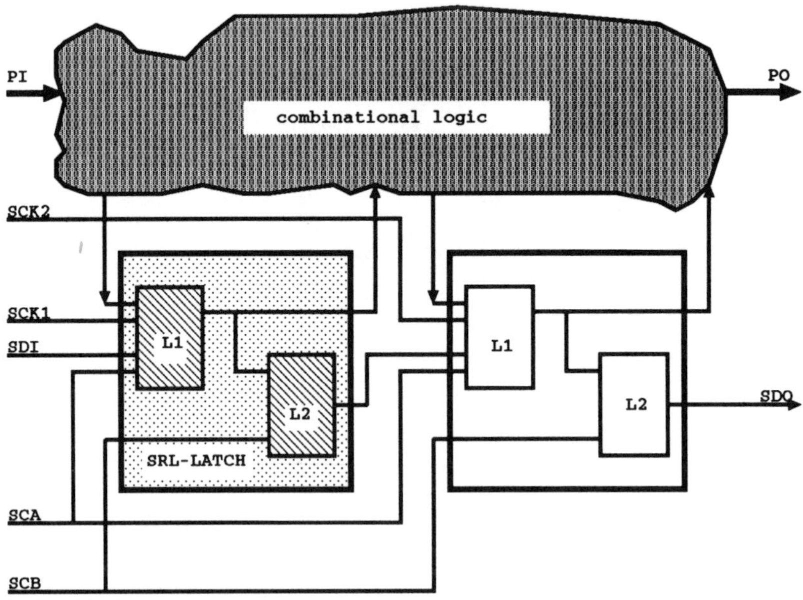

Figure 10.3: LSSD SRL and principle of LSSD.

The LSSD technique normally requires four additional primary inputs/outputs:

SDI	serial data input
SDO	serial data output
SCA	master clock for shift
SCB	slave clock for shift

LSSD allows the *L2*-Latches to be used for normal system functions as well. By doing this, the area overhead of LSSD can be reduced to nearly 0%. In this case the *SCB* clock is also used functionally. Note that the additional *SDO*-output can be omitted if a data output of any of the latches directly feeds a primary output. In this case this latch is inserted as the last latch in the scan chain.

This design technique can be extended to support designs where the system clock inputs of the *L1*-latches are not fed directly from primary inputs (*gated clocks*). In this case the test pattern generator has to consider the gating logic also. Due to this requirement automatic test pattern generation becomes more complicated but is still based on a combinational model. Advantages include more freedom for the designer and often a smaller area overhead. This aspect is of specific interest in synthesized circuits where the clocks controlling the registers are gated by signals from the control-FSM.

Scan Path. Whereas LSSD is best suited for latch-based designs, the Scan Path technique is dedicated to designs based on master-slave flip-flops [GeNe84]. In this design technique all the storage elements are implemented using scannable flip-flops as shown in Figure 10.4. The data input of the D-flip-flops is

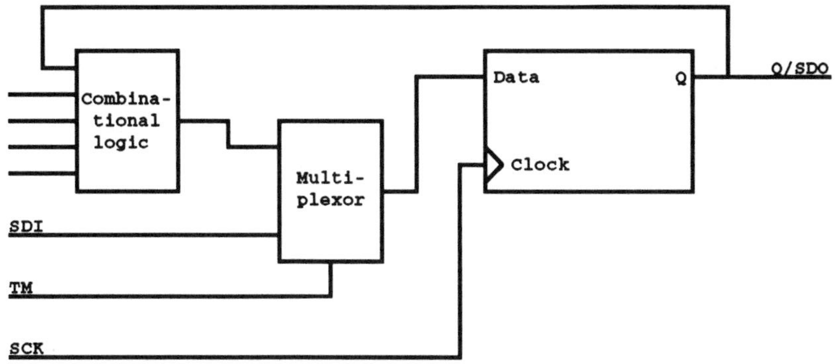

Figure 10.4: Universal flip-flop for Scan Path.

now fed by the output of a multiplexor which allows for the selection between two different data sources. One is the functional data input, the other the serial test input. Therefore these flip-flops are often called *multiplexed-data flip-flops* (MD-flip-flop) [McCl86]. The scan chain is built by connecting the serial data inputs and the Q-outputs in series. The Scan Path technique requires only three additional pins:

SDI	serial data input
SDO	serial data output
TM	test mode

The SDI input can be omitted if the data input of a flip-flop is already fed by a primary input. The SDO output can be omitted if the Q-output of a flip-flop directly feeds a primary output.

During test first all the flip-flops are checked by shifting in and out a "0011..." pattern as described earlier. Then for each pattern for the combinational logic the following procedure is repeated:

1. Set test mode TM to shift mode

2. Shift in the test data by clocking the system clock CK and applying data to the SDI input

3. Observe the primary outputs
4. Switch test mode TM to functional mode
5. Apply one system clock cycle
6. Switch test mode back to shift mode
7. Shift out test responses and observe them at the SDO output.

In a synthesis environment the Scan Path technique is also easy to integrate. The only problem is that the Scan Path technique uses the same clock both in shift and functional mode. This does not allow the clock to be gated, i.e., inhibited by a control signal, which often occurs in synthesized circuits; for instance, registers are only loaded in specific states of the controller.

Partial scan. Whereas the basic idea of full scan design was to include all storage elements of a circuit into the scan chain, the partial scan approaches attempt to reduce the number of storage elements included in the scan chain. This can be achieved using both the LSSD or the Scan Path technique. The advantage of partial scan is the lower area overhead. The disadvantage is the more difficult test pattern generation. However under certain circumstances test patterns can still be generated using methods for combinational logic. The basic requirements are:

- All feed back loops are broken up in test mode by including appropriate storage elements into the scan chain.
- If all registers that are included in the scan chain are viewed as new primary inputs/outputs, the resulting circuit structure must expose a *pipeline structure*. In pipeline structures combinational logic is followed by registers and vice versa (see Figure 10.5).

In a synthesis environment these basic requirements can often be met. For example, synthesis tools often provide registers for the circuit inputs/outputs. Since these registers are directly controllable/observable they do not need to be included in the scan chain, without increasing the complexity of test pattern generation.

For a more detailed discussion of various partial scan approaches see [Kunz88, Tris80, MaDN88, GuGB89, ChAg89a, GuMu90, KuWu90].

Boundary scan. The idea of scan design has been extended to support board level testing. The main idea is to include every input and output of a chip into a special scan chain on the chip—the boundary scan. The boundary scan chains of the individual chips on a board are again chained on board level

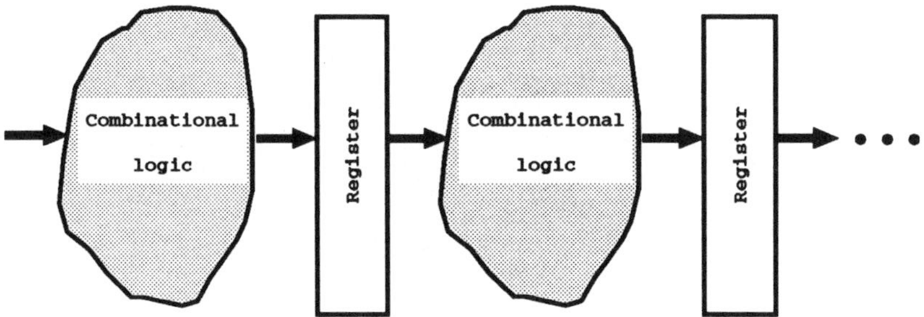

Figure 10.5: Pipeline structure.

through a special *test access port* (TAP) on each chip. Through the TAP also all operations of the Boundary Scan are controlled. The TAP consists of four pins:

TDI	Test data input: This pin is used to shift in test data and instructions.
TDO	Test data output: This pin is used to shift out the test data.
TMS	Test mode select: This pin controls the different tests performed on the chip.
TCK	Test clock: This clock allows the usage of the boundary scan independently of the system clock.

The boundary scan can be used for both *internal* and *external* test. The internal test examines the chip itself and is necessary, e.g., for fault diagnosis on board. Various internal test methods such as BIST or Scan Path can be controlled through the TAP without any additional pin. The external test tests the interconnections between the individual components on a board. Thus in-circuit-test can be performed easily even if advanced technologies like surface mounted devices (SMD) are used. A third test mode is the *sample* mode. In this mode the input and output signals of a chip are sampled and observed independently of the normal operation of the chip.

The basic principle of boundary scan on a printed circuit board (PCB) is shown in Figure 10.6. Boundary scan has been accepted as IEEE standard P1149.1 [IEEE90].

10.4.2 Modular Test Methods

The complexity of electronic devices, measured by the number of gates, has increased in the last decade by a factor of nearly 100. Devices frequently have

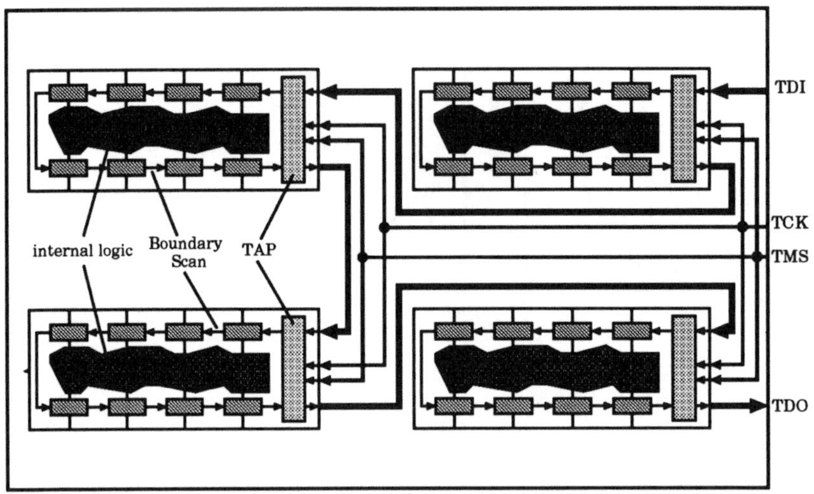

Figure 10.6: Boundary scan principle.

a complexity of some millions of gates. On the other hand, the computational expense of test generation grows more than linearly with the number of gates, provided the number of accessible pins does not change significantly. Hence, the continuing rapid increase of circuit complexity results in an explosion of computational expense, if test engineers go on handling a circuit as a large flat logic-level structure rather than as a hierarchical structure consisting of several separately treatable partitions.

Moreover, modern electronic circuits, especially integrated circuits, usually consist of a number of heterogeneous modules. The heterogeneity refers to the function (e.g., multiplier, controller, RAM), the design style (e.g., random logic, PLA, pipeline style), or the test methodology applicable to a module. Further, it is more often the case that modules are used in a design for which test patterns already exist (from a previous design or from external libraries if standardized modules are employed).

For the reasons given above, modular (i.e., hierarchical) design for testability and test pattern generation appear indispensable [MuHa88, BDSS89, CSIM89, RoJG89, BDSS90]. In a modular representation from the testing point of view, a circuit is considered as a combination of *modules*. A module is defined as an instance of a *module type*. A module type is an abstract entity characterized by a *function* attribute, a *design style* attribute, and an *interface* attribute. Each module may, in turn, be decomposed into a modular structure built up from lower-level modules. If during this top-down decomposition a module is reached for which test methods of manageable complexity are known, further decomposition can be stopped. This latter type of module is called a *primitive testable module*.

The basic idea in modular testing is to generate test patterns first for primitive testable modules, and then to transform them bottom-up to the boundary of higher-layer modules until the system boundary is reached. For simplicity, it is assumed in the following that the circuit is modeled in a two-layer hierarchy: the layer of the complete circuit, called the *system layer*, and the layer of primitive testable modules, called the *module layer*. The procedure of modular testing is roughly outlined in Figure 10.7.

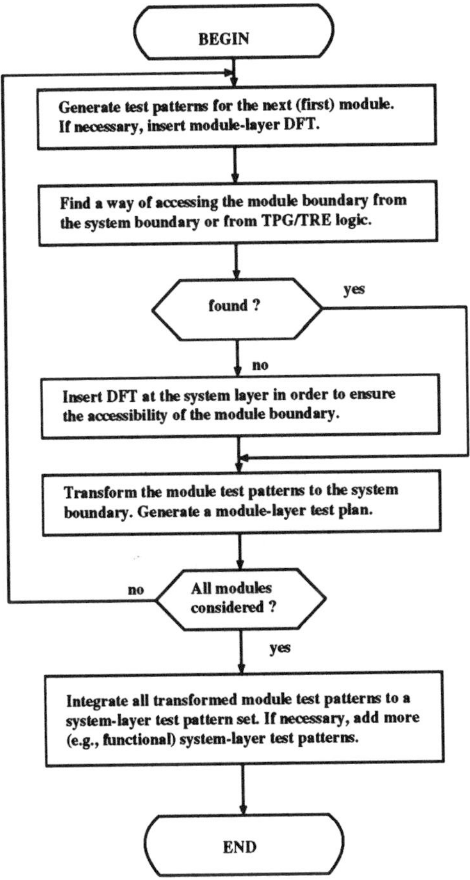

Figure 10.7: Simplified flowchart of modular testing.

First, test patterns are generated for a module by means of a test generation approach dedicated to the specific module type. In general, the module structure must obey certain DFT rules or DFT logic must be inserted in order to make the module-specific test approach applicable. If the module-layer test patterns are already known or if they are to be generated in-circuit, the step of test pattern generation is omitted for this module.

Next, the module-layer test patterns and test responses have to be transformed to the system boundary or—in case of a self-test—to an internal test pattern generation/test response evaluation logic (see Section 10.4.3). This problem is almost as complex as the general test pattern generation problem. However, this task can be simplified by restricting it to a search of paths through modules with *a priori* known simple transfer properties (e.g., transparency mode). If no such path can be found, DFT aids are inserted at the system layer which enable a test pattern transfer to/from the module, e.g., via multiplexors, busses, scan chains, or extra pins.

If the test pattern transfer is guaranteed, a sequence of activations, the so-called test plan, has to be determined which stimulates the generation (in case of in-circuit generation), transfer, and evaluation of the module test patterns during test application. The test plan normally describes symbolic or real signal values and clock pulses to be applied to the system boundary for executing the above task. An example is given later in this section.

After this procedure has been performed for each module, all transformed module test patterns are merged into a complete and compact system-layer test pattern set. In this operation, parallelism in test application can be exploited to minimize test time. The problem of minimizing the test time can be solved using graph-theoretical methods, e.g., [KiSa82, AbBr86, SaKi88].

Additional test patterns have to be added for testing the interconnections of modules, the operation of the entire system in the functional mode, and the timing of the system (see Section 10.5).

The modular test approach is illustrated using the example circuit shown in Figure 10.8a. The circuit consists of a ROM, a register, an adder, and a multiplexor.

According to the flowchart of Figure 10.7, first, test patterns have to be generated for each module. Assume the adder is selected as the first module and test patterns are generated for the adder by means of a dedicated approach.

Then, a set of paths must be found each connecting an input or output of the adder with an input or output of the system, respectively. Consider first the input ADD.IN1. The preceding module is the ROM. A ROM generally has no transfer ability; hence, there is no transfer path feeding to ADD.IN1. Therefore, the no-branch of the flowchart (Figure 10.7) is entered and transfer logic has to be inserted. A multiplexor is chosen and inserted for this purpose,

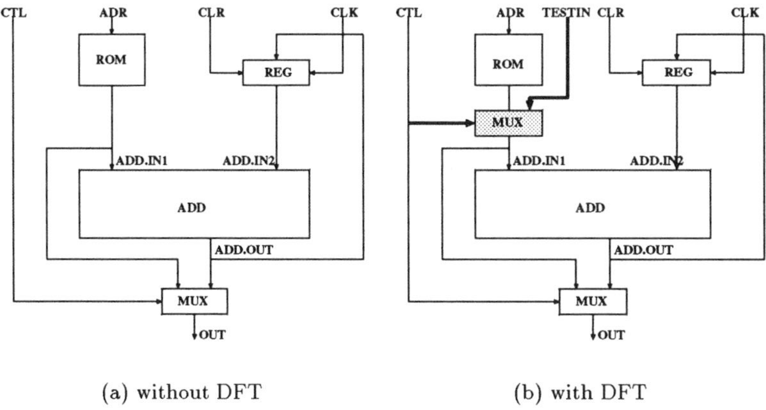

(a) without DFT

(b) with DFT

Figure 10.8: Circuit without and with DFT for modular test.

as shown in Figure 10.8b. An additional system-layer input TESTIN feeds to ADD.IN1 through the multiplexor in the test mode controlled by the signal CTL.

Next, the input ADD.IN2 of the adder is considered. Is there a path from the system boundary to ADD.IN2 ? ADD.IN2 is fed from the register. A register is transparent with a delay of one clock cycle. The register is fed from the adder. An adder is transparent for one of its input signals, if zero is applied to the other input (addition of zero). Thus, if a zero can be applied to ADD.IN2, then the adder is transparent for ADD.IN1. A zero can be applied to ADD.IN2 by applying a pulse to the system-layer input CLR (i.e., by clearing the register). On the other hand, ADD.IN1 is accessible from TESTIN, as pointed out earlier. Consequently, a transfer path exists from the system-layer input TESTIN to ADD.IN2.

Finally, the output ADD.OUT of the adder is considered. ADD.OUT is observable at the system-layer output OUT via the output multiplexor.

The above steps ensure that the boundary of the adder is accessible from the system boundary. The next task, according to the flowchart, is to generate a test plan for the adder. The resulting test plan is:

```
For each pattern
   apply a pulse to CLR;
   apply '1' to CTL;
   apply pattern for ADD.IN2 to TESTIN;
   apply a pulse to CLK;
   apply pattern for ADD.IN1 to TESTIN;
   observe ADD.OUT at OUT;
```

The other modules of the circuit can be treated similarly and so a complete test pattern set for the entire circuit can be achieved.

The procedure of modular testing can be automated within a synthesis system, provided that dedicated test methods for the module types supported by the synthesis system are available. The synthesis algorithm has to take care of generating architectures with easy accessibility of all modules. A synthesis system that supports modular testing is the PIRAMID system [BDSS89].

10.4.3 Built-In Self-Test (BIST)

10.4.3.1 Benefits of Built-In Self-Test

In conventional test approaches, test patterns are selected manually or generated by CAD tools. These patterns, along with the corresponding circuit responses, are then transferred to an automated test equipment (ATE) and are applied to the circuit (chip, board, system) under test. The signal values at the circuit outputs are measured by the ATE and compared with the expected responses. This procedure is referred to as an *external test.*

External testing exhibits, however, some drawbacks:

1. ATEs are expensive. The price of ATEs that incorporate major features may climb up to some millions of dollars. Therefore, test time is dearly paid. Moreover, for different package levels (chip, board, system) different ATE environments are needed.

2. External test is subject to restrictions due to the architecture and technology of ATEs:

   - The timing capabilities and especially the test rate are limited. Obviously, it is not easy to test the timing behavior of a fast new-generation circuit with an ATE of a previous technology generation.
   - The number of pins addressable by an ATE is limited. Though average devices usually have less pins than the ATE can address, it happens that some special-purpose circuits have more pins, and, therefore, tedious tricks must be applied in order to test them.

- Reloading the pin memories of ATEs may corrupt the test of dynamic circuits because of the long reload time. During this time, the state of a dynamic circuit may change.
- The throughput of production test is limited by the fact that conventional ATEs do not support concurrent testing of more than one circuit.

3. For a maintenance test, the circuit has to be set out of operation and to be transferred to a test equipment.

All these problems can be avoided or, at least, diminished if the circuit provides the capability to test itself. An expensive ATE is not required for a logic test. Merely low-cost equipment is needed for the parametric DC and AC test or for a short additional logic test session. The limitations pointed out above do not apply. Moreover, the circuit can be tested under the real-operation clock rate. Finally, the self-test used in production test can also be used in the field as a maintenance test.

Especially in synthesis, where low-cost design and short time-to-market are the targets, automatic BIST is attractive because of the characteristics mentioned above. If the synthesized circuit includes BIST capabilities, the designer does not have to trouble about tedious and time-consuming test generation.

10.4.3.2 A Survey of BIST Techniques

The basic configuration of a BIST structure is outlined in Figure 10.9. The block denoted *module under test* may be the entire circuit (except the BIST logic) or only a part of the circuit. In the latter case, only this part of the circuit is tested by this configuration.

The *test pattern generator* usually is either a register-based structure (e.g., linear feedback shift register, counter), a memory in which the test patterns are stored, or a processor which generates test patterns according to some stored program.

The *test pattern path* is the path traversed by the test patterns from the test pattern generator to the module under test (in one or more clock cycles). In the simplest case, this path is a wire (or a bundle of wires); in the general case, it consists of a sequence of building blocks with certain transfer properties. The transfer properties must obey some requirements implied by the test method applied. If, for instance, an exhaustive test is used (see next paragraph), the transfer function of the path must be bijective.[4] If a random test is used, the transfer function must not affect the statistical distribution of the generated

[4]The function must provide all possible outputs, and for each output there must be a unique input combination that generates this output.

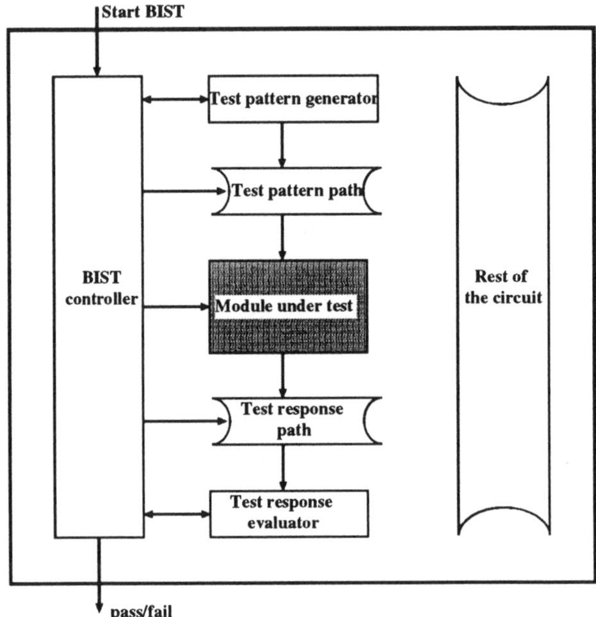

Figure 10.9: Basic configuration of BIST.

patterns. The simplest transfer property is the *identity* property where the test patterns are passed without alteration.

The *test response path* transfers the test responses from the output of the module under test to the test response evaluator. By analogy to the test pattern path, it must possess certain transfer properties depending on the response evaluation method used.

The *test response evaluator* may be a data-compressing register (e.g., signature analyzer), a memory containing the expected responses in conjunction with a comparator, or a more intelligent software-driven unit.

The *BIST controller* controls the complete BIST session. It sets the circuit to the BIST mode, when a "start BIST" signal is activated, initializes the BIST logic and the module under test, activates the test pattern generator, the test pattern path, the test response path, and the test response evaluator, controls the clocking, and, when the BIST process is finished, it provides the pass/fail signal (or other diagnostic information) and switches off the BIST mode. The BIST controller is an implementation of the state transition graph of the BIST process. It can be synthesized by means of well-known controller synthesis approaches, or a parameterized library module can be used.

In synthesizing a BIST configuration, functional features already existing can be exploited for BIST purposes. If, for instance, a counter is needed for the normal circuit function, and the BIST method to be implemented requires a counter as test pattern generator, the same counter can be used. Of course, an appropriate test pattern path has to be provided between the counter and the module under test. In many cases, an existing functional component can easily be extended to a BIST component. If, for example, a functional register is augmented to a signature analyzer, the additional logic is less than if a completely new signature analyzer were inserted. Furthermore, if the circuit has "built-in intelligence" (e.g., a processor system), this intelligence may be exploited to save test-logic overhead.

A well-known problem in BIST is how to test the BIST logic. To a certain degree, the BIST logic is also tested during the BIST operation. Some BIST components can be tested in an extra BIST session (the BIST components being treated as modules under test), others (e.g., the controller) must be tested externally. For example, registers in the BIST logic can be configured to be scannable, thus enabling the use of a scan-based test.

As pointed out earlier, the module under test is not necessarily the entire functional circuit, but it may be a part of it. If the circuit consists of several heterogeneous modules or if it has been partitioned to reduce complexity, each module can be tested separately (*modular BIST*). In this case, a configuration according to Figure 10.9 exists for each module. Though the module tests are logically separated, BIST components such as test pattern generators and test response evaluators can be shared by the tests of different modules in order to reduce hardware overhead. Moreover, one BIST component can be provided with different BIST operations; for instance, a register may be designed in such a way that it can operate in two different modes: a test pattern generator mode and a test response evaluator mode. But in most of the BIST techniques it cannot operate in both modes at the same time. In Figure 10.10a a circuit consisting of two modules which have to be tested by BIST is shown. In Figure 10.10b a BIST configuration is shown in which the test pattern generators and the test response evaluators are different for both modules. For simplicity, the BIST controller is not shown; the test pattern generators and the test response evaluators are assumed to be transparent in the functional mode. In this case, both modules can be tested simultaneously. The configuration shown in Figure 10.10c uses the same component as test response evaluator for module 1 and as test pattern generator for module 2; thus, the logic overhead is lower than in Figure 10.10b. However, the tests of both modules cannot be carried out simultaneously.

Saving hardware overhead results, in general, in an increased test time. Therefore, optimal BIST synthesis is a trade-off problem between minimal

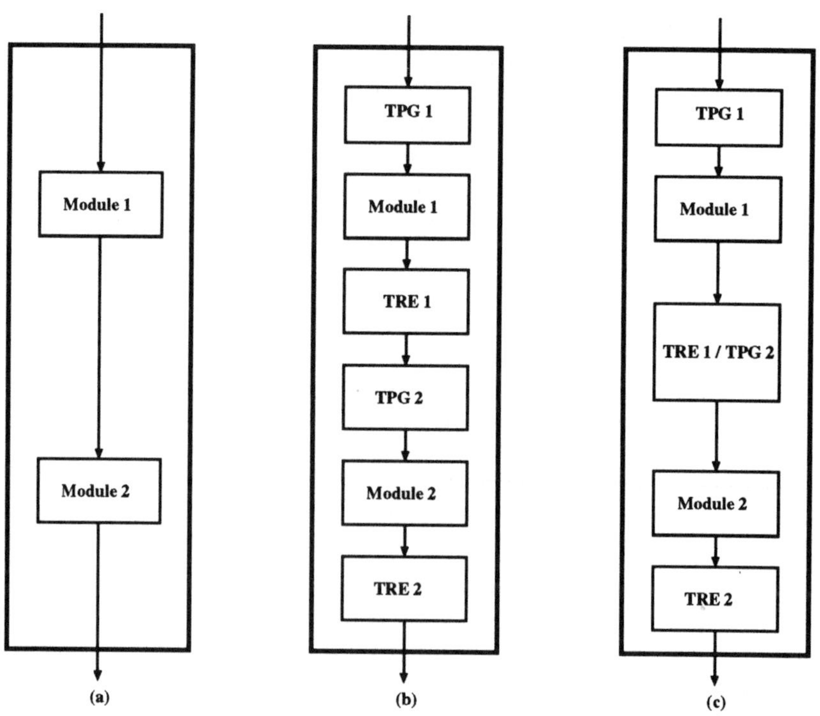

Figure 10.10: Modular BIST for minimal test time (b) and minimal hardware overhead (c). TPG: Test pattern generator; TRE: Test response evaluator.

hardware overhead and minimal test time, in analogy to synthesis of the normal circuit function, where a trade-off between hardware and performance is common.

In analogy to high-level synthesis, an appropriate BIST scheduling and allocation of BIST components to BIST operations (test pattern generation, test response evaluation) has to be performed during BIST synthesis. Furthermore, the test pattern path and the test response path have to be established; the insertion of multiplexors, scan chains or busses may turn out to be necessary. Finally, the BIST controller is synthesized and the overall structure is optimized. BIST synthesis can take place either in common with synthesis of the normal circuit function or in an alternate procedure.

Conventional BIST techniques can be roughly divided into three major classes: exhaustive BIST, random BIST, and special BIST methods for special structures. Often, different BIST techniques are combined into mixed BIST approaches.

Exhaustive BIST. In exhaustive BIST all possible patterns are applied to the inputs of a module. If the module is combinational, the exhaustive BIST is *complete* for static faults,[5] i.e., all irredundant static faults are detected. The number of patterns for an exhaustive BIST is 2^n where n is the number of module inputs. For large n, this number of patterns results in a prohibitive test time. Consequently, exhaustive BIST is applicable to modules with a limited number of inputs (typically n $\leq$ 20) only.

For test pattern generation, an n-bit counter or an n-bit *linear feedback shift register (LFSR)* can be used (Figure 10.11) which—if properly configured—generates 2^n -1 different patterns in a "random" order (the all-0-pattern is excluded; however, it can be applied in a separate mode, e.g., by clearing the LFSR). The test response evaluation usually is performed by compression of

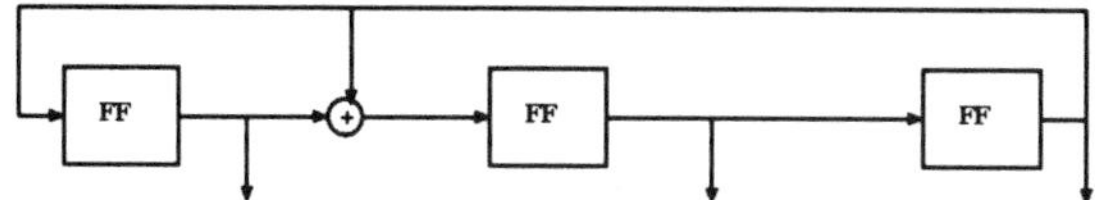

Figure 10.11: Example of a LFSR as test pattern generator.

all responses to a *signature*. A circuit commonly used for signature calculation is a *multiple input signature register* (MISR). A MISR compresses a series of data at its inputs to a single signature. An example of an MISR is shown in Figure 10.12. For a study of properties and the theoretical foundations of

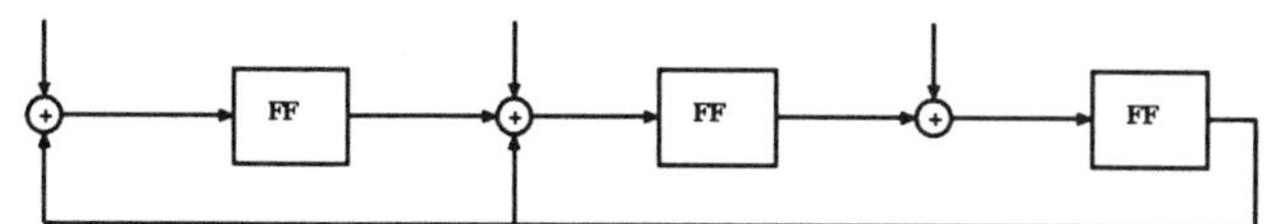

Figure 10.12: Example of a multiple input signature register.

LFSRs and MISRs see, for example, [BaAS87].

A modified form of exhaustive BIST is *pseudo-exhaustive* BIST [McCl84, Udel86, WaMc86, WuHe89]. The basic idea in pseudo-exhaustive BIST is to divide the set of module inputs into subsets (not necessarily disjoint), each influencing only a limited part of the module up to some outputs, and to apply exhaustive patterns to each subset. These subsets can be determined by segmenting the circuit graph into *cones* (Figure 10.13). Each cone begins at an output (vertex of the cone) and contains all predecessors of the output. The inputs of the cones define the subsets mentioned above.

[5]A fault is called *static* if it can be stimulated by a single pattern applied to the location of the fault.

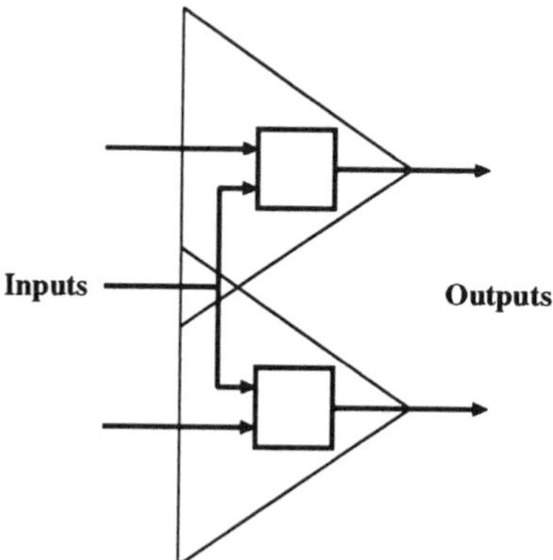

Figure 10.13: Cones for pseudo-exhaustive test.

In the case of combinational modules, pseudo-exhaustive test is complete for static faults, provided that faults outside a segment do not affect the interior of the segment. The total number of patterns is less than in exhaustive BIST, if the overlaps of different segments are sufficiently small.

Various implementations of test pattern generators for pseudo-exhaustive BIST have been proposed in the literature aiming at minimizing hardware overhead and test time [WuHe88, WaMc86, Udel86]. Their automatic generation within a synthesis system appears feasible.

Random BIST. Random BIST is the most popular BIST technique [BaAS87, McCl85, OhWM87, Wund89]. It is based on randomly generating test patterns. Since random test pattern generators in reality are deterministic structures, the patterns generated by them are—strictly speaking—pseudo-random. However, they provide a good approximation for random patterns, and so the term "random" is used here.

Obviously, a random test is, in general, not complete, i.e., it does not test every fault. Nevertheless, a relationship between the number of patterns and the corresponding fault coverage can be estimated statistically from the circuit structure; thus, a number of random patterns can be determined which yield a required fault coverage (with a statistical confidence level).

The simplest form of random test is to randomly apply to each module input 0's and 1's with a probability of 0.5 *(uniform random test)*. However, depending on the circuit structure, the number of uniform patterns needed to detect a given fault may be huge. For example, the probability of detecting a stuck-at-0 fault on the output of a 10-bit AND gate with uniform random patterns is 2^{-10} because only the all-1-pattern detects this fault. A more sophisticated approach uses *weighted random patterns* which are, in general, distributed with some bias rather than uniformly [Wund87, BrGK89]. An optimal distribution is determined by taking the properties of the current circuit structure into account in order to minimize the number of patterns needed. For this procedure, the patterns at different inputs are assumed to be statistically independent.

There are circuit structures which are easily random testable (with few patterns) and others which are not easily random testable, even with weighted patterns. Especially, many sequential structures are very hard to test with random patterns because the probability that a special pattern *sequence* occurs, which is necessary to test a fault, is very low. A number of approaches have been suggested in the literature to modify the circuit structure such that it becomes random testable. This is often achieved by connecting internal points of the circuit with the test pattern generator or the test response evaluator, or by partitioning the circuit into random-testable modules and applying modular BIST.

Random BIST normally uses LFSRs as test pattern generators and as test response evaluators. A frequently used LFSR-based structure is the *built-in logic block observer (BILBO)* which combines the operations of a normal register, a shift register (scan path), a random pattern generator, and a signature register [KoMZ79]. Weighted random patterns can be generated by a combination of a LFSR (e.g., BILBO) which generates uniform patterns and a subsequent weight logic.

If random BIST is incorporated in a synthesis system, algorithms of random-pattern testability analysis must be included in the system to compute the number of random patterns (or the fault coverage if the number of patterns is given), the weights (in case of a weighted random test), and, if necessary, to determine where and which DFT aids have to be inserted in order to obtain random-testable structures.

Two extended forms of random BIST are sketched in the following as examples out of a wide range of random-BIST applications: the *LOCST (LSSD On-Chip Self-Test)* approach [LeBl84] and the *circular BIST* [KrPi87, KrPi89, Stro88b, Stro88a, PrBL88]. These two approaches are selected because they seem to be best suited to integration into a synthesis environment.

LOCST. The LOCST approach consists of LFSR-based random testing combined with scan design and boundary-scan [LeBl84], see also [Komo82]. The input boundary scan cells, the internal memory elements, and the output boundary-scan cells form a scan chain. An appropriate number of memory elements at the beginning of the scan chain, which are assumed to be on the circuit boundary, are configured into a LFSR for random pattern generation. Similarly, a sequence of memory elements at the end of the scan chain, which also have to be on the circuit boundary, are configured into a signature register. The signature register has a parallel input port for signals coming directly from the combinational part of the circuit, as well as a serial input connected with the internal scan chain.

After initializing the scan chain, the following procedure is performed:

1. Apply scan clocks until the scan chain up to the signature register is filled with random patterns generated by the LFSR. Simultaneously, the previous content of the scan chain is shifted into the signature register for compression (serial mode). At the end of this step, the random pattern shifted into the scan chain is applied to the combinational part of the circuit and the corresponding response is present at the parallel inputs of the internal scan chain and of the signature register.

2. Apply a system clock to capture the response of the combinational part into the internal scan chain and the signature register (parallel mode).

3. Repeat 1 and 2 for a specified number of patterns.

Finally, the resulting signature is read out and compared with the expected signature.

A very low hardware overhead of 2% is observed. This overhead corresponds to the test pattern generator, the test response evaluator, and the BIST controller, but not to the scan path. Including scan path, an overhead of 10–15% is expected.

A drawback of LOCST is the long test time due to the serial shifting of each test pattern. This time can be reduced by using several parallel scan chains [BaAS87]. Another drawback is that some faults may never be detected by this method because certain patterns are never applied to the combinational part *(limit cycling)*. If n is the length of the test pattern generator and m the length of the internal scan chain, the number of unique test patterns applied to the combinational part is $(LCM(2^n - 1, m))/m$.[6] Depending on n and m, this number has a minimum of $(2^n - 1)/m$ and a maximum of $2^n - 1$ patterns. The condition for maximal pattern size is: $LCM(2^n - 1, m) = (2^n - 1)m$. To meet this maximum condition, dummy memory elements can be added. However, limit cycling can still occur for $m > n$ since not all 2^m possible patterns can be generated. For the faults not covered by LOCST, an external scan-based test can be applied.

Circular BIST. Circular BIST is a special form of random BIST in which some selected functional memory elements are replaced by an augmented

[6] LCM: least common multiple.

flip-flop cell and are configured to a circular path [KrPi87, KrPi89, Stro88b, Stro88a, PrBL88]. The basic idea is to use the compressed test response after each clock cycle as the next "random" test pattern.

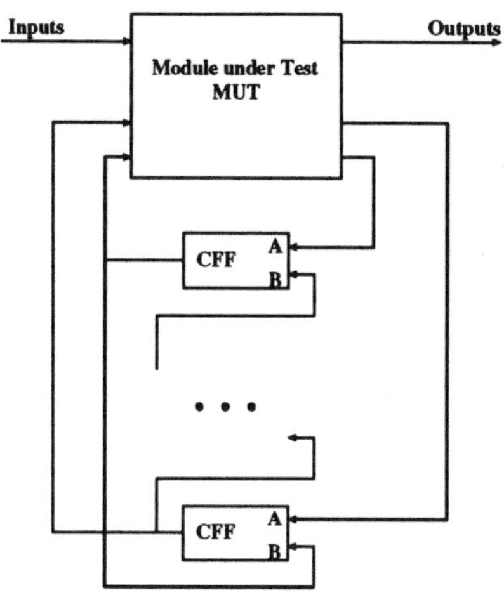

Figure 10.14: Simplified configuration of circular BIST.

A simplified configuration of circular BIST is shown in Figure 10.14. For simplicity, the controller and the control lines are not shown. The block denoted MUT (module under test) is the circuit without the memory elements selected for the circular path. The blocks CFF are the augmented flip-flops of the

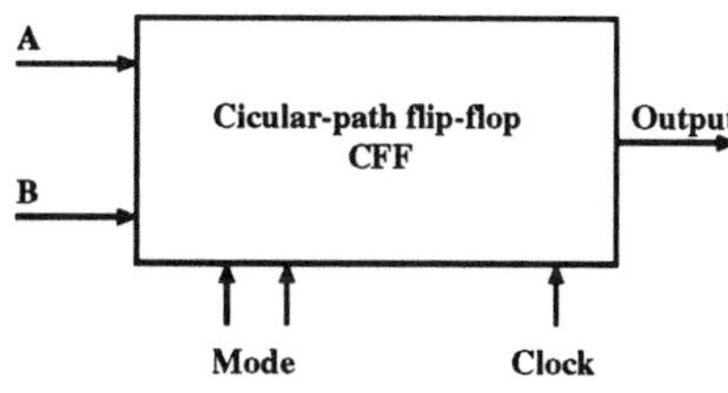

Mode	Output
Functional mode	A
BIST mode	$A \oplus B$
Reset mode	0

Figure 10.15: The circular-path flip-flop.

circular path. They have two data inputs and at least three modes of operation: the functional mode, the BIST mode, and the reset mode (Figure 10.15). The circular BIST proceeds as follows:

First, all CFFs are reset and the MUT is initialized. Next, the CFF chain is set to the BIST mode. The content of the CFFs is applied to the inputs of the MUT. Then, the CFF chain is clocked and each MUT output connected with a CFF is XOR-ed with the value of the previous CFF to produce a signature. This signature is applied to the inputs of the MUT as the next "random" pattern and the process continues for a predetermined number of clock cycles. The final signature is read by an ATE or a diagnostic unit.

The MUT inputs coming from the primary inputs of the circuit also have to be fed with some patterns. If a boundary scan is used, the boundary-scan flip-flops are included to the CFF chain and, thus, the primary inputs feed directly to the CFF chain and do not need to be treated separately (they can be set to a fixed logic value during BIST). Analogously, the primary outputs come directly from the CFF chain (boundary scan) so that the entire MUT logic contributes to the signature.

An extended approach [Stro88b] uses an extra signature register inserted into the CFF chain in order to hold the signature after the BIST session outside the functional logic until it is read by an external diagnostic unit (which possibly runs asynchronously with respect to the circuit under test). Furthermore, it exploits the fact that with the two bits used to control the three modes of a CFF, an additional fourth mode can be controlled as well. Hence, a shift mode is inserted which provides the ability of using the CFF chain also as a (partial) scan path, independently of the circular BIST. The addition of this capability has no significant effect on the logic overhead of the CFFs. However, it requires an additional circuit pin feeding to the controller, since the modes of operation for the *entire* circuit (not for the CFFs) are increased from two (functional mode, circular BIST mode) to three (addition of the scan-test mode).

A benefit of circular BIST in comparison with other BIST techniques is—provided the circuit structure is suitable—the short test time in combination with a relatively low hardware overhead. Each test pattern needs exactly one clock cycle, whereas existing functional flip-flops and boundary-scan flip-flops are used for test pattern generation and test response evaluation.

A major drawback of circular BIST is that the random properties of the test patterns applied depend on the structure of the circuit under test, since the test patterns are identical to the signatures. In case of an inconvenient circuit structure, the random properties are poor and much more clock cycles are necessary, compared with a conventional random test, in order to obtain a high fault coverage. It may even happen that the fault coverage never exceeds some limit because certain patterns never occur due to the circuit structure (limit cycling).

The question which functional flip-flops must be included in the circular path cannot be answered easily. It depends on how random-pattern resistant the remaining module under test (MUT) is and to what extent the flip-flops of the MUT corrupt the random properties of the pattern generation. From the overhead point of view, as few flip-flops as possible should be included in the circular path, whereas from the testability point of view as many as possible are asked for. Some heuristic criteria are proposed in [PrBL88] for the selection of flip-flops. Another problem is to find a convenient topological order of the CFFs within the chain with respect to the randomness of the patterns. It seems to be indispensable to run a fault simulation in order to ensure the circular-BIST testability of a circuit with a given flip-flop selection and a given CFF order. For the integration of circular BIST in a synthesis system, an algorithm for flip-flop selection and ordering, and a fault simulator have, therefore, to be included into the system.

Special BIST methods for special structures. For special module structures, especially for regular structures such as RAMs, PLAs, multipliers, etc., a large variety of special BIST methods exists. These methods are based on specific fault models rather than the classical stuck-at model.

For RAMs, various fault models and test techniques are reported [SuWa84, FuDC86, JaSt86, MaPF88, DeBT89, FrSK89, SaAV89]. The fault models concern faults in the cell array, the decoder logic, and the read/write logic. With respect to their effect, the faults can be roughly grouped into stuck-at faults, stuck-open faults preventing the access to a cell, transition faults, coupling faults, and data retention faults [DeBT89]. Many RAM BIST approaches use up/down counters as address generators, special register structures as data generators, a signature register as test response evaluator, and a more or less complex test controller. The simplest test procedure consists in successively writing 0 and 1 to each cell and reading the content of the cell and of the adjacent cells.

For the test of PLAs, crosspoint faults (extra or missing crosspoints in the AND or OR array) are of primary importance. For a comprehensive study of the numerous PLA test techniques, the reader is referred to the literature, e.g., [DaMu81, SaMU85, TrFA85, HaRe86, Fuji88, LiMC88, UpSa88].

Mixed BIST approaches. Among the numerous implementations of mixed BIST approaches reported, two that are well supported by CAD tools are roughly sketched out in the following: *ASTA* and *TIGER*.

ASTA. The BIST system ASTA [HTHI90, Illm85, IlCl89] combines a variety of different approaches for different parts of a circuit: exhaustive test, quasi-exhaustive test, subset test, and memory test. In a circuit designed with ASTA, all functional registers exhibit a special structure which allows many modes of operation, e.g., load mode, hold mode, shift mode, random test pattern generation mode (LFSR), signature analysis mode. This special register is approximately 50% bigger than a simple scan-path register. Additionally, a second type of LFSR in combination with a multiplexor, a so-called fence, is used for partitioning the circuit during the BIST mode. For each test session, a maximum number of 2^{20} patterns is allowed. Exhaustive test is applied to modules with ≤ 20 inputs provided that all module inputs are fed from a single register.

Quasi-exhaustive test is a special type of random test. It is used for modules with a number of inputs $n \leq 17$. It applies 2^{n+3} random patterns which are reported to yield a fault detection probability of more than 99.9% for irredundant combinational faults. The pattern size 2^{n+3} is larger than the pattern size 2^n necessary for an exhaustive test. This is due to the fact that an exhaustive test requires a *unique* LFSR to generate all *different* patterns in a 2^n-session, whereas a quasi-exhaustive test can be performed with more than one LFSRs used in parallel, or with a LFSR with more than n stages. This allows a shared use of LFSRs—each for testing several modules of different sizes—without reconfiguring the LFSRs. Since such a configuration does not necessarily generate different patterns within a 2^n-session, a longer session is needed to achieve a high fault coverage.

Subset test compounds a number of special manually specifiable test techniques for regular structures. Finally, a special random-like test method was developed for testing memories.

In the ASTA system, fault coverage seems to have a higher priority than hardware overhead.

TIGER. TIGER is an expert system for automatic testability insertion [Abad89]. It supports not only pure BIST but any test methodology described in a knowledge base. It is selected as a representative of a class of knowledge-based DFT/BIST systems, see e.g., [AbBr85, JoBa86, KiTH88, ZhBr88, BhPa89]. These systems rely on the modular test approach. A set of test methods, each applicable to a number of module types, are formally described in a knowledge base. This formal description contains the configuration of the logic necessary for testing the module, and the test plan. Given a circuit description at the module level (e.g., register-transfer level), and a set of constraints (e.g., required fault coverage, limits of hardware overhead, test time, etc.), an "optimal" combination of feasible test methods for all modules is selected from the knowledge base. These test methods are instantiated and the test controller is synthesized according to their test plans.

This approach has the benefit that a circuit-specific "optimal" trade-off between high fault coverage, low hardware overhead and short test time can be achieved by exploiting the degrees of freedom inherent in the selection and combination of test methods from a pool of test methods.

10.4.3.3 When to Use BIST?

As pointed out, BIST has a variety of benefits in comparison with external testing: no expensive ATE is needed; test is performed "at speed"; the same technique can be used for production and maintenance test. The price one has to pay for these benefits is a hardware overhead and, in certain cases, a performance degradation due to the additional logic. Whether this price is too high or not, depends on the case at hand. The benefits and costs have to be estimated carefully before coming to a decision about the use of BIST. Assuming that currently synthesis is mainly used in the case of application-specific circuits and systems where a low-cost design and a short time-to-market has higher priority than optimized area and performance, BIST is an attractive test strategy for this class of products.

10.4.4 Easily Testable Controllers

The result of algorithmic synthesis is often a processor-like structure consisting of a data path and a controller. A data path can be represented by a data flow graph, whereas a controller is specified as a finite state machine by means of a state transition graph. The primary task in synthesis for testability is to transform the specification to structures which are not only optimized in terms of hardware overhead and performance but also completely and easily testable.

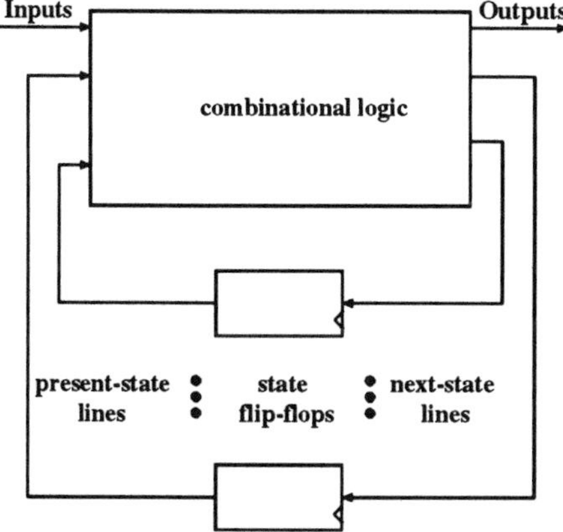

Figure 10.16: General controller model.

The controller is implemented either as micro-coded memory or as a sequential structure consisting of a combinational logic block (PLA or multi-level logic) and state registers. In this section, the second implementation type (Figure 10.16) is considered and the problem of synthesizing testable controllers of this type is addressed.

Optimal controller synthesis (cf. Chapter 5.3), independent of testability, consists of three steps targeting minimal hardware: state minimization, state assignment (encoding), and combinational logic minimization. In this context, the following questions are relevant if testability has to be taken into account: Is a controller synthesized this way always completely and easily testable? If not, is it possible to modify the above three steps such that a completely/easily testable controller results without additional hardware? If not, how can a trade-off between low hardware overhead and complete/easy testability be gained?

First, the requirement of complete testability is considered. A circuit is *completely testable* if for each fault according to a specified fault model, there exists at least one sequence of test patterns such that the output vector after applying the sequence to the faulty circuit is different from the corresponding output vector of the fault-free circuit. If no such sequence exists for a fault, the fault is called *redundant* (see also Section 10.4.5). Since the controller-synthesis procedure mentioned above produces a minimum-logic controller, it intuitively seems to be rather unlikely that such a controller has "many" redundant faults. Moreover, it has been shown that it is possible with a modified iterative approach to obtain a completely testable controller with respect to single stuck-at faults without additional hardware [DMNS90, DeMN89a]. This can be achieved by exploiting degrees of freedom (so-called *don't care sets*). However, the test patterns must be generated by a sequential test pattern generation technique which can be extremely time-consuming.

Next, the question is posed whether the controller can be synthesized in such a manner that a *low* test pattern generation expense is needed while retaining complete testability. One possibility is, of course, to use a scan path. It has been shown, however, that the above goal can also be achieved with a constrained state assignment technique without requiring a scan path [DMNS89a]. State assignment is performed in such a way that within a prespecified number (e.g., 1) of clock cycles after fault excitation, the controller comes to a state that is in the faulty case distinguishable at the controller outputs from that in the fault-free case. In contrast to the first approach, controllers synthesized with this technique exhibit increased area. Test pattern generation essentially consists of (a) combinational test pattern generation for determining, for each fault of the combinational logic, the input and present-state pattern which excites the fault and propagates its effect to the output or the next-state lines, plus (b) search in the state transition graph to justify the state determined in (a) (all states are assumed to be reachable from a reset state). This procedure is less time-consuming than in the previous case. The synthesis procedure is less time-consuming, too.

Other controller synthesis approaches also aiming at a reduction of computational expense are presented in [DeKe89, ChAg89b]. The approach proposed in [ChAg89b] relies on synthesizing pipeline structures, where possible. For pipeline structures, the expense of sequential test pattern generation is lower than in the general case (see also Section 10.4.1). However, a pipeline type implementation of an arbitrary state transition graph requires, in general, more flip-flops compared to the minimized implementation (where feedbacks occur instead of pipeline stages).

In the above considerations the controller is assumed to have accessible inputs and outputs. In reality, however, certain inputs and outputs are connected with the data path and are not directly accessible from the outside. In this case, DFT aids have to be inserted in order to make them accessible (see Section 10.4.2).

Finally, the intensive research currently directed to the complex task of synthesizing delay testable controllers has to be mentioned, e.g., [DeKe90].

10.4.5 Elimination of Redundancies and Logic Optimization

Redundancies impose large problems on test data generation. A node is called redundant if the output function of the circuit is independent of the value of that node in all states and for all possible inputs of the circuit. A stuck-at fault at a redundant node however can change the functionality of the circuit if other faults are also present. A fault at a redundant node can cause a previously detectable fault to become undetectable or another undetectable fault to become detectable. During test pattern generation faults at redundant nodes cannot be stimulated and/or observed and thus not be tested. The detection of redundancies or the proof that a node is redundant is known to be NP-complete. Therefore, circuits must contain the least possible number of redundancies because a test pattern generator spends a lot of time in trying to generate a test pattern for these faults without success. If a circuit is synthesized automatically, this has to be considered by the logic optimization tool. Ideally the circuits generated by logic optimization tools are free of redundancies and therefore 100% testable. In practice this cannot always be obtained within reasonable CPU time (except for 2-level logic) but some heuristics in logic synthesis tools can provide acceptable solutions. The removal of redundancies is not only important in order to achieve good testability but also to obtain a minimal logic. An example for this is shown in Figure 10.17.

Several approaches to this problem are known. In [Bran83] an algorithm is proposed for the fast detection of local redundancies. The algorithm performs local topological searches for redundancies. Because of this local view, the algorithm does not detect *all* redundancies. It does however never declare an irredundant node as redundant. The algorithm is used inside a logic synthesis system for further logic optimizations.

The approach published in [Bart88] guarantees totally irredundant and therefore 100% testable logic. The test patterns for the circuit are produced as a by-product. However, the CPU time requirements for this approach are very high.

In [DMNS88a] an approach for the synthesis and optimization of 100% testable sequential machines is presented. The basic idea is to perform a constrained state assignment combined with logic optimizations (see also Section 10.4.4).

Since the identification of redundancies is a by-product of automatic test pattern generation (ATPG), some approaches are published to use ATPG systems for redundancy elimination and further optimization of the logic [Brgl89]. ATPG algorithms have been developed that manage to identify all redundant faults in all of the ISCAS [BrFu85] benchmarks with a limit of only ten backtracking steps [ScAu88, ScAu89]. Schulz applied the following three step strategy to test pattern generation and redundancy identification:

- Generation of random test patterns
- Fault simulation
- Deterministic test pattern generation with integrated fault simulation for new generated test patterns. During test pattern generation use of contrapositive implications[7] is made. These implications have to be "learned" from simulations. The use of this "learning" technique drastically reduces the number of necessary backtracks.

Removal of redundancies. Each time a redundant node is found by any of the above methods the redundant logic can be eliminated. This must be done node by node and cannot be done globally for all nodes that are redundant. The reason for this is that a removed redundant node may make another redundant node irredundant. The elimination of the redundancy can be described as follows (see also Figure 10.17).

- Remove all gates and nodes in the part of the circuit driving *only*[8] the redundant node.
- If the redundant node is not testable for stuck-at-0 faults bind the node to 0 otherwise to 1.

[7] $(P \Rightarrow Q) \Leftrightarrow (\neg Q \Rightarrow \neg P)$

[8] Commonly called the *fan out free region*.

- Propagate implications of the fixed value through the fan out of the node. This means:
  - Eliminate the corresponding input at a gate if the value is not dominant (e.g., 0 at an OR-gate)
  - Eliminate the gate if the value is dominant (e.g., 0 at an AND-gate) and propagate the corresponding output value (e.g., 0 at an AND-gate).

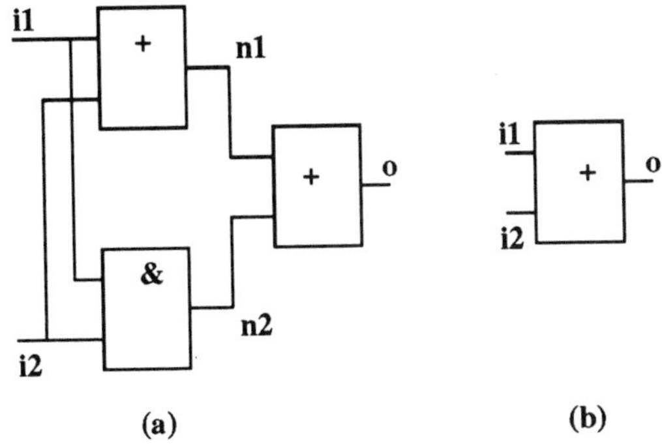

Figure 10.17: Optimization through redundancy removal: (a) before and (b) after optimization. The fault "node n2 stuck-at-0" is not detectable.

10.4.6 Modification of Data and Control Flow Graph

The test methods like scan design or BIST are inserted into the circuit *after* synthesis at the logic level. Some approaches are published to insert testability already at an earlier step in the synthesis process.

10.4.6.1 Modification of Data Flow

In [Marh89] an approach is shown that allows for consideration of testability after the data flow graph is constructed. The basic idea is to enhance the data flow graph prior to synthesis in the following way.

A test data flow is added to each operation (e.g., "+") in the data flow graph. The test data flow describes the flow of test patterns from a *test data source* (e.g., LFSR, scan chain or primary input) to a *test data sink* (e.g., MISR or primary output). This extension allows a unified allocation of functional and test operations to functional blocks, registers and interconnect. In order to reduce the area overhead, transparency properties of operations can be exploited to share test data sources and sinks by several operations. Registers can be used both as functional registers and as test registers with one or more of the following functions:

- Access function: serial read/write in test mode
- Test pattern generation (e.g., exhaustive or random)
- Test response analysis (e.g., signature computation)

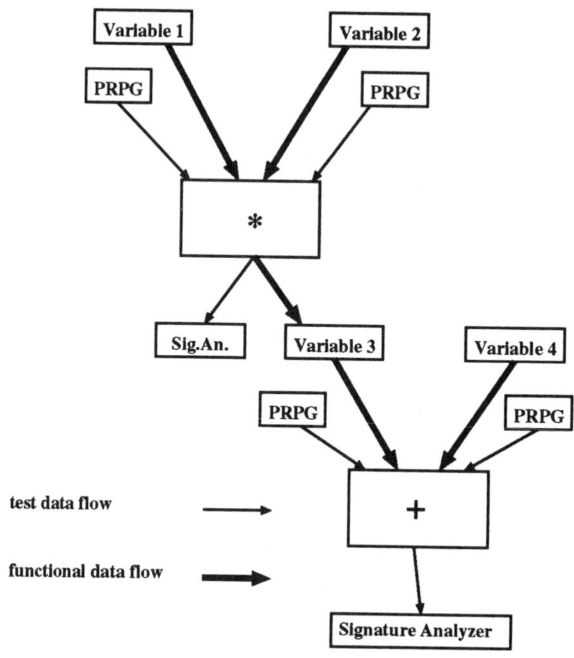

Figure 10.18: Augmented flow graph.

In the example of Figure 10.18 the augmented flow graph is shown for a self test strategy. Pseudo random pattern generators (PRPGs) are added as test data source to each operator. Signature analysis registers are used as test data sinks. In Figure 10.19 the circuit after allocation is presented. For the operation "$*$" a multiplier was allocated, for "+" an adder. The variables 1, 2, and 4 were allocated to the registers that are used as PRPG in test mode. Because the multiplier is transparent to random patterns, one of the PRPGs for the adder can be omitted. In this example 4 multifunctional registers are used to implement 10 data flow/test data flow functions.

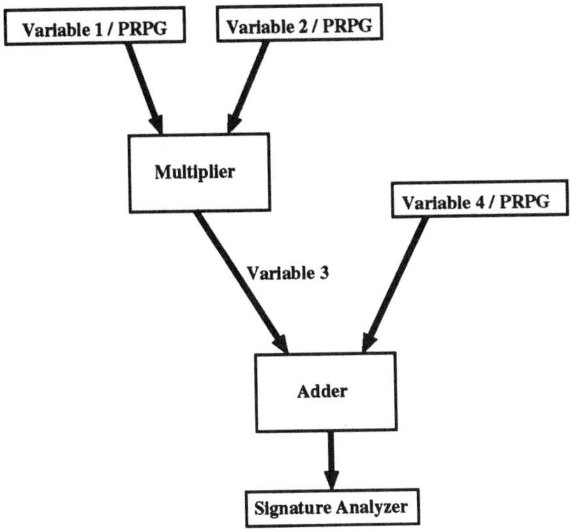

Figure 10.19: Data path after allocation.

The main advantage of this approach is that it leads directly to a register transfer-level circuit, including all test hardware. But because this approach is applied prior to scheduling and allocation, it requires a *single* test method, e.g., BIST, to be applied for the whole circuit (*homogeneous test*). This is due to the fact that it is not possible to apply a functional unit specific test method to an operation when it is not known yet what type of functional unit it is mapped onto later on. This standardization can make early DFT more expensive than if it is applied at a later stage of synthesis.

Therefore, [Marh89] proposed a slightly modified approach. The insertion of test data sources and sinks is done *after* scheduling and allocation. Now it is possible to apply a specific test method for each of the functional units. Transparency properties can also be exploited in order to reduce the complexity of the test hardware. Also, already allocated functional registers can be converted to test registers. Similar approaches are presented in [AbBr85, KiTH88].

10.4.6.2 Modification of Control Flow

The control part of a synthesized circuit can be represented at an abstract level using state diagrams. In [AgCh90] a method is proposed how testability of a controller can be incorporated already at the state transition diagram level. The basic idea is to incorporate a test function into the state transition diagram of the original machine. The test function is built in such a way that each state of the FSM can be set and observed easily. The test state machine uses the same number of state variables as the functional machine. After construction of the test machine the two state transition diagrams are superimposed such that a minimum number of new transitions is needed. Because all states of the test machine are easily controllable and observable the same holds true for the states of the functional state machine. The state assignment and logic minimization is done on the combined machine, thus optimizing both functional and test hardware.

10.5 Test Data Generation

In the previous sections of this chapter, the main focus is on DFT and the synthesis of testable circuits. However, a complete synthesis system must also generate the test data for the synthesized circuit (at least, if an external test is to be applied, rather than BIST).

General test pattern generation techniques for arbitrary random-logic structures, such as PODEM [Goel81], FAN [FuSh83], or SOCRATES [ScTS87], [ScAu89], as well as a number of known test pattern generation techniques for regular structures (RAMs, PLAs, etc.) can be applied to synthesized circuits. However, most of these techniques are established for more or less arbitrary structures. In contrast, the modules and circuits generated by a synthesis system do not have an arbitrary structure but a well-defined structural style inherent in the synthesis methodology used. Thus, test patterns can be generated in a more efficient manner using methods specially optimized for the structural styles supported by the given synthesis system, rather than general methods. For instance, the PLA generator of a synthesis system generates not only the PLA structure but in addition to that, the corresponding test patterns.

The patterns generated for the modules are transformed to the system level as described in Section 10.4.2. As also mentioned in Section 10.4.2, the test patterns generated for each module according to a fault model must be extended by functional test patterns at the system level in order to examine the functional mode of the overall system. Usually, these functional patterns are identical to the simulation patterns. They are chosen manually in such a way that each "function" (operation) of the system is stimulated at least once. In a synthesis environment where "correctness by construction" of the circuit structure is assumed, and therefore—at least in the ideal case—no simulation is needed, the patterns for a functional test session must be generated automatically. This can be done with the aid of the control-flow representation (e.g., state transition graph or Petri net) available during synthesis. Under the assumption that each "function" of the circuit corresponds to an edge in the control flow graph, functional test pattern generation can be modeled as the generation of a set of pattern sequences which sensitize all edges of the control flow graph. The inputs of the data path not affecting the control flow may be arbitrary. However, they must ensure that the output patterns allow for detecting whether the desired edge or—in the faulty case—another edge has been sensitized.

In addition to test pattern generation for static faults, delay-test pattern generation is also significant [Smit85, LiRe87, ScFF89]. Unfortunately, a satisfactory quality assessment of a delay-test pattern set is very difficult due to the intractable large number of structural paths in a large circuit and to the lack of information about functional path sensitizable for a signal transition. Functional and structural information—available during synthesis—about critical paths and critical timing conditions of single modules or of the entire circuit are useful for a delay test.

After test pattern generation for static and delay faults has been performed, a complete test program is integrated which contains the logic test (i.e., the test of the logic behavior by means of the generated patterns) and the parametric test (i.e., the test of electrical parameters such as voltage levels, leakage currents, DC and AC current consumption, etc.). Finally, the test program is converted to the language of the ATE to be used. Since this stage of test data generation is merely an integration and conversion procedure and does not depend on the specific synthesis methodology, conventional stand-alone post-processors can be used.

10.6 Outlook

The aim of this chapter is to consider the test problem from the synthesis point of view and to show some possible solutions. Since the main objective of synthesis is to minimize time-to-market, one cannot afford to neglect testability and test data generation which are bottlenecks even in conventional design. Therefore, the parameter testability has to be included as a steering criterion of synthesis, together with hardware overhead and performance.

Testability must be taken into account at all stages of the synthesis process, from specification to low-level module generation. Testability considerations at early synthesis stages prevent unnecessary time-consuming feedbacks in the synthesis process. On the other hand, completely and easily testable modules are a prerequisite for the overall testability of a circuit.

Built-in self-test is expected to play a prominent role in the synthesis area, though it is in many cases necessary to supplement BIST with a short external test session to test the few faults not covered by BIST. A combination of dedicated module-specific test methodologies appears to be more efficient than a "universal" methodology.

As mentioned earlier, the test techniques outlined in this section are only exemplarily chosen and do not form a complete test strategy for a synthesis system. The test strategy supported by a synthesis system has, of course, to be selected in accordance to the overall synthesis strategy.

10.7 Problems for the Reader

1. The circuit shown in Figure 10.20 consists of an adder ADD and two registers REG1 and REG2. Both registers can be cleared asynchronously. Write a test plan for testing externally all single stuck-at faults on the data inputs and outputs of REG1 and REG2.

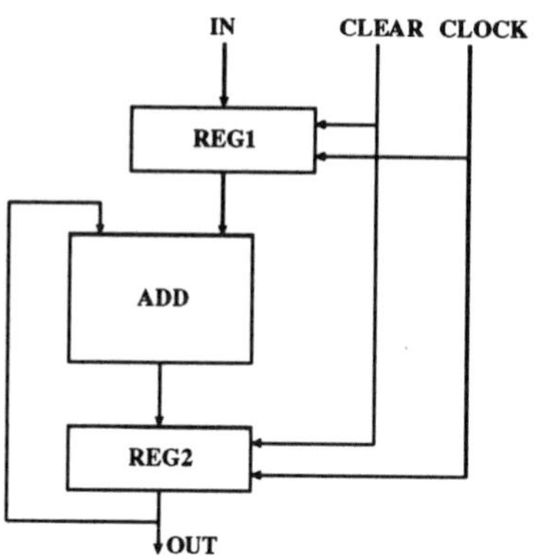

Figure 10.20: Test plan.

2. In circular BIST the so-called adjacency problem is well-known. It arises if in the functional configuration the output of a memory element feeds directly to the input of another memory element, and in the circular chain the corresponding augmented flip-flops are also directly connected in the same order. Why does this situation impair the random properties of circular BIST? How can the problem be avoided?

Bibliography

[Abad89] M.S. Abadir. TIGER: Testability Insertion Guidance Expert System. In *Proceedings of the International Conference on Computer Aided Design (ICCAD)*, 1989.

[AbBr85] M.S. Abadir and M.A. Breuer. A Knowledge-Based System for Designing Testable VLSI Chips. *IEEE Design & Test of Computers*, August 1985.

[AbBr86] M.S. Abadir and M.A. Breuer. Test Schedules for VLSI Circuits Having Built-In Test Hardware. *IEEE Transactions on Computers*, 1986.

[ABCo86] M. Ajmone Marson, G. Balbo, and G. Conte. *Performance Models of Multiprocessor Systems.* MIT Press, Cambridge MS, 1986.

[ADA83] American National Standards Institute, 1430 Broadway, New York, New York 10018. *ANSI Reference Manual for the ADA Programming Language*, 1983.

[AdPo86] H.H. Adelsberger and U.W. Pooch, et al. Rule Based Object Oriented Simulation Systems. In *[LuAd86]*, 1986.

[AgCh90] V.D. Agrawal and K-T. Cheng. Test Function Specification in Synthesis. In *Proceedings of the 27th ACM/IEEE Design Automation Conference (DAC)*, pages 235–240, 1990.

[AHL91] Abstract Hardware Ltd., Brunel Science Park, Uxbridge UB8 3PH, UK. *LAMBDA Version 4.0 Reference Manual and User Guide*, July 1991.

[AhSU86] A.V. Aho, R. Sethi, and J.D. Ullman. *Compilers—Principles, Techniques, and Tools.* Addison-Wesley Publishing Company, 1986.

[Aker78] S.B. Akers. Binary Decision Diagrams. *IEEE Transactions on Computers*, Vol. C-27, June 1978.

[AmBa89] R. Amann and U.G. Baitinger. Optimal State Chains and State Codes in Finite State Machines. *IEEE Transactions on Computer Aided Design*, CAD–8(2), Febr. 1989.

[AmBW89] T. Amon, G. Borriello, and W. Winder. A Unified Behavioral/Structural Representation for Simulation and Synthesis. In *Proceedings of the Fourth International Workshop on High-Level Synthesis*, 1989.

[Apos89] E. Aposporidis, F. Hoppe, F. Lohnert, P. Mehring, and H.U. Post. PMLS—A Parallel Multi-Level VLSI Simulator. In *Proceedings of the EUROMICRO 89*, Sept. 1989.

[AsDN91] P. Ashar, S. Devadas, and R. Newton. Optimum and Heuristic Algorithms for an Approach to Finite State Machine Decomposition. *IEEE Transactions on Computer Aided Design*, CAD–10(3), March 1991.

[ASSP90] P. Abouzeid, K. Sakouti, G. Saucier, and F. Poirot. Multilevel Synthesis Minimizing the Routing Factor. In *Proceedings of the IFIP Working Conference on Logic and Architecture Synthesis*, 1990.

[BaAS87] P.H. Bardell, W.H. McAnney, and J. Savir. *Built-In Test for VLSI: Pseudorandom Techniques.* John Wiley & Sons, 1987.

[Back78] J. Backus. Can Programming Be Liberated from the von Neumann Style? A Functional Style and Its Algebra of Programs. *Communications of the ACM*, 21(8), August 1978.

[BaMa90] M. Balakrishnan and P. Marwedel. Integrated Scheduling and Binding: A Synthesis Approach for Design Space Exploration. In *Proceedings of the 26th ACM/IEEE Design Automation Conference (DAC)*, pages 68–74, 1989.

[Barb81] M.R. Barbacci. Instruction Set Processor Specifications (ISPS): The Notation and Its Applications. *IEEE Transactions on CAD*, pages 24–40, Jan. 1981.

[Bart88] K.A. Bartlett, R.K. Brayton, G.D. Hachtel, R.M. Jacoby, C.R. Morrison, R.L. Rudell, A. Sangiovanni-Vincentelli, and A.R. Wang. Multilevel Logic Minimization Using Implicit Don't Cares. *IEEE Transactions on CAD*, 7(6), June 1988.

[BaSK91] H. Bauer, C. Sporrer, and T.H. Krodel. On Distributed Logic Simulation Using Time Warp. In *Proceedings of the IFIP VLSI 91*. North Holland, 1991.

[BDPV88] D.K. Beece, G. Deibert, G. Papp, and F. Villante. The IBM Engineering Verification Engine. In *Proceedings of the 25th ACM/IEEE Design Automation Conference (DAC)*, 1988.

[BDSS89] F. Beenker, R. Dekker, R. Stans, and M. van der Star. A Testability Strategy for Silicon Compilers. In *Proceedings of the International Test Conference (ITC)*, 1989.

[BDSS90] F. Beenker, R. Dekker, R. Stans, and M. van der Star. Implementing Macro Test in Silicon Compiler Design. *IEEE Design & Test of Computers*, April 1990.

[BeAl88] R.A. Bergamaschi and D.J. Allerton. A Graph-based Silicon Compiler for Concurrent VLSI Systems. In *Proceedings of the CompEuro 88*, pages 36–47, 1988.

[BeCa89] R. Bergamaschi and R. Camposano. Synthesis Using Path-Based Scheduling: Algorithms and Exercises. In *Proceedings of the Fourth International Workshop on High-Level Synthesis*, 1989.

[Beck85] B. Becker. An Easily Testable Optimal-time VLSI Multiplier. In *Proceedings of the Euromicro*, 1985.

[Beck88] B. Becker. Efficient Testing of Optimal Time Adders. *IEEE Transactions on Computers*, 37(9):1113 ff, September 1988.

[BeCP91] R. Bergamaschi, R. Camposano, and M. Payer. Area and Performance Optimizations in Path-Based Scheduling. In *Proceedings of the European Design Automation Conference (EDAC)*, 1991.

[BeHS90] R.K. Brayton, G.D. Hachtel, and A.L. Sangiovanni-Vincentelli. Multilevel Logic Synthesis. In *Proceedings of the IEEE*, volume Vol. 78, February 1990.

[BeLW91] M. Bechtold, T. Leyendecker, and I. Wich. A Dynamic Framework for Simulator Tool Integration. In F.J. Rammig and R. Waxman, editors, *Electronic Design Automation Frameworks*. North Holland, 1991.

[Berg88] R.A. Bergamaschi. Automatic Synthesis and Technology Mapping of Cominational Logic. In *Proceedings of the International Conference on Computer Aided Design (ICCAD)*, pages 466–473, 1988.

[BhLe90] J. Bhasker and H.-C. Lee. An Optimizer for Hardware Synthesis. *IEEE Design & Test for Computers*, pages 20–36, Oct. 1990.

[BHMS84] R. Brayton, G. Hachtel, C. McMullen, and A. Sangiovanni-Vincentelli. *Logic Minimization Algorithms for VLSI Synthesis*. Kluwer Academic Publishers, 1984.

[BhPa89] S. Bhawmik and P. Palchaudhuri. DFT EXPERT: Designing Testable VLSI Circuits. *IEEE Design & Test of Computers*, October 1989.

[BiGS89] W.P. Birmingham, A.P. Gupta, and D.P. Siewiorek. The MICON System for Computer Design. In *Proceedings of the 26th ACM/IEEE Design Automation Conference (DAC)*, pages 135–140, 1989.

[BiSu89] G. Birtwistle and P.A. Subrahmanyam, editors. *VLSI Specification, Verification, and Synthesis.* Kluwer Academic Publishers, 1989.

[BKKR86] R. Brück, B. Kleinjohann, T. Kathöfer, and F.J. Rammig. Synthesis of Concurrent Modular Controllers from Algorithmic Descriptions. In *Proceedings of the 23rd ACM/IEEE Design Automation Conference (DAC)*, pages 285–292, 1986.

[BoMo88] R.S. Boyer and J.S. Moore. The User's Manual for A Computational Logic. Technical Report 18, Computational Logic Inc., Austin, TX, February 1988.

[Bori88] G. Boriello. *A New Interface Specification Methodology and its Application to Transducer Synthesis.* PhD thesis, U.C. Berkeley, 1988.

[Borr81] D. Borrione. Langue de description des systemes logiques—Proposition pour une methode formelle de definition. These d'Etat, INPG Grenoble, 1981.

[Bran83] D. Brand. Redundancies and Don't Cares in Logic Synthesis. *IEEE Transactions on Computers*, C-32(10):947–952, October 1983.

[Bray88] R.K. Brayton, R. Camposano, G. De Micheli, R.H.J.M. Otten, and J. van Eijndhoven. The Yorktown Silicon Compiler System. In D.D. Gajski, editor, *Silicon Compilation*, pages 204–310. Addison-Wesley, 1988.

[BrFr76] M.A. Breuer and A.D. Friedman. *Diagnosis and Reliable Design of Digital Systems.* Computer Science Press, 1976.

[BrFu85] F. Brglez and H. Fujiwara. A Neutral Netlist of 10 Combinational Benchmark Circuits and A Target Translator in FORTRAN. In *Proceedings of the IEEE International Symposium On Circuits and Systems*, pages 663–698, June 1985.

[BrGK89] F. Brglez, G. Gloster, and G. Kedem. Hardware-Based Weighted Random Pattern Generation for Boundary Scan. In *Proceedings of the International Test Conference (ITC)*, 1989.

[Brgl89] F. Brglez, D. Bryan, J. Calhoun, G. Kedem, and R. Lisanke. Automated Synthesis for Testability. *IEEE Transactions on Industrial Electronics*, 36(2):263–277, May 1989.

[BrHS81] R.K. Brayton, G.D. Hachtel, and A. Sangiovanni-Vincentelli. A Taxonomy of CAD for VLSI . In *Proceedings of the 1981 European Conference on Circuit Theory and Design*, pages 34–57, 1981.

[BrHu89] B.C. Brock and W.A. Hunt Jr. The Formalization of a Simple Hardware Description Language. In L. Claesen, editor, *Proceed-*

ings of the IMEC/IFIP International Workshop on Applied Formal Methods for Correct VLSI Design, Houthalen, Belgium, 1989.

[BrHu90] B.C. Brock and W.A. Hunt Jr. A Formal Introduction to a Simple HDL. In J. Staunstrup, editor, *Formal Methods for VLSI Design*, IFIP WG 10.5 Lecture Notes. North-Holland, 1990.

[Brow81] D.W. Brown. A State-Machine Synthesizer - SMS. In *Proceedings of the 28th ACM/IEEE Design Automation Conference (DAC)*, 1981.

[BRSW87] R. Brayton, R. Rudell, A. Sangiovanni-Vincentelli, and A. Wang. MIS: A Multiple-Level Logic Optimization System. *IEEE Transactions on Computer Aided Design*, Vol. 6, November 1987.

[Brya86] R.E. Bryant. Graph-Based Algorithms for Boolean Function Manipulation. *IEEE Transactions on Computers*, August 1986.

[BuLa87] R. Buschke and K. Lagemann. SIMUEVA, a CAD Program Package for Efficient Analysis of Simulation Results. In *Proceedings of the IEEE COMPEURO 87*, 1987.

[Busc90a] H. Busch. Proof-Based Transformation of Formal Hardware Models. In G. Jones and M. Sheeran, editors, *Proceedings of the Workshop on Designing Correct Circuits, Oxford.* Springer, September 1990.

[Busc90b] R. Buschke. Grundlagen und Methoden zur computerunterstützten Auswertung von Simulationsergebnissen im VLSI-Design. Dissertation, FB Informatik, Univ. Hamburg, 1990.

[CaBe90] R. Camposano and R. Bergamaschi. Synthesis Using Path-Based Scheduling: Algorithms and Exercises. In *Proceedings of the 27th ACM/IEEE Design Automation Conference (DAC)*, pages 450–455, 1990.

[CaBr87] R. Camposano and R.K. Brayton. Partitioning Before Logic Synthesis. In *Proceedings of the International Conference on Computer Aided Design (ICCAD)*, pages 324–326, 1987.

[CADA91] Racal Redac Inc. *CADAT 2000 System Reference Manual*, 1991.

[Camp91a] R. Camposano. Path-Based Scheduling For Synthesis. *IEEE Transactions on Computer Aided Design*, 10(1):85–93, January 1991.

[Camp91b] R. Camposano, R.A. Bergamaschi, C.E. Haynes, M. Payer, and S.M. Wu. The IBM High-Level Synthesis System. In *Trends in High-Level Synthesis.* Kluwer Academic Publishers, 1991.

[CaRo89] R. Camposano and W. Rosenstiel. Synthesizing Circuits From Behavioral Descriptions. *IEEE Transactions on CAD*, 8:171–180, Feb. 1989.

[CBBG87] E. Clarke, S. Bose, M. Browne, and O. Grumberg. The Design and Verification of Finite State Hardware Controllers. Technical Report CMU–CS–87–145, Carnegie Mellon University, CS Dept, July 1987.

[ChAg89a] K.-T. Cheng and V.D. Agrawal. An Economical Scan Design for Sequential Logic Test Generation. In *Proceedings of the International Symposium on Fault-Tolerant Computing (FTCS)*, pages 28–35, 1989.

[ChAg89b] K.-T. Cheng and V.D. Agrawal. Design of Sequential Machines for Efficient Test Generation. In *Proceedings of the International Conference on Computer Aided Design (ICCAD)*, 1989.

[Chin90] S.-K. Chin. Synthesis of Arithmetic Hardware Using Hardware Metafunctions. *IEEE Transactions on Computer-Aided Design of Integrated Circuits and Systems*, 9(8), August 1990.

[Chu74] Y. Chu. Introducing CDL. *IEEE Computer*, 7(12), 12 1974.

[Clae90] L. Claesen, editor. *Formal VLSI Specification and Synthesis—VLSI Design Methods, Volumes I and II.* North-Holland, 1990.

[ClTh90] R.J. Cloutier and D.E. Thomas. The Combination of Scheduling, Allocation, and Mapping in a Single Algorithm. In *Proceedings of the 27th ACM/IEEE Design Automation Conference (DAC)*, pages 71–76, 1990.

[Coel89] D.R. Coelho. *The VHDL Handbook.* Kluwer Academic Publishers, 1989.

[Coff76] E.G. Coffman Jr., editor. *Computer and Job Scheduling Theory.* J. Wiley, New York, 1976.

[Comp74] Various papers in IEEE Computer. *IEEE Computer*, 7(12), 12 1974.

[Comp77] Various papers in IEEE Computer. *IEEE Computer*, 10(6), 6 1977.

[CSIM89] F. Catthor, J. van Sas, L. Inze, and H. De Man. A Testability Strategy for Multiprocessor Architecture. *IEEE Design & Test of Computers*, April 1989.

[DACA90] Dosis GmbH, Dortmund. *DACAPO III System User Manual*, 1990.

[DaLH89] N. Daeche, M. Longley, and K. Hanna. Overview of the Interactive VERITAS$^+$ Environment. Technical Report, Faculty of Information Technology, University of Kent, Kent CT2 7NT, U.K., November 1989.

[DaMu81] W. Dähn and J. Mucha. A Hardware Approach to Self-Testing of Large Programmable Logic Arrays. *IEEE Transactions on Computers*, November 1981.

[Dani70] S.F. Daniels. A Technique For the Analysis of Digital Systems Dynamics. Ph.D. Thesis, Case Western Reserve University, CS, University Microfilms, 1970.

[DeBr85] G. De Micheli, R.K. Brayton, and A. Sangiovanni-Vincentelli. Optimal State Assignment for Finite State Machines. *IEEE Transactions on Computer Aided Design*, CAD–4(3):269–284, July 1985.

[DeBT89] R. Dekker, F. Beenker, and L. Thijssen. Realistic Built-In Self-Test for Static RAM's. *IEEE Design & Test of Computers*, February 1989.

[DeDN85] P. Dewilde, E. Deprettere, and R. Nouta. Parallel and Pipelined VLSI Implementations of Signal Processing Algorithms. In T. Kailath S.Y. Kung, H.J. Whitehouse, editor, *VLSI and Modern Signal Processing*. Prentice Hall, 1985.

[DeKe89] S. Devadas and K. Keutzer. Boolean Minimization and Algebraic Factorization Procedures for Fully Testable Sequential Machines. In *Proceedings of the International Conference on Computer Aided Design (ICCAD)*, 1989.

[DeKe90] S. Devadas and K. Keutzer. Synthesis and Optimization Procedures for Robustly Delay-Fault Testable Combinational Logic Circuits. In *Proceedings of the 27th ACM/IEEE Design Automation Conference (DAC)*, 1990.

[DeKu88] G. De Micheli and D.C. Ku. HERCULES - A System for High-Level Synthesis. In *Proceedings of the 25th ACM/IEEE Design Automation Conference (DAC)*, pages 483–488, 1988.

[DeMa87] H. De Man. Evolution Towards Third Generation Custom VLSI Design. *Revue de Physique appliquee*, 22, 1987.

[DeMa90] H. De Man, F. Catthoor, G. Goossens, J. Vanhoof, J. van Meerbergen, S. Note, and J. Huisken. Architecture-Driven Synthesis Techniques for VLSI Implementation of DSP Algorithms. *Proceedings of the IEEE*, 78(2), February 1990.

[Demi86a] G. De Micheli. Symbolic Design of Combinational and Sequential Logic Circuits Implemented by Two-Level Logic Macros. *IEEE Transactions on Computer Aided Design*, CAD–5(4):597–616, October 1986.

[Demi86b] G. De Micheli. Symbolic Design of Combinational and Sequential Logic Circuits Implemented by Two-Level Logic Macros. Research Report RC 11672, IBM-T.J. Watson Research Center, January 1986.

[DeMN89a] S. Devadas, H.-K.T. Ma, and A.R. Newton. Redundancies and Don't Cares in Sequential Logic Synthesis. In *Proceedings of the International Test Conference (ITC)*, 1989.

[DeNe88] S. Devadas and A.R. Newton. Decomposition and Factorization of Sequential Finite State Machines. In *Proceedings of the International Conference on Computer Aided Design (ICCAD)*, pages 148–151, 1988.

[DeNe89] S. Devadas and A.R. Newton. Algorithms for Hardware Allocation in Data Path Synthesis. *IEEE Transactions on CAD*, pages 768–781, July 1989.

[Denn84] J.B. Dennis. Models of Data Flow Computation. In M. Broy, editor, *International Summer School: Control Flow and Data Flow - Concepts of Distributed Programming*, pages 346–354. Springer Verlag, 1985.

[Detj87] E. Detjeus, G. Gannot, R. Rudell, A. Sangiovanni-Vincentelli, and A. Wang. Technology Mapping in MIS. In *International Conference on Computer Aided Design (ICCAD)*, pages 116–119. IEEE, November 1987.

[Deva88] S. Devadas. Approaches to Multi-Level Sequential Logic Synthesis. In *Proceedings of the 26th ACM/IEEE Design Automation Conference (DAC)*, 1988.

[Diet74] D.L. Dietmeyer. Introducing DDL. *IEEE Computer*, 7(12), 12 1974.

[DISI91] W. Jud. Simulation Digitaler Systeme - Das DISIM-3-System. *ASIM Mitteilungen aus den Arbeitskreisen*, (5):1–4, 1987.

[DKMT90] G. De Micheli, D. Ku, F. Mailhot, and T. Truong. The Olympus Synthesis System. In *IEEE Design and Test of Computer, 7(5)*, pages 37–53, 1990.

[DMNS88a] S. Devadas, H.-K.T. Ma, A.R. Newton, and A. Sangiovanni-Vincentelli. Synthesis and Optimization Procedures for Fully and Easily Testable Sequential Machines. In *Proceedings of the International Test Conference (ITC)*, pages 621–630, 1988.

[DMNS88b] S. Devadas, H.K. MA, A.R. Newton, and A. Sangiovanni-Vincentelli. MUSTANG: State Assignment of Finite State Machines Targeting Multilevel Logic Implementations. *IEEE Transactions on Computer Aided Design*, CAD–7(12):1290–1300, December 1988.

[DMNS89a] S. Devadas, H.-K.T. Ma, A.R. Newton, and A. Sangiovanni-Vincentelli. A Synthesis and Optimization Procedure for Fully and Easily Testable Sequential Machines. *IEEE Transactions on CAD*, October 1989.

[DMNS89b] S. Devadas, H.-K.T. Ma, A.R. Newton, and A. Sangiovanni-Vincentelli. The Relationship between Logic Synthesis and Test. In *Proceedings of VLSI*, 1989.

[DMNS90] S. Devadas, H.K.T. Ma, A.R. Newton, and A. Sangiovanni-Vincentelli. Irredundant Sequential Machines Via Optimal Logic Synthesis. *IEEE Transactions on Computer Aided Design*, 9(1):8–18, January 1990.

[DrHa89] D. Drusinsky and D. Harel. Using StateCharts for Hardware Description and Synthesis. *IEEE Transactions on CAD*, 8(7):798–807, July 1989.

[DuDi75] J.R. Duley and D.L. Dietmeyer. A Digital System Design Language (DDL). *IEEE ToC*, C-24(2), 1975.

[Duzy89] P. Duzy, H. Krämer, M. Neher, M. Pilsl, W. Rosenstiel, and T. Wecker. CALLAS—Conversion of Algorithms to Library Adaptable Structures. In *Proceedings of the IFIP VLSI Conference, Munich*, pages 197–208, 1989.

[EdFo87] M.D. Edwards and J. Forrest. Initial Architecture for the BEAST Mapping Process. ESPRIT Project AIDA report: AIDA/UMIST/004-1, University of Manchester, Institute of Science and Technology, 1987.

[EdFo88] M.D. Edwards and J. Forrest. Alternative Architectures for the BEAST Mapping Process. ESPRIT Project AIDA report: AIDA/UMIST/007-1, University of Manchester, Institute of Science and Technology, 1988.

[EgRo88] H.W. Egdorf and D.D. Robert. Discrete Event Simulation in the Artificial Intellegence Environment. In *Proceedings of the Conference on AI and Simulation, AI Papers*, 1988.

[EiWi77] E.B. Eichelberger and T.W. Williams. A Logic Design Structure for LSI Testability. In *Proceedings of the 14th ACM/IEEE Design Automation Conference (DAC)*, pages 462–468, 1977.

[ElÖZ86] M.S. Elzas, T.I. Ören, and B.P. Zeigler. *Modelling and Simulation Methodology in the Artificial Intellegence Era.* North Holland, 1986.

[ErDe91] S. Ercolani and G. De Micheli. Technology Mapping for Electrically Programmable Gate Arrays. In *Proceedings of the 28th ACM/IEEE Design Automation Conference (DAC)*, pages 234–239, June 1991.

[ErLa85] M.D. Ercegovac and T. Lang. *Digital Systems and Hardware/Firmware Algorithms.* John Wiley & Sons, Inc., 1985.

[FFFH90] S. Finn, M. Fourman, M. Francis, and R. Harris. Formally Based System Design—Interactive Synthesis Based on Computer-Assisted Formal Reasoning. In L. Claesen, editor, *Formal VLSI Specification and Synthesis—VLSI Design Methods, Volume I*, pages 139–152. North-Holland, 1990.

[FHPZ87] M.P. Fourman, R.C. Holte, W.J. Palmer, and R.M. Zimmer. Top-Down Design as Bottom-Up Proof. In *Proceedings of the Electronic Design Automation Conference*, pages 617–628, Wembley, 1987.

[FiHa86] A.J. Field and P.G. Harrison. *Functional Programming.* International Computer Science Series. Addison-Wesley, 1986.

[Fogg89] D.C. Fogg. Operator Selection: Two Approaches. In *Proceedings of the Fourth International High-Level Synthesis Workshop*, Kennebunkport, March 1989.

[FoMa89] M.P. Fourman and E. Mayger. Formally Based System Design—Interactive Hardware Scheduling. In G. Musgrave and U. Lauther, editors, *Proceedings of the International Conference on VLSI.* Elsevier Science Publishers, 1989.

[Forg81] C.L. Forgy. *OPS5 User's Manual.* Carnegie Mellon University, 1981.

[Four90] M.P. Fourman. Formal System Design. In J. Staunstrup, editor, *Formal Methods for VLSI Design*, Chapter 5, pages 191–236. North-Holland, 1990.

[FrRC90] R.J. Francis, J. Rose, and K. Chung. Chortle: A Technology Mapping Program for Lookup Table-Based FPGAs. In *Proceedings of the 27th ACM/IEEE Design Automation Conference (DAC)*, pages 613–619, 1990.

[FrRV91] R. J. Francis, J. Rose, and Z. Vranesic. Technology Mapping for Delay Optimization of Lookup Table-Based FPGAs. In *International Workshop on Logic Synthesis*, 1991.

[FrSK89] M. Franklin, K.K. Saluja, and K. Kinoshita. Design of a BIST RAM with Row/Column Pattern Sensitive Fault Detection Capability. In *Proceedings of the International Test Conference (ITC)*, 1989.

[FuDC86] A. Fuentes, R. David, and B. Courtois. Random Testing versus Deterministic Testing of RAM's. In *Proceedings of the International Symposium on Fault-Tolerant Computing (FTCS)*, 1986.

[Fuji85] H. Fujiwara. *Logic Testing and Design for Testability.* MIT Press, 1985.

[Fuji88] H. Fujiwara. A Design of Programmable Logic Arrays with Random Pattern Testability. *IEEE Transactions on CAD*, January 1988.

[FuSh83] H. Fujiwara and T. Shimono. On the Acceleration of Test Generation Algorithms. *IEEE Transactions on Computers*, December 1983.

[GaJo79] M.R. Garey and D.S. Johnson. *Computers and Intractability.* W.H. Freeman and Company, New York, 1979.

[GaKu83] D. Gajski and R. Kuhn. Guest Editors' Introduction: New VLSI Tools. *IEEE Computer*, 6(12):11–14, 12 1983.

[GBDH86] D. Gregory, K. Bartlett, A. De Geus, and G. Hachtel. SOCRATES: A System for Automatically Synthesizing and Optimizing Combinational Logic. In *Proceedings of the 23rd ACM/IEEE Design Automation Conference (DAC)*, 1986.

[GeMu91] M. Geiger and T. Mueller-Wipperfuerth. FSM Decomposition revisited: Algebraic Structure theory applied to MCNC benchmark FSMs. In *Proceedings of the 28th ACM/IEEE Design Automation Conference (DAC)*, 1991.

[GeNe84] M. Gerner and H. Nertinger. Scan Path in CMOS Semicustom LSI Chips? In *Proceedings of the International Test Conference (ITC)*, 1984.

[Geni90] D. Genin, P. Hilfinger, J. Rabaey, C. Scheers, and H. De Man. DSP Specification Using The Silage Language. In *Proceedings of the International Conference on Acoustics, Speech, and Signal Processing (ICASSP 90)*, pages 1057–1060, 1990.

[Girc87] E.F. Girczyc. Loop Winding - A Data Flow Approach to Functional Pipelining. In *Proceedings of the ISCAS, Philadelphia*, pages 382–385, 1987.

[GlUm90] W. Glunz and G. Umbreit. VHDL for High-Level Synthesis of Digital Systems. In *Proceedings of the 1st European Conference on VHDL*, 1990.

[Goel81] P. Goel. An Implicit Enumeration Algorithm to Generate Tests for Combinational Logic Circuits. *IEEE Transactions on Computers*, March 1981.

[Golu80] M.C. Golumbic. *Algorithmic Graph Theory and Perfect Graphs.* Academic Press, 1980.

[Goor89] A.J. van de Goor. *Computer Architecture and Design.* Addison-Wesley, 1989.

[Goos87] Gert Goosens and et. al. An Efficient Microcode Compiler for Custom DSP Processors. In *Proceedings of the International Conference on Computer Aided Design (ICCAD)*, pages 24–27, Santa Clara, November 1987.

[Goos88] G. Goossens, D. Lanneer, J. Vanhoof, J. Rabaey, J. van Meerbergen, and H. De Man. Optimization-Based Synthesis of Multiprocessor Chips for Digital Signal Processing with Cathedral-II. In G. Saucier and P.M. McLellan, editor, *Logic and Architecture Synthesis*, 1988.

[Goos89] G. Goossens. *Optimisation Techniques for Automated Synthesis of Application-Specific Signal-Processing Architectures.* PhD thesis, Katholieke Universiteit Leuven, 1989.

[Gord86] M.J. Gordon. Why Higher-Order Logic is a Good Formalism for Specifying and Verifiying Hardware. In G.J. Milne and P.A. Subrahmanyam, editors, *Formal Aspects of VLSI Design.* North-Holland, 1986.

[Gord87] M.J. Gordon. A Proof Generating System for Higher-Order Logic. Technical Report 103, University of Cambridge, January 1987.

[GrMu89] W. Grass and M. Mutz. Modular Implementation of Finite State Machines Observing Topological Constraints. In *Proceedings of CHDL and their Application*, 1989.

[GrRo91] M. Groh and W. Rosenstiel. Technology Mapping for Table-Look-Up Programmable Gate Arrays. In *Proceedings MCNC International Workshop on Logic Synthesis*, May 1991.

[GuBi89] B. Gurunath and N.N. Biswas. An Algorithm for Multiple Output Minimization. *IEEE Transactions on Computer Aided Design*, Vol. 8, November 1989.

[GuGB89] R. Gupta, R. Gupta, and M. Breuer. An Efficent Implementation of the BALLAST Partial Scan Architecture. In *Proceedings of the IFIP Conference VLSI89*, pages 133–142, 1989.

[Guib89] F. El Guibaly. Design and Analysis of Arbitration Protocols. *IEEE Transactions on Computers*, 2(38):161–171, 1989.

[GuKR91] P. Gutberlet, H. Krämer, and W. Rosenstiel. CASCH - a Scheduling Algorithm for High Level Synthesis. In *Proceedings of the European Design Automation Conference (EDAC)*, pages 311–315, Feb. 1991.

[GuMi90] R. Gupta and G. De Micheli. Partitioning of Functional Models of Synchronous Digital Systems. In *Proceedings of the International Conference on Computer Aided Design (ICCAD)*, pages 216–219, 1990.

[GuMu90] H. Gundlach and K.-D. Müller-Glaser. On Automatic Insertion of Testpoints into Sequential Circuits. In *Proceedings of the International Test Conference (ITC)*, 1990.

[HaDL89] F.K. Hanna, N. Daeche, and M. Longley. VERITAS$^+$: A Specification Language Based on Type Theory. In M. Leeser and G. Brown, editors, *Hardware Specification, Verification, and Synthesis: Mathematical Aspects (Lecture Notes in Computer Science 408)*, pages 358–379. Springer, 1989.

[HaEl89] B.S. Haroun and M.I. Elmasy. Architectural Synthesis for DSP Silicon Compilers. *IEEE Transactions on CAD* , pages 431–447, April 1989.

[HaFi85] W. Hahn and K. Fischer. High Performance Computing for Digital Design Simulation. In *Proceedings of the IFIP VLSI 85.* North Holland, 1985.

[HaHE90] W. Hahn, A. Hagerer, and M. Eisenhut. Timing Models for Compiler-Driven Logic Simulation. In *Proceedings of the SCS European Simulation Multiconference*, 1990.

[HaLD90] F.K. Hanna, M. Longley, and N. Daeche. Formal Synthesis of Digital Systems. In L. Claesen, editor, *Formal VLSI Specification and Synthesis—VLSI Design Methods, Volume I.* North-Holland, 1990.

[Hard91] W. Hardt. Implementation hierarchischer Steuerwerke beschrieben durch modifizierte Statecharts. Diploma Thesis, University of Paderborn, Department of Computer Science, 1991.

[HaRe86] D.S. Ha and S.M. Reddy. On the Design of Random Pattern Testable PLA's. In *Proceedings of the International Test Conference (ITC)*, 1986.

[Hare87] D. Harel. StateCharts: A Visual Formalism for Complex Systems. *Science of Computer Programming*, 8:231–274, 1987.

[Hart77] R. Hartenstein. *Fundamentals of Structured Hardware Design.* North Holland, 1977.

[HaSt66] J. Hartmanis and R.E. Stearns. *Algebraic Structure Theory of Sequential Machines.* Prentice Hall, Englewood Cliffs, 1966.

[HaSt71] A. Hashimoto and J. Stevens. Wire Routing by Optimizing Channel Assignment within Large Apertures. In *Proceedings of the 8th Design Automation Workshop*, pages 155–159, 1971.

[Haye86] J.P. Hayes. Digital Simulation with Multiple Logic Values. *IEEE ToCAD*, 5(2), April 1986.

[HCLH90] C.-Y. Huang, Y.-S. Chen, Y.-L. Lin, and Y.-C. Hsu. Data Path Allocation Based on Bipartite Weighted Matching. In *Proceedings of the 27th ACM/IEEE Design Automation Conference (DAC)*, pages 499–504, 1990.

[HePo89] A. Hemani and A. Postula. NISCHE: A Neural Net Inspired Scheduling Algorithm. In *Proceedings of the Fourth International Workshop on High-Level Synthesis*, Oct. 1989.

[Herb89] J. Herbert. Temporal Abstraction of Digital Designs. In G.J. Milne, editor, *The Fusion of Hardware Design and Verification.* North-Holland, 1989.

[Hilf85] P. Hilfinger. SILAGE, A High Level Language and Silicon Compiler for Digital Signal Processing. In *Proceedings of the Custom Integrated Circuits Conference (CICC)*, May 1985.

[Hill74] F.J. Hill. Introducing AHPL. *IEEE Computer*, 7(12), 12 1974.

[HILO91] Genrad Inc. *System HILO 4.0 System Reference Manual*, 1991.

[Hoar78] C.A.R. Hoare. Communicating Sequential Processes. *Comm. ACM*, 21(8), 1978.

[Hoar85] C.A.R. Hoare. *Communicating Sequential Processes.* Prentice-Hall, 1985.

[HoCO74] S.J. Hong, R.G. Cain, and D.L. Ostapko. MINI: A Heuristic Approach for Logic Minimization. *IBM Journal of Research and Development*, Vol. 18, September 1974.

[Hoer89] E. Hörbst et al. Synthese - die Entwurfsmethode der Zukunft. *Elektronik*, (23), 1989.

[Hohl90] A. Hohl. HDL for System Design. In H. Schwärtzel and I. Mizin, editors, *Advanced Information Processing, Proc. of a Joint Symposium Information Processing and Software Systems Design Automation.* Siemens AG, Academy of Science of the USSR, Springer-Verlag, 6 1990.

[Hort89] P.D. Hortensius, R.D. McLoead, M. Pries, D.M. Miller, and H.C. Card. Cellular Automata-Based Pseudorandom Number Generators for Built-In Self-Test. *IEEE Transactions on CAD*, August 1989.

[HTHI90] S. Hodgson, L. Theobald, W.B. Hughes, and R.J. Illman. ASTA - An Integrated System for BIST Analysis and Automatic Test Data Generation. In *Proceedings of the European Design Automation Conference (EDAC)*, 1990.

[Hu61] T.C. Hu. Parallel Sequencing and Assembly Line Problems. *Operations Research*, 9, A5.2:841–848, 1961.

[Huet75] G. Huet. A Unification Algorithm for Typed λ-Calculus. *Theoretical Computer Science*, 1:27–57, 1975.

[HwHL90] C.-T. Hwang, Y.-C. Hsu, and Y.-L. Lin. Optimum and Heuristic Data Path Scheduling under Resource Constraints. In *Proceedings of the 27th ACM/IEEE Design Automation Conference (DAC)*, pages 65–70, 1990.

[HwHL91] C.-T. Hwang, Y.-C. Hsu, and Y.-L. Lin. Scheduling for Functional Pipelining and Loop Winding. In *Proceedings of the 28th ACM/IEEE Design Automation Conference (DAC)*, pages 764–769, 1991.

[HwLH91] C.-T. Hwang, J.-H. Lee, and Y.-C. Hsu. A Formal Approach to the Scheduling Problem in High Level Synthesis. *IEEE Transactions on CAD*, 10:464–475, April 1991.

[IEEE90] The Institute of Electrical and Electronics Engineers, Inc, 345 East 47th Street, New York, NY 10017, USA. *IEEE Standard Test Access Port and Boundary Scan (IEEE Std 1149.1-1990)*, 1990.

[IlCl89] R. Illman and S. Clarke. Built-In Self Test of the MACROLAN Chip. In *Proceedings of the International Test Conference (ITC)*, 1989.

[Illm85] R. Illman. Self-Tested Data Flow Logic: A New Approach. *IEEE Design & Test of Computers*, April 1985.

[IsYY90] N. Ishiura, H. Yasuura, and S. Yajima. NES: The Behavioral Model for the Formal Semantics of a Hardware Design Language UDL/I. In *Proceedings of the 27th ACM/IEEE Design Automation Conference (DAC)*, 7 1990.

[JaSt86] S.K. Jain and C.E. Stroud. Built-In Self Testing of Embedded Memories. *IEEE Design & Test of Computers*, October 1986.

[Jeff85a] Jefferson. D.R. Implementation of Time Warp on the Caltech Hypercube. In *Proceedings of the SCS Distributed Simulation Conference*, Jan. 1985.

[Jeff85b] Jefferson. D.R. Virtual Time. *ACM TOPLAS*, 7(3), July 1985.

[JePa91] A. Jerraya and P.G. Paulin. SIF: An Interchange Format for the Design and Synthesis of High-Level Controllers. In *High Level Synthesis Workshop 91*, 1991.

[JePC91] A. Jerraya, P.G. Paulin, and S. Curry. Meta VHDL for Higher Level Controller Modelling and Synthesis. In *Proceedings of the IFIP VLSI'91 Conference Edinburgh*, 1991.

[JeSo82] D. Jefferson and H. Sowizral. Fast Concurrent Simulation Using the Time Warp Mechanism, Part I: Local Control Rand Corporation. In *Rand Note N-1906-AF*, 1982.

[JoBa86] N.A. Jones and K. Baker. An Intelligent Knowledge-Based System Tool for High-Level BIST Design. In *Proceedings of the International Test Conference (ITC)*, 1986.

[John67] S.C. Johnson. Hierarchical Clustering Schemes. In *Psychometrika*, pages 241–254, September 1967.

[John86] S.D. Johnson. Digital Design in a Functional Calculus. In G.J. Milne and P.A. Subrahmanyam, editors, *Formal Aspects of VLSI Design*. North-Holland, 1986.

[Joyc89] J. Joyce. Generic Structures in the Formal Specification and Verification of Digital Circuits. In G.J. Milne, editor, *The Fusion of Hardware Design and Verification*. North-Holland, 1989.

[Joyc91] J. Joyce. More Reasons Why Higher-Order Logic is a Good Formalism for Specifying and Verifying Hardware. In P.A. Subrahmanyam, editor, *Proceedings of the ACM International Workshop on Formal Methods in VLSI Design, Miami*, 1991.

[Karm84] N. Karmakar. A new Polynomial Algorithm for Linear Programming. *Combinatorica*, 4(4):373–395, 1984.

[Karp89] K. Karplus. Using If-Then-Else DAG's for multi-level Minimization. In *Decennial Caltech Conference on VLSI*, May 1989.

[Karp91a] K. Karplus. Amap: A Technology Mapper for Selector-Based Field-Programmable Gate Arrays. In *Proceedings of the 28th ACM/IEEE Design Automation Conference (DAC)*, pages 244–247, June 1991.

[Karp91b] K. Karplus. Xmap: A Technology Mapper for Table-Lookup Field-Programmable Gate Arrays. In *Proceedings of the 28th ACM/IEEE Design Automation Conference (DAC)*, pages 240–243, June 1991.

[KeLi70] B.W. Kernighan and S. Lin. An Efficient Heuristic Procedure for Partitioning Graphs. In *The Bell System Technical Journal, 49(2)*, 1970.

[Keut87] K. Keutzer. DAGON: Technology Binding and Local Optimization by DAG Matching. In *Proceedings of 24th ACM/IEEE Design Automation Conference (DAC)*, pages 341–347, June 1987.

[KiSa82] C.R. Kime and K.K. Saluja. Test Scheduling in Testable VLSI Circuits. In *Proceedings of the International Symposium on Fault-Tolerant Computing (FTCS)*, 1982.

[KiTH88] K. Kim, J.G. Tront, and D.S. Ha. Automatic Insertion of BIST Hardware Using VHDL. In *Proceedings of the 27th ACM/IEEE Design Automation Conference (DAC)*, pages 9–15, 1990.

[KlKu91] B. Kleinjohann and E. Kupitz. Tight Integration in a Hardware Synthesis System. In *[RaWa91]*, 1991.

[KNRR88] H. Krämer, M. Neher, G. Rietsche, and W. Rosenstiel. Data Path and Control Synthesis in the CADDY System. In *Proceedings of the International Workshop on Logic and Architecture Synthesis*, pages 229–242, 1988.

[KoGD90] M. Koster, M. Geiger, and P. Duzy. ASIC Design Using the CALLAS Synthesis System: A Case Study. In *Proceedings of the International Conference on Computer Design (ICCD)*, 1990.

[Kogg81] P.M. Kogge. *The Architecture of Pipelined Computers.* Hemisphere Publishing Corporation, 1981.

[Komo82] D. Komonytsky. LSI Self-Test Using Level Sensitive Scan Design and Signature Analysis. In *Proceedings of the International Test Conference (ITC)*, 1982.

[KoMZ79] B. Könemann, J. Mucha, B. Zwiehoff. Built-In Logic Block Observation Technique. In *Proceedings of the International Test Conference (ITC)*, pages 37–41, 1979.

[KoNT90] N. Koike, T. Nakata, and S. Takasaki. Integrated CAD Environments for Simulation Engines. In *Proceedings of the SCS European Simulation Multiconference*, 1990.

[KrAn90] T.H. Krodel and K.J. Antreich. An Accurate Model for Ambiguity Delay Simulation. In *Proceedings of the European Design Automation Conference (EDAC)*, 1990.

[KrPi87] A. Krasniewski and S. Pilarski. Circular Self-Test Path: A Low-Cost BIST Technique. In *Proceedings of the 24th ACM/IEEE Design Automation Conference (DAC)*, 1987.

[KrPi89] A. Krasniewski and S. Pilarski. Circular Self-Test Path: A Low-Cost BIST Technique for VLSI Circuits. *IEEE Transactions on CAD*, January 1989.

[KrRo89] H. Krämer and W. Rosenstiel. Synthesis of Multi-Processor Architectures from Behavioral Descriptions. In *Proceedings of the Fourth International Workshop on High-Level Synthesis*, October 1989.

[KrRo90] H. Krämer and W. Rosenstiel. System Synthesis Using Behavioral Descriptions. In *Proceedings of the European Design Automation Conference (EDAC)*, pages 277–282, 1990.

[KuDe89] D.C. Ku and G. De Micheli. Optimal Synthesis of Control Logic from Behavioral Specifications. Technical Report CSL-TR-89-402, Computer Systems Laboratory, Stanford University, November 1989.

[KuDe90] D. Ku and G. De Micheli. Relative Scheduling under Timing Constraints. In *Proceedings of the 27th ACM/IEEE Design Automation Conference (DAC)*, pages 59–64, 1990.

[KuFD91] D. Ku, D. Filo and G. De Micheli. Control Optimization Based on Resynchronization of Operations. In *Proceedings of the 28th ACM/IEEE Design Automation Conference (DAC)*, pages 366–371, 1991.

[Kunz88] A. Kunzmann. Produktionstest synchroner Schaltwerke auf der Basis von Pipelinestrukturen. *Proceedings of the 18. GI-Jahrestagung*, pages 92–105, 1988.

[KuPa87] F.J. Kurdahi and A.C. Parker. REAL: A Program for REgister ALlocation. In *Proceedings of the 24th ACM/IEEE Design Automation Conference (DAC)*, pages 210–215, 1987.

[KuPa90a] K. Küçükçakar and A.C. Parker. MABAL: A Software Package for Module and Bus Allocation. *International Journal of Computer Aided VLSI Design*, pages 419–436, 1990.

[KuPa90b] K. Küçükçakar and A. Parker. CHOP: A Constraint-Driven System Level Partitioner. Report CEng 90-26, University of Southern California, Los Angeles, 1990.

[KuWu90] A. Kunzmann and H.-J. Wunderlich. An Analytical Approach to the Partial Scan Problem. *Journal of Electronic Testing: Theory and Applications*, 1:163–174, 1990.

[LaTh89] E.D. Lagnese and D.E. Thomas. Architectural Partitioning for System Level Desing. In *Proceedings of the 26th ACM/IEEE Design Automation Conference (DAC)*, pages 62–67, 1989.

[LeBl84] J.J. LeBlanc. LOCST: A Built-In Self-Test Technique. *IEEE Design & Test of Computers*, November 1984.

[Lee59] C.Y. Lee. Representation of Switching Circuits by Binary-decision Programs. *Bell Systems Technical Journal*, Vol. 38, July 1959.

[Lees90] M. Leeser and G. Brown, editors. *Hardware Specification, Verification, and Synthesis: Mathematical Aspects (Lecture Notes in Computer Science 408)*. Springer, 1990.

[LeHL89] J.-H. Lee, Y.-C. Hsu, and Y.-L. Lin. A New Integer Linear Programming Formulation for the Scheduling Problem in Data Path Synthesis. In *Proceedings of the International Conference on Computer Aided Design (ICCAD)*, pages 20–23, 1989.

[Leng90] T. Lengauer. *Combinatorial Algorithms for Integrated Circuit Layout.* JOHN WILEY and SONS, New York, 1990.

[LeRa83] K.D. Lewke and F.J. Rammig. Description and Simulation of MOS Devices in Register Transfer Languages. In *Proceedings of the IFIP VLSI 83.* North Holland, 1983.

[LeRS83] C.E. Leiserson, F.M. Rose, and J.B. Saxe. Optimizing Synchronous Circuits by Retiming. In *Proceedings of the Third Caltech Conf. on VLSI*, pages 87–116, 1983.

[LeSa81] C.E. Leiserson and J.B. Saxe. Optimizing Synchronous Systems. In *Proceedings of the Twenty-Second Ann. Symp. on Foundations of Computer Science*, pages 23–26, 1981.

[LeSa83] C.E. Leiserson and J.B. Saxe. Optimizing Synchronous Systems. *Journal of VLSI and Computer Systems*, 1(1), 1983.

[LeSa89a] R. Leveugle and G. Saucier. Optimized Synthesis of Dedicated Controllers with Convenient Checking Capabilities. In *Proceedings of the International Test Conference (ITC)*, 1989.

[LeSa89b] R. Leveugle and G. Saucier. Synthesis of Dedicated Controllers for Concurrent Checking. In *Proceedings of the VLSI 89 international conference*, pages 123–132. IFIP, August 1989.

[LeWB91] T.G.R. van Leuken, P. van der Wolf, and P. Bingley. Standardization Concepts in the Nelsis CAD Framework. In F.J. Rammig and R. Waxman, editors, *Electronic Design Automation Frameworks*. North Holland, 1991.

[LiBK88] R. Lisanke, F. Brglez, and Gershon Kedem. McMAP: A Fast Technology Mapping Procedure for Multi-Level Logic Synthesis. In *ICCD*, pages 257–261, October 1988.

[LiGa89] J.S. Lis and D.D. Gajski. VHDL Synthesis Using Structured Modelling. In *Proceedings of the 26th ACM/IEEE Design Automation Conference (DAC)*, pages 606–609, 1989.

[LiMC88] L. Liu and E.J. McCluskey. Design of Large Embedded CMOS PLA's for Built-In Self-Test. *IEEE Transactions on CAD*, January 1988.

[LiNe89a] B. Lin and A.R. Newton. Restructuring State Machines and State Assignment: Relationship to Minimizing Logic Across Latch Boundaries. In *Proceedings of the International Workshop on Logic Synthesis*, Amsterdam, 1989. North-Holland.

[LiNe89b] B. Lin and A.R. Newton. Synthesis of Multiple Level Logic from Symbolic High–Level Description Languages. In *Proceedings of the VLSI 89 international conference*, pages 187–196. IFIP, August 1989.

[LiRe87] C.J. Lin and S.M. Reddy. On Delay Fault Testing in Logic Circuits. *IEEE Transactions on CAD*, September 1987.

[LMOI89] S.P. Levitan, A.R. Martello, R.M. Owens, and M.J. Irwin. Using VHDL as a Language for Synthesis of CMOS VLSI Circuits. In J.A. Darringer and F.J. Rammig, editors, *Computer Hardware Description Languages*, pages 331–343, 1989.

[LMWV91] P.E.R. Lippens, J.L. van Meerbergen, A. van der Werf, and W.F.J. Verhaegh. PHIDEO: A Silicon Compiler for High Speed Algorithms. In *Proceedings of the European Design Automation Conference (EDAC)*, pages 436–441, 1991.

[LuAd86] P.A. Luker and H.H. Adelsberger. Intelligent Simulation Environments. In *SCS Simulation Series, 17:1*, 1986.

[LyEG90] T.A. Ly, W.L. Elwood, and E.F. Girczyc. A Generalized Interconnect Model for Data Path Synthesis. In *Proceedings of the*

27th ACM/IEEE Design Automation Conference (DAC), pages 168–173, 1990.

[MaBa88] H.-J. Mathony and U.G. Baitinger. CARLOS: An Automated Multilevel Logic Design System for CMOS Semi-Custom Integrated Circuits. *IEEE Transactions on Computer Aided Design*, March 1988.

[MaDN88] H.-K.T. Ma, S. Devadas, and A.R. Newton. An Incomplete Scan Design Approach to Test Generation for Sequential Machines. In *Proceedings of the International Test Conference (ITC)*, pages 730–734, 1988.

[MaFu89] Y. Matsunaga and M. Fujita. Multi-Level Logic Optimization Using Binary Decision Diagrams. In *Proceedings of the International Conference on Computer Aided Design*, 1989.

[MaPF88] P. Mazumder, J.H. Patel, and W.K. Fuchs. Methodologies for Testing Embedded Content Addressable Memories. *IEEE Transactions on CAD*, January 1988.

[Marc87] M. Marchesi, P. Antognetti, G.M. Bisio, D.D. Caviglia, P. Traverso, and F.J. Rammig. Object-oriented Programming for VLSI CAD Tool Integration, Tool Integration and Environments. In *Proceedings of the International Federation on Information Processing (IFIP) Working Group (WG) 10.2 Workshop*, 11 1987.

[Marh89] M. Marhöfer. Allocation of Test Hardware Within High-Level Synthesis. In *12th Annual IEEE Workshop on Design for Testability, Vail(CO)*, April 1989.

[März89] S. März, K. Buchenrieder, P. Duzy, R. Kumar, and T. Wecker. CALLAS - A System for Automatic Synthesis of Digital Circuits from Algorithmic Behavioral Descriptions. In *Proceedings of the EuroASIC Conference*, Jan. 1989.

[März90] S. März. High-Level Synthesis - A Step Towards System Design Automation. In H. Schwärtzel and I. Mizin, editors, *Advanced Information Processing, Proceedings of a Joint Symposium Information Processing and Software Systems Design Automation.* Siemens AG, Academy of Science of the USSR, Springer-Verlag, 6 1990.

[Math88] H.-J. Mathony. *Algorithmische Entwurfsverfahren für zwei- und mehrstufige Schaltnetze.* VDI-Verlag GmbH, 1988.

[May83] M.D. May. OCCAM. *ACM Sigplan Notices*, 18(4), April 1983.

[McCl56] E.J. McCluskey. Minimization of Boolean Functions. *Bell Systems Technical Journal*, Vol. 35, April 1956.

[McCl84] E.J. McCluskey. Verification Testing - A Pseudoexhaustive Test Technique. *IEEE Transactions Computers*, June 1984.

[McCl85] E.J. McCluskey. Built-In Self Test Structures. *IEEE Design & Test of Computers*, April 1985.

[McCl86] E.J. McCluskey. *Logic Design Principles, With Emphasis on Testable Semicustom Circuits.* Prentice-Hall International, Inc, 1986.

[McFa83] M.C. McFarland. Computer-Aided Partitioning of Behavioral Hardware Description . In *Proceedings of the 20th ACM/IEEE Design Automation Conference (DAC)*, pages 472–478, 1983.

[McFa86] M.C. McFarland. Using Bottom-Up Design Techniques in the Synthesis of Digital Hardware from Abstract Behavioral Descriptions. In *Proceedings of the 23rd ACM/IEEE Design Automation Conference (DAC)*, pages 474–480, 1986.

[McKo90] M.C. McFarland and T.J. Kowalski. Incorporating Bottom-Up Design into Hardware Synthesis. *IEEE Transactions on CAD*, pages 938–950, Sept. 1990.

[McPC90] M.C. McFarland, A.C. Parker, and R. Camposano. The High-Level Synthesis of Digital Systems. *Proceedings of the IEEE*, 78(2):301–318, Feb. 1990.

[Meer88] J. Van Meerbergen. Efficient Controller Architectures for DSP Compilers. In *Proceedings of the IEEE Custom Integrated Circuits Conference (CICC)*, 1988.

[Melh89] T. Melham. Abstraction Mechanisms for Hardware Verification. In G. Birtwistle and P.A. Subrahmanyam, editors, *VLSI Specification, Verification, and Synthesis*, pages 27–72. Kluwer Academic Publishers, 1989.

[Merm85] J. Mermet. Several Steps Towards a Circuit Integrated CAD System: CASCADE. In *CHDL 85 - Computer Hardware Description Languages and their Application.* North Holland, 1985.

[Meye89] E. Meyer. VHDL Strives to Cover Both Synthesis and Modeling. *Computer Design*, 10 1989.

[Mich91] G. DeMicheli. Synchronous Logic Synthesis: Algorithms for Cycle-Time Minimization. *IEEE Transaction on Computer Aided Design*, 10(1):63–73, January 1991.

[Micz86] A. Miczo. Digital Logic Testing and Simulation. In *Harper & Row, New York*, 1986.

[Miln89] G.J. Milne, editor. *The Fusion of Hardware Design and Verification.* North-Holland, 1989.

[MiSa91a] A. Mignotte and G. Saucier. Control Flowchart Synthesis: A Practical Approach between High Level and Low Level Synthesis. In *Proceedings of the Fifth International High-Level Synthesis Workshop*, pages 126–133, Germany, March 1991.

[MiSa91b] A. Mignotte and G. Saucier. A Generalized Model for Resource Assignment. In *Proceedings of the Fifth International Workshop on High-Level Synthesis*, March 1991.

[MiTH90] R. Milner, M. Tofte, and R. Harper. *The Definition of ML*. MIT Press, Cambridge, MA, 1990.

[MLSB89] C.W. Moon, B. Lin, H. Savoj, and R.K. Brayton. Technology Mapping for Sequential Logic Synthesis. In *Logic Synthesis*. MCNC, November 1989.

[MoPT85] J.D. Morrison, N.E. Peeling, and T.L. Thorp. The Design Relationale of ELLA, a Hardware Design and Description Language. In *Proceedings of 7th International Conference on Computer Hardware Description Languages and their Applications*. North Holland, 1985.

[MSBS89] S. Malik, E.M. Sentovich, R.K. Brayton, and A. Sangiovanni-Vincentelli. Retiming and Resynthesis: Optimizing Sequential Networks with Combinational Techniques. In *Proceedings of the International Workshop on Logic Synthesis*, Amsterdam, 1989. North-Holland.

[MSBS91] S. Malik, E.M. Sentovich, R.K. Brayton, and A. Sangiovanni-Vincentelli. Optimizing Sequential Networks with Combinational Technique. *IEEE Transaction on Computer Aided Design*, 10(1):74–84, January 1991.

[MuGe91] T. Mueller-Wipperfuerth and M. Geiger. Algebraic Decomposition of MCNC Benchmark FSMs for Logic Synthesis. In *Proceedings of the Conference EUROASIC 91, Paris*, 1991.

[MuHa88] B. Murray and J. Hayes. Hierarchical Test Generation Using Precomputed Tests for Modules. In *Proceedings of the International Test Conference (ITC)*, 1988.

[MuRa90] W. Müller and F. Rammig. ODICE: Object-Oriented Hardware Description in CAD Environment. In *Proceedings of the IFIP Ninth International Symposium on Computer Hardware Description Languages and their Applications*, pages 19–34, 1990.

[Nage87] L.W. Nagel. SPICE2: A Computer Program to Simulate Semiconductor Circuits. Memo No. RL-M-520, May 1975, University of California, Berkeley, May 1975.

[NaSp90] Z. Navabi and J. Spillane. Hardware Generation from a Synthesis Subset of VHDL. In *Proceedings of the VHDL Users' Group*, 4 1990.

[NaVG91] S. Narayan, F. Vahid, and D.D. Gajski. Translating System Specifications to VHDL. In *Proceedings of the European Design Automation Conference (EDAC)*, pages 390–394, February 1991.

[Nest87] J.A. Nestor. *Specification and Synthesis of Digital Systems with Interfaces.* PhD thesis, Carnegie Mellon University, April 1987.

[Newp86] J.R. Newport. An Introduction to OCCAM and the Development of Parallel Software. *Software Engineering Journal*, July 1986.

[Newt91] A.R. Newton. Has CAD for VLSI Reached a Dead End? In *Proceedings of the International Federation on Information Processing (IFIP) VLSI '91*, 1991.

[NGCD91] S. Note, W. Geurts, F. Catthor, and H. De Man. Cathedral-III: Architecture-Driven High-Level Synthesis for High Throughput DSP Applications. In *Proceedings of the 28th ACM/IEEE Design Automation Conference (DAC)*, pages 597–602, 1991.

[Niem91] M. Niemeyer. Simulation of Heterogeneous Models With a Simulator Coupling System. In *Proceedings of the SCS 1991 European Simulation Multiconference*, June 1991.

[Nomu85] H. Nomura. Current Status, Future Trends, and Impact of VLSI. In *Proceedings of the IFIP VLSI Conference, Tokyo*, pages 3–11, 1985.

[Oczk90] A. Oczko. Hardware Design with VHDL at a Very High Level of Abstration. In *Proceedings of the 1st European Conference on VHDL*, 1990.

[OhWM87] M.J. Ohletz, T.W. Williams, and J.P. Mucha. Overhead in Scan and Self-Testing Design. In *Proceedings of the International Test Conference (ITC)*, 1987.

[PaGa87] B.M. Pangrle and D.D. Gajski. Design Tools for Intelligent Silicon Compilation. *IEEE Transactions on CAD*, 6(6), Nov. 1987.

[PaKn89a] P.G. Paulin and J.P. Knight. Scheduling and Binding Algorithms for High-Level Synthesis. In *Proceedings of the 26th ACM/IEEE Design Automation Conference (DAC)*, pages 1–6, 1989.

[PaKn89b] P.G. Paulin and J.P. Knight. Force-Directed Scheduling for the Behavioral Synthesis of ASICs. *IEEE Transactions on CAD*, pages 661–679, June 1989.

[PaPa88] N. Park and A.C. Parker. Sehwa: A Software Package for Synthesis of Pipelines from Behavioral Specifications. *IEEE Transactions on CAD*, 7(3):356–370, March 1988.

[PaPM86] M. Mlinar A.C. Parker, J.T. Pizarro. MAHA: A Program for Data Path Synthesis. In *Proceedings of the 23rd ACM/IEEE Design Automation Conference (DAC)*, pages 235–240, 1986.

[PaSt82] C.H. Papadimitriou and K. Steiglitz. *Combinatorial Optimization: Algorithms and Complexity.* Prentice Hall, Englewood Cliffs N.J., 1982.

[Paul86] L.C. Paulson. Natural Deduction as Higher-Order Resolution. *Journal of Logic Programming*, 3:237–258, 1986.

[Paul88] P.G. Paulin. *High-Level Synthesis of Digital Circuits Using Global Scheduling and Binding Algorithms*. PhD thesis, Carleton University, 1988.

[PaWl87] A. Pawlak and W. Wlodzimierz. Modern Object-Oriented Programming Language as a HDL. In *Proceedings of the IFIP Eight International Symposium on Computer Hardware Description Languages and their Applications*, pages 343–362, 1987.

[PeKL88] Z. Peng, K. Kuchcinski, and B. Lyles. CAMAD: A Unified Data Path / Control Synthesis Environment. In *Proceedings of the IFIP Working Conf. on Design Methodologies for VLSI and Computer Architecture*, pages 53–67, 1988.

[Pete77] J.L. Peterson. Petri Nets. *ACM Computing Surveys*, 9(3):223–252, Sept. 1977.

[Pfaf90] H. Pfaffhausen. Realisierung eines regelbasierten Expertensystems durch einen Logiksimulator. In *it 4/90*. R. Oldenbourg Verlag, August 1990.

[Pfaf91] H. Pfaffhausen. Ein wissensbasierter Ansatz zur automatischen Durchführung von Experimenten in der Logiksimulation. Dissertation, Universität-GH-Paderborn, 1991.

[Pfah87] P. Pfahler. Automated Datapath Synthesis: A Compilation Approach. *Microprocessing and Microprogramming*, 21:577–584, 1987.

[Pilo83] R. Piloty, M. Barbacci, D. Borrione, D. Dietmeyer, F. Hill, and P. Skelly. CONLAN Report. In *Lecture Notes in Computer Science No. 151*. Springer, 1983.

[PoRa89] M. Potkonjak and J. Rabaey. A Scheduling and Resource Allocation Algorithm for Hierarchical Signal Flow Graphs. In *Proceedings of the 26th ACM/IEEE Design Automation Conference (DAC)*, pages 7–12, 1989.

[PrBL88] M.M. Pradhan, E.J. O'Brien, and S.L. Lam. Circular BIST with Partial Scan. In *Proceedings of the International Test Conference (ITC)*, 1988.

[Quin55] W.V. Quine. A Way to Simplify Truth Functions. *Amer. Math Mon.*, Vol. 62, November 1955.

[Raba88] J. Rabaey, H. De Man, J. Vanhoof, G. Goossens, and F. Catthoor. CATHEDRAL-II: A Synthesis System for Multiprocessor DSP Systems. In D.D. Gajski, editor, *Silicon Compilation*, pages 311–360. Addison-Wesley, 1988.

[RaLi75] C.V. Ramamoorthy and H.F. Li. Some Problems in Parallel and Pipeline Processing. In *Proceedings COMPCON*, pages 177–180, 1975.

[Ramm80a] F.J. Rammig. Five Valued Quasi Boolean Functions. In *Proceedings of the 5th European Meeting of Cybernetics and Systems Research*, 1980.

[Ramm80b] F.J. Rammig. Preliminary CAP/DSDL Language Reference Manual. Technical Report 129, University of Dortmund, Dept. Computer Science, 1980.

[Ramm82] F.J. Rammig. Preliminary CAP/DSDL Language Reference Manual, Forschungsbericht 131. Technical Report, Abteilung Informatik, Universität Dortmund, 1982.

[Ramm89] F.J. Rammig. *Systematischer Entwurf digitaler Systeme.* Teubner Stuttgart, 1989.

[RaWa91] F.J. Rammig and R. Waxman, editors. *Electronic Design Automation Frameworks.* North Holland, 1991.

[RCHP91] J.M. Rabaey, C. Chu, P. Hoang, and M. Potkonjak. Fast Prototyping of Datapath-Intensive Architectures. *IEEE Design & Test of Computers*, pages 40–51, 1991.

[RiNe90] G. Rietsche and M. Neher. CASTOR: State Assignment in a Finite State Machine Synthesis System. In *Proceedings of the IFIP Working Conference on Logic and Architecture Synthesis*, pages 15–19, Paris, May 30th-June 1st 1990. IFIP Working Group 10.5.

[RoBT85] C.W. Rose, M. Buchner, and Y. Trivedi. Integrating Stochastic Performance Analysis with System Design Tools. In *Proceedings of the 22nd ACM/IEEE Design Automation Conference (DAC)*, pages 482–488, 1985.

[RoJG89] W. Roth, M. Johansson, and W. Glunz. The BED Concept - A Method and a Language for Modular Test Generation. In *International Conference on VLSI (VLSI'89)*, 1989.

[RoKr91] W. Rosenstiel and H. Krämer. Scheduling and Assignment in High Level Synthesis. In R. Camposano and W. Wolf, editors, *High-Level VLSI Synthesis*, pages 355–381. Kluwer Academic Publishers, 1991.

[SaAV89] J. Savir, W.H. McAnney, and S.R. Vecchio. Testing for Coupled Cells in Random-Access Memories. In *Proceedings of the International Test Conference (ITC)*, 1989.

[SaBo91] A. Salem and D. Borrione. Formal Semantics of VHDL Timing Constructs. In *Proceedings of the 2nd European Conference on VHDL.* Swedish Institute of Microelectronics, 1991.

[SaDP90] G. Saucier, C. Duff, and F. Poirot. State Assignment of Controllers for Optimal Area Implementation. In *Proceedings of the 1st European Design Automation Conference (EDAC)*, pages 546–551, 1990.

[SaDS87] G. Saucier, M.C. Depaulet, and P. Sicard. ASYL: A Rule–based System for Controller Synthesis. *IEEE Transactions on Computer Aided Design*, CAD–6:1088–1098, November 1987.

[SaKi88] J. Sayah and C.R. Kime. Test Scheduling for High Performance VLSI System Implementation. In *Proceedings of the International Test Conference (ITC)*, 1988.

[SaMU85] J. Salick, M.R. Mercer, and B. Underwood. Built-In Self Test Input Generator for Programmable Logic Arrays. In *Proceedings of the International Test Conference (ITC)*, 1985.

[SaPo89] G. Saucier and F. Poirot. State Assignment Using a New Embedding Method Based On an Intersecting Cube Theory. In *Proceedings of the 26th ACM/IEEE Design Automation Conference (DAC)*, pages 321–326, 1989.

[ScAu88] M.H. Schulz and E. Auth. Advanced Automatic Test Pattern Generation and Redundancy Identification Techniques. In *18th International Symposium on Fault-Tolerant Computing*, June 1988.

[ScAu89] M. Schulz and E. Auth. Essential: An Efficient Self-Learning Test Pattern Generation Algorithm for Sequential Circuits. In *Proceedings of the International Test Conference (ITC)*, 1989.

[ScFF89] M.H. Schulz, K. Fuchs, and F. Fink. Advanced Automatic Test Pattern Generation Techniques for Path Delay Faults. In *19th International Symposium on Fault-Tolerant Computing*, 1989.

[Sche91] J. Scheichenzuber. Hardware Specification and Representation Using a Dataflow Notation. In *Proceedings of the Fifth International Workshop on High-Level Synthesis*, March 1991.

[ScTS87] M. Schulz, E. Trischler, and T. Sarfert. SOCRATES: A Highly Efficient Automatic Test Pattern Generation System. In *Proceedings of the International Test Conference (ITC)*, 1987.

[SGLM90] J. Scheichenzuber, W. Grass, U. Lauther, and S. März. Global Hardware Synthesis from Behavioral Dataflow Descriptions. In *Proceedings of the 27th ACM/IEEE Design Automation Conference (DAC)*, pages 456–461, 1990.

[Shan48] C.E. Shannon. The Synthesis of Two-terminal Switching Circuits. *Bell Systems Technical Journal*, 1948.

[Shee86] M. Sheeran. Design and Verification of Regular Synchronous Circuits. *IEE Proceedings, Partition E (Computers and Digital Technology)*, 133(5):295–304, September 1986.

[ShMA85] R.E. Shannon, R. Mayer, and H.H. Adelsberger. Expert Systems and Simulation. *SIMULATION*, 44, June 1985.

[Siew74a] D. Siewiorek. Introducing ISP. *IEEE Computer*, 7(12), 12 1974.

[Siew74b] D. Siewiorek. Introducing PMS. *IEEE Computer*, 7(12), 12 1974.

[Smit85] G.L. Smith. Model for Delay Faults Based upon Paths. In *Proceedings of the International Test Conference (ITC)*, 1985.

[Snep85] J.L.A. van de Snepscheut. *Trace Theory and VLSI Design*. Springer, 1985.

[Snow78] E.A. Snow, D.P. Siewiorek, and D.E. Thomas. A Technology-Relative Computer-Aided Design System: Abstract Representations, Transformations, and Design Tradeoffs. In *Proceedings of the 15th ACM/IEEE Design Automation Conference (DAC)*, pages 220–226, 1978.

[SpTh90] D.L. Springer and D.E. Thomas. Exploiting the Special Structure of Conflict and Compatibility Graphs in High-Level Synthesis. In *Proceedings of the International Conference on Computer Aided Design (ICCAD)*, pages 254–257, 1990.

[STAT90] I-Logix Inc, Burlington, MA 01803. The STATEMATE Approach to Complex Systems.

[StBo88] L. Stok and R. van den Born. EASY: Multiprocessor Architecture Optimization. In G. Saucier and P.M. McLellan, editor, *Logic and Architecture Synthesis*, pages 313–328. Elsevier Science Publishers, 1988.

[Stok91] L. Stok. *Architectural Synthesis and Optimization of Digital Systems*. PhD thesis, Technische Universiteit Eindhoven, 1991.

[Stro86] B. Stroustrup. *The C++ Programming Language*. Addison-Wesley Reading, 1986.

[Stro88a] C.E. Stroud. An Automated BIST Approach for General Sequential Logic Synthesis. In *Proceedings of the 25th ACM/IEEE Design Automation Conference (DAC)*, 1988.

[Stro88b] C.E. Stroud. Automated BIST for Sequential Logic Synthesis. *IEEE Design & Test of Computers*, December 1988.

[Subr91] P.A. Subrahmanyam, editor. *Proceedings of the ACM International Workshop on Formal Methods in VLSI Design, Miami*, 1991.

[SuCh72] S.Y.H. Su and P.T. Cheung. Computer Minimization of Multiple Valued Switching Functions. *IEEE Transactions on Computers*, C-21:995–1003, December 1972.

[SuLP88] L.F. Sun, J.M. Liaw, and T.M. Parng. Automated Synthesis of Microprogrammed Controllers in Digital Systems. *IEEE Proceedings*, 135(4):231–240, July 1988.

[SuWa84] Z. Sun and L.T. Wang. Self-Testing of Embedded RAM's. In *Proceedings of the International Test Conference (ITC)*, 1984.

[SvWh79] A. Svoboda and D.E. White. *Advanced Logical Circuit Design Techniques.* Garland Press, New York, 1979.

[SWBS88] A. Saldanha, A.R. Wang, R.K. Brayton, and A. Sangiovanni-Vincentelli. Multi-Level Logic Simplification Using Don't Cares and Filters. In *Proceedings of the 26th ACM/IEEE Design Automation Conference (DAC)*, 1989.

[Take81] A. Takeuchi. Object Oriented Description Environment for Computer Hardware. In *Proceedings of the IFIP Fifth International Symposium on Computer Hardware Description Languages and their Applications*, pages 197–210, 1981.

[Tarj83] R.E. Tarjan. *Data Structures and Network Algorithms.* Society for Industrial and Applied Mathematics, Philadelphia, PA, 1983.

[Temm89] K.-H. Temme. CHARME: A Synthesis Tool for High Level Chip Architecture Planning. In *Proceedings of the Custom Integrated Circuits Conference (CICC)*, 1989.

[Thom88] D.E. Thomas, E.M. Dirkes, R.A. Walker, J.V. Rajan, J.A. Nestor, and R.L. Blackburn. The System Architect's Workbench. In *Proceedings of the 25th ACM/IEEE Design Automation Conference (DAC)*, pages 337–343, 1988.

[Thom90] D.E. Thomas, E.D. Lagnese, R.A. Walker, J.A. Nestor, J.V. Rajan, and R.L. Blackburn. *Algorithmic and Register-Transfer Level Synthesis: The System Architect's Workbench.* Kluwer Academic Publishers, 1990.

[Thom91] D. Thomas. *The Verilog Hardware Description Language.* Kluwer Academic Publisher, 1991.

[TrFA85] R. Treuer, H. Fujiwara, and V.K. Agarwal. Implementing a Built-In Self-Test PLA Design. *IEEE Design & Test of Computers*, April 1985.

[Tric87] H. Trickey. Flamel: A High-Level Hardware Compiler. In *IEEE Transactions on CAD*, volume 6, pages 259–269, March 1987.

[Tris80] E. Trischler. Incomplete Scan Path with an Automatic Test Generation Methodology. In *Proceedings of the International Test Conference (ITC)*, pages 153–162, 1980.

[TsSi86] C.-J. Tseng and D.P. Siewiorek. Automated Synthesis of Data Paths in Digital Systems. *IEEE Transactions on CAD*, 5(3), July 1986.

[Udel86] J.G. Udell. Test Set Generation for Pseudo-Exhaustive BIST. In *Proceedings of the International Conference on Computer Aided Design (ICCAD)*, 1986.

[Umbr90] G. Umbreit. VHDL-Subset for the CALLAS Synthesis System. Technical Report, ZFE IS EA 11, Siemens AG, 1990.

[UpSa88] S.J. Upadhyaya and K.K. Saluja. A New Approach to the Design of Built-In Self Testing PLA's for High Fault Coverage. *IEEE Transactions on CAD*, January 1988.

[VAKL91] W.F.J. Verhaegh, E.H.L. Aarts, J.H.M. Korst, and P.E.R. Lippens. Improved Force-Directed Scheduling. In *Proceedings of the European Design Automation Conference (EDAC)*, pages 430–435, 1991.

[VaNG91] F. Vahid, S. Narayan, and D.D. Gajski. SpecCharts : A Language for System Level Synthesis. In *Proceedings of the IFIP Tenth International Symposium on Computer Hardware Description Languages and their Applications*, pages 145–154, April 1991.

[VdWe91] A. van der Werft, B.T. McSweeney, and J.L. van Meerbergen. Flexible Data Path Compilation for PHIDEO. In *Proceedings of EuroASIC'91*, May 1991.

[VeCD89] D. Verkest, L. Claesen, and H. De Man. On the Use of the Boyer-Moore Theorem Prover for Correctness Proofs of Parameterized Hardware Modules. In L. Claesen, editor, *Formal VLSI Specification and Synthesis—VLSI Design Methods, Volume I*, pages 99–116. North-Holland, 1990.

[VHDL87] The Institute of Electrical and Electronics Engineers, Inc., 345 East 47th Street, New York, NY 10017, USA. *IEEE Standard VHDL Language Reference Manual (IEEE Std 1076-1987)*, 1987.

[ViSa89] T. Villa and A. Sangiovanni-Vincentelli. NOVA: State Assignment of Finite State Machines for Optimal Two-Level Logic Implementations. In *Proceedings of the 26th ACM/IEEE Design Automation Conference (DAC)*, pages 327–332, 1989.

[Wake90] J.F. Wakerly. *Digital Design Principles and Practices.* Prentice-Hall, New Jersey, 1990.

[WaMc86] L.-T. Wang and E.J. McCluskey. Circuits for Pseudo-Exhaustive Test Pattern Generation. In *Proceedings of the International Test Conference (ITC)*, 1986.

[WaNe89] S.H. Wang and A.R. Newton. BEAVER: A Behavioral Formal Verifier for VLSI Design. In L. Claesen, editor, *Proceedings of the IMEC/IFIP International Workshop on Applied Formal Methods for Correct VLSI Design*, 1989.

[WaTh85] R.A. Walker and D.E. Thomas. A Model of Design Representation and Synthesis. In *Proceedings of the 22nd ACM/IEEE Design Automation Conference (DAC)*. IEEE, 1985.

[WaTh87] R.A. Walker and D.E. Thomas. Design Representation and Transformation in the System Architect's Workbench. In *Proceedings of the International Conference on Computer Aided Design (ICCAD), New York*, pages 166 – 169, 1987.

[WaTh89] R.A. Walker and D.E. Thomas. Behavioral Transformations for Algorithmic Level IC Design. *IEEE Transactions on CAD* , pages 1115–1127, Oct. 1989.

[WDGS88] T.W. Williams, W. Dähn, M. Grützner, and C.W. Starke. Bounds and Analysis of Aliasing Errors in Linear Feedback Shift Registers. *IEEE Transactions on CAD*, January 1988.

[WeBP91] N. Wehn, J. Biesenack, and M. Pilsl. A New Approach to Multiplexer Minimization in the CALLAS Synthesis Environment. In *Proceedings of the IFIP VLSI Conference, Edinburgh*, 1991.

[Weis87] D. Weise. Functional Verification of MOS Circuits. In *Proceedings of the 24th ACM/IEEE Design Automation Conference (DAC)*, 1987.

[WePa91] J.-P. Weng and A.C. Parker. 3D Scheduling: High-Level Synthesis with Floorplanning. In *Proceedings of the 28th ACM/IEEE Design Automation Conference (DAC)*, pages 668–673, 1991.

[WhNe90] G.S. Whitcomb and A.R. Newton. Abstract Data Types and High-Level Synthesis. In *Proceedings of the 27th ACM/IEEE Design Automation Conference (DAC)*, pages 680–685, 1990.

[Wind90] P. Windley. A Hierarchical Methodology for Verifying Microprogrammed Microprocessors. In *1990 IEEE Symposium on Research in Security and Privacy*, 1990.

[Wins84] P.H. Winston. *Artifical Intelligence.* Addison-Wesley Publishing Company, 1984.

[Wolf89] W.H. Wolf. How to Build a Hardware Description and Measurement System on an Object Oriented Programming Language. *IEEE Transactions on CAD*, 8(3):288–301, 1989.

[WoTL91] W. Wolf, A. Takach, and T.-C. Lee. Architectural Optimization Methods for Control-Dominated Machines. In R. Camposano and W. Wolf, editors, *High-Level VLSI Synthesis*, pages 231–254. Kluwer Academic Publishers, 1991.

[WuHe88] H.-J. Wunderlich and S. Hellebrand. Generating Pattern Sequences for the Pseudo-Exhaustive Test of MOS-Circuits. In *Proceedings of the International Symposium on Fault-Tolerant Computing (FTCS)*, 1988.

[WuHe89] H.-J. Wunderlich and S. Hellebrand. The Pseudo-Exhaustive Test of Sequential Circuits. In *Proceedings of the International Test Conference (ITC)*, 1989.

[Wund85] H.-J. Wunderlich. PROTEST: A Tool for Probabilistic Testability Analysis. In *Proceedings of the 22nd ACM/IEEE Design Automation Conference (DAC)*, 1985.

[Wund87] H.-J. Wunderlich. Self Test Using Unequiprobable Random Patterns. In *Proceedings of the International Symposium on Fault-Tolerant Computing (FTCS)*, 1987.

[Wund89] H.-J. Wunderlich. The Design of Random-Testable Sequential Circuits. In *Proceedings of the International Symposium on Fault-Tolerant Computing (FTCS)*, 1989.

[YaIs89] H. Yasuura and N. Ishiura. Semantics of a Hardware Design Language for Japanese Standardization. In *Proceedings of the 26th ACM/IEEE Design Automation Conference (DAC)*, 1989.

[ZhBr88] X.-A. Zhu and M.A. Breuer. Selecting Test Methodologies. *IEEE Design & Test of Computers*, October 1988.

Index